Hair Test Interpretation: Finding Hidden Toxicities

Andrew Hall Cutler, Ph.D., P. E.

www.noamalgam.com

Publisher's Cataloging-in-Publication data
Cutler, Andrew Hall 1956-2017
Hair Test Interpretations: Finding Hidden Toxicities /Andrew Hall Cutler.
Includes index.
ISBN 978-0-9676168-2-7

1. Hair--Diseases.2. Hair--Analysis 3. Hair manifestations of general diseases. 4.Trace elements--Diagnostic use. I. Title

RL151 C88 2004
613.26
LCCN 2005281997
OCLC 62187958

Disclaimer

I have written this book to provide information on a controversial subject - the use of hair testing to identify heavy metal toxicities from various exposures, many of which are not only nonoccupational but are iatrogenic.

The purpose of this book is to educate any interested reader. This book as complete and as accurate as I could make it. The author assumes no liability for errors or omissions. The book is provided in an "as is" condition, on an "as available" basis, with no express or implied warranties of accuracy or suitability for any particular purpose. Much of this book is a statement of opinion in areas where the facts are controversial or do not exist. The contents can not reasonably be presumed to be error free.

I do not advocate the use of any particular form of health care. I am not offering a diagnosis or suggesting treatment for any specific individual. This book is not a substitute for medical advice or counseling and is not intended as such. Each person's own particular situation is unique.

This book relies heavily on statistical concepts and analysis. Due to the very nature of statistics, only probabilities can be determined. Nothing is certain. Also, the probabilities are no better than the data used to calculate them, which is imperfect to start with. Due to both the statistical nature of this work and to the imperfection of the available data, which is beyond the control of the author, exceptions to the rules stated herein can reasonably be expected to occur. The author can accept no responsibility or legal liability if you turn out to be one of those exceptions.

The author assumes no liability or responsibility for any loss, damage, or injury caused or alleged to be caused directly or indirectly by the information contained in this book or information omitted from it. This book is sold with the understanding that the author is not liable for misunderstanding, misinterpretation, misuse or misapplication of the information in it. Use of the information in this book is at the sole risk of the reader. There is always some risk involved in health care. Please do not use this book unless you are willing to assume that risk.

Table of Contents

Acknowledgments

I would like to thank Darrel Hickok. MD and *Doctor's Data, Inc.*, for their kind permission to use their hair test report. I would also like to thank Lori Knowles for facilitating the hair test form permission. I would also like to thank Jerry Deutsch, LN, Mukesh Suthar, MD, Gordon Downie, MD, PhD, Carol Wester, CNS and Julia Trudeau, DC for thoughtful reviews of the book in draft form. I am grateful to Jim Adams, Ph. D. for sharing unpublished data with me. I also very much appreciate the efforts of Craig Barth, Jessica Kerger, JD and Alicia Spear, MS who assisted with selection and clarity of content and conveyal of ideas, each in their own way. Very special thanks and appreciation are due to Kitt Carlton-Wippern, Ph.D. for both great assistance with the statistical discussion as well as thorough and thoughtful review of the draft manuscript. Thanks are also due Julie Anderson, ARNP, Linda Lacy and Moria Merriwhether for special contributions to this work. Of course I am also very grateful to all the people who were willing to give me permission to publish their hair test results and backgrounds as the examples in this book. These people's kindness and generosity in sharing their stories and test results made it possible to write this book. As confidentiality is promised, customary, and expected they can not be thanked by name but I am sure the reader can appreciate my deep gratitude to them.

Preface

This book explains how to properly interpret a hair test and gives specific examples of how to do this using *Doctor's Data Laboratories* and *Great Plains Laboratories* tests.

This book is intended for a broad audience, including physicians, other kinds of licensed health care practitioners, and laymen. Physicians have their own special language to describe health care concepts. While this language is clear, precise, and standardized, it is also not understood by laymen. In order to make this book accessible to as many people as possible, I have used common laymen's terms for health care concepts and conditions in most cases. I have mentioned the proper medical terminology in many cases but it simply isn't practical to do this throughout the book.

This book is a practical book. It provides detailed academic style information only in those areas where that is important so readers can understand things in order to be able to use the methods described in this book. For example, I do go through the details of how to calculate probabilities so that people who routinely use tests with a different format can come up with their own counting rules for the labs they use. In other areas little information is provided in the interest of brevity. I cite references only when you might actually want to verify them or check out what else they say, not simply to produce a long important looking academic style bibliography that nobody will ever read.

Because the book is intended to be useful in a practical sense, some material is repeated if it is relevant in several sections so that people don't have to go flipping through the book when they are trying to look up one specific thing. For example, the effects of low magnesium are repeated under all the toxic elements that can cause magnesium levels to be reduced, and how to determine the amount of vitamin B-6 to use is repeated wherever relevant.

The book starts out telling you how to interpret a hair test for mercury and other heavy metals. I provide a step by step discussion of this with figures to make it easy to do.

Next, I actually interpret some example hair tests that are particularly compelling in terms of them being real people diagnosed with horrible diseases that have no good treatments who turned out to have a heavy metal problem instead. This should provide some motivation to study the rest of the book carefully.

Following the introductory examples are some graphs giving my best estimates[*] regarding what fraction of sick people have heavy metal poisoning or other specific problems underlying their conditions. This is very useful in understanding why a hair test is worth doing early on.

Having explained how mercury affects hair test results, I discuss the other toxic element, how useful hair testing is for checking them, how likely the test is to be falsely normal[†] or falsely positive[‡] (and why this might happen), and give a brief description of what kinds of things happen with people who have too much of each toxic metal in them.

Once I have discussed the toxic elements, I discuss the meaning of the essential element levels as measured in hair. High hair levels do not always mean high body levels – for example, hair zinc levels go up when the body can't hold it in. Some of the essential elements can be toxic if they build up to a high enough level. I discuss briefly what the essential elements do for us, and what happens to people who end up with toxic levels of them.

Then I provide information on what other kinds of information you can derive from a hair test regarding how someone's body is working, such as whether they have too much stress, poor digestion, have adrenal issues or thyroid problems.

Following this I provide many, many example hair tests – actual case studies – with interpretation and background information as to what is going on with the person whose test it is. All of the tests used in this book are real tests from real people and accurate information is provided about what was going on with them, what they did, and what happened.

After the examples of useful, informative hair tests I provide examples of why you need to make sure the right test is ordered and the right information written on the test order form by showing an assortment of hair tests for a family of four where some tests were ordered improperly and the interpretation changed when a corrected test report based on accurate age, sex, and type of test desired was issued.

Following this, I provide a very brief discussion of what to do based on the test results, e. g. how to chelate heavy metals. Hopefully this discussion will help you avoid the use of harmful or ineffective protocols, but it is BRIEF and necessarily omits much detail. Anyone wishing to actually pursue any of these would be wise to consider researching other materials[§].

Finally, I provide an appendix with the equations and numerical probabilities that underlie the counting rules approach, as well as an appendix providing information on how to deal with tests from other laboratories if you cannot possibly get a doctor's data or great plains laboratory test.

Peer review is considered an important part of modern medicine. This book has been peer reviewed by relevant experts and I have addressed all the reviewers' concerns.

[**] Made by comparing published information from which underlying causes can be determined with patient reports regarding success or failure of proper detoxification to bring significant lasting improvement. Since outcome studies for alternative medicine treatments based on politically incorrect diagnoses (for example, how many autistic children are cured by mercury chelation) are not yet available in the mainstream medical literature I rely on patient or caregiver reports when these are numerous and informative enough to generate good statistical data.

[†] This is referred to as beta error in statistics.

[‡] This is referred to as alpha error in statistics.

[§] e. g. *Amalgam Illness: Diagnosis and Treatment*, see www.noamalgam.com.

Introduction

For historical reasons, health care providers believe that poisoning occurs on the job and all they have to do is ask about people's work to make sure their patients aren't poisoned. In one way modern medicine is a victim of its own success – through intensive occupational and public health efforts, very few people get poisoned at work any more. Today most people get poisoned at home.

Unfortunately most health care providers are only familiar with occupational poisonings. They are not familiar with what kinds of tests to order for other kinds of exposures or how to interpret those tests. This wouldn't be a problem if the test results were obvious and straightforward. Unfortunately they are not. Blood and urine tests often are normal even in very poisoned people. Hair tests are hard to interpret because one of the poisons – mercury – actually damages how the body moves minerals around and makes the test results wild. Crazy. Out of control

In fact, saying that mercury makes hair mineral test results go out of control is a good way to view it. "In control" and "out of control" are important engineering concepts in statistical process control of the sort most popularly associated with Deming. Process engineers derive some rules to draw process control charts so that someone at a production facility can measure each widget as it comes off the line and quickly determine whether an off size part is just a fluke or if there is a manufacturing problem that requires the line to be shut down and everything recalibrated[1]. I derived some statistical rules for checking whether mineral transport is in control or out of control on a hair test in the same manner that the process engineer does this for his facility. In this engineering analogy, each human being is a "production lot" of assorted mineral transport proteins, and I am able to use a hair test measure the performance of 23 of these mineral transport proteins in that particular person/production lot to determine whether the mineral transport process in that "production lot" is in control or out of control.

I applied these rules to many tests and found them to correlate very well with what was going on with the people. I also found them to correlate quite well with whether people would improve on chelation or not.

I didn't derive all these statistical rules and then go looking at hair tests. Instead, I compiled information obtained from

[1] Statistical process control is discussed in the *Review for the Professional Engineers' Examination for Industrial Engineers*, revised edition, Donovan Young, PhD, PE, Engineering and Management Press, Norcross, GA in section 4.3.1 on page 126. I derive different rules since I am not taking serial measurements but rather 23 measurements at once. The counting rules apply the same criteria used in the process control charts presented therein, of checking for how many points are outside two sigmas (in the red), how many are outside one sigma (not in the middle band) and how many are on one side of the mean. Given the way that hair test results are reported there is no convenient way to check whether there is a measurement outside three sigma, and in a hair test a three sigma result means that the "out of control" issue is that there is a genetic variant in one particular mineral transport protein rather than that there is the generalized derangement of mineral transport mercury induces.

several alternative doctors (or their patients' renditions of their comments) regarding which hair tests were "toxic" and which weren't. I noticed that indeed, the tests did somehow look different. Clearly these tests were indicating something was wrong and a few talented alternative physicians had learned to intuitively recognize the mineral patterns diagnostic of the underlying problem. What I did was use equations to reduce this intuitive art to a science, where a few simple rules allow you to learn how to check for mercury problems in minutes.

I have seen a very large number of tests of sick people who thought they might be toxic by consulting to people's health care providers on their cases, and by having tests sent to me by health care providers who wanted to know what I thought, and by seeing tests posted on various internet listservers, as well as by having people just share their tests with me.

I have also seen a great number of hair tests from normal healthy people – something most health care providers don't get the opportunity to do[2]. I've reviewed hundreds of tests that the students at a local medical school did on themselves and discussed health histories with many of the students whose test results did not come back completely normal (about 80% did come back normal). I've seen 100 tests with detailed health histories for students, staff and others at a different medical school which were specifically gathered to include mostly normal healthy people[3]. I've also seen the hair tests for a research study on results from autistic children versus normal control children. Most of the tests that are supposed to come out normal actually do according to the counting rules. The ones the counting rules say are "out of control" are usually associated with a health history that is also out of control.

In addition to this I have also seen a large number of cases where someone is sick, their hair test says they are poisoned, and they get better when they use an appropriate treatment to remove the poison. In addition to getting better, the people who repeat the hair test afterwards also find it is normal. People who don't detoxify properly usually find their hair test continues to show toxicity every time they repeat it. Several examples of hair tests improving with chelation (pages 182 to 192) and not improving without chelation (pages 175 to 181) are included in this book.

The interesting part of this is how often it turns out that people have the wrong diagnosis. A lot of times their health care provider tells them they have some or other horrible and incurable condition, but their hair test tells them that they are poisoned. The kinds of poisoning that hair tests can pick up are usually treatable.

For example, the schizophrenic boy whose case is discussed on pages 47-50 is greatly improved from chelation. Mercury poisoning is a well known cause of psychosis. It is seldom checked for once the doctor finds out the schizophrenic person doesn't work in a mercury using factory. This boy was quite lucky that his parents knew to check for mercury when he became schizophrenic at age 16. The schizophrenic woman discussed on pages 50-52 was not so lucky. She also became schizophrenic at age 16. She spent the next 40 years insane, in and out of institutions, on medicine with horrible side effects, her life totally disrupted. Someone finally thought to do a hair test on her at age 56.

Parkinson's disease is a slow, degenerative neurological disease that ends in death. There is no medication that slows its progression[4]. It is well known that mercury poisoning can be misdiagnosed as Parkinson's disease.

---

[2] Normal healthy people don't make doctor's appointments and then pay for hair tests.

[3] This information being gathered for a research project on hair test analysis.

[4] A recent study does suggest that the over the counter nutritional supplement coenzyme Q-10 can slow the progression of Parkinson's disease.

When the gentleman whose situation is discussed on pages 34-36 got a hair test he found out he had mercury poisoning, not Parkinson's disease. Unlike Parkinson's disease, mercury poisoning can be cured and this gentleman is now fine.

Mercury is known to cause autoimmune disease in laboratory animals and to accelerate the disease in animal models of lupus used for research purposes[5]. Despite this scientific knowledge, most health care practitioners do not consider mercury relevant to lupus. The woman whose situation is discussed on pages 205-206 was lucky enough to see someone who did a hair test – and found out she has mercury poisoning, not lupus.

There is also much evidence that multiple sclerosis can be due to mercury[6]. The story on pages 201-204is about a gentleman who experienced 20 years of misery before he discovered he really had a mercury problem, not MS.

This kind of story is repeated many times in this book. Autism, cancer, attention deficit, chronic fatigue, fibromyalgia, and many other conditions sometimes turn out to be heavy metal problems.

Another story I run into repeatedly is desperate sick people and their frustrated health care providers having gotten a hair test they don't know how to interpret. When mineral transport goes out of control and becomes deranged hair tests become quite difficult to figure out. After talking to frustrated people enough times I learned how to explain hair test interpretation and teach them how to do it.

I've also learned how confused people get themselves by reading journal paper abstracts they find on PubMed via the internet. Many journal papers actually say something quite different than what the one paragraph abstract when you read the entire paper, and that many papers simply got things wrong[7]. I know that it is a lot more helpful to read textbooks when trying to understand basic concepts and well established facts. For example, there are a very large number of papers on the effects of, say, lead or mercury on each of the thousands and thousands of different biochemical systems and metabolic intermediates in our bodies, but you don't find descriptions of what poisoned people are like in these papers.

---

[5] Low-Dose Exposure to Inorganic Mercury Accelerates Disease and Mortality in Acquired Murine Lupus, Via *et al.*, *Environmental Health Perspectives* volume **111**, pages 1273-7 , 2003.

[6] Evidence that Mercury from Silver Dental Fillings may be an Etiological Factor in Multiple Sclerosis, Siblerud and Kienholz, *Science of the Total Environment*, volume **142**, pages 191-205, 1994; and Cerebrospinal Fluid Protein Changes in Multiple Sclerosis After Dental Amalgam Removal, Huggins and Levy, *Alternative Medicine Review*, volume **3**, pages 295-300, 1998.

[7] I can absolutely attest from personal experience (and years of frustration) as a Ph. D. scientist performing research in chemistry, materials science and space engineering how easy it is to obtain misleading, wrong or uninterpretable data in trying to do experimental research. The frustration and torment of doing real research is something health care providers don't experience in school so they tend to fall into two camps – they either believe everything they hear, or they only believe politically correct things that are indexed in PubMed (that is they are alternative or mainstream in their views). When interpreting real data in other fields, reasonable people may and often do differ. Science depends on evaluation of ALL facts, which specifically includes your own experiences and experiments, experiences and experiments friends and colleagues tell you about, and everything in the literature that you think is accurately reported. Because everyone has different experiences and different views on which papers are right and which wrong when the published reports don't agree, all scientific advances are controversial for many years before becoming generally accepted. Unfortunately some health care providers aren't sophisticated enough in their understanding of science to be aware of how uncertain and limited current knowledge can be even when it is widely accepted as accurate and true.

The symptoms of poisoning are described in textbooks and review papers. I also know that the most recent textbook is not necessarily the clearest source. Often older textbooks contain a section written by someone who didn't have preconceptions about what he was or wasn't seeing, and who turned out to not only be an observant clinician but also a lucid writer. Most modern textbook sections are written by authors who simply read this older literature rather than relying on their own observations – you'll get a lot more out of reading the original description yourself than you will out of reading someone's "Cliff Notes®" version of it. Medicine, like science, is not about fashion and fad. If something was true 50 or 100 years ago it is still true today.

People also don't appreciate the limitations of PubMed (or any other literature tool). In addition to using a lot of literature tools (*Chemical Abstracts*, *Gmelin*, *Beilstein*, *Science Abstracts A*, *Science Citation Index*, *Dialog's* various databases, assorted patent databases, *Bilogical Abstracts*, *Index Medicus* – the paper precursor to PubMed) I worked as a library assistant during graduate school and the librarian took an interest in explaining how these literature tools were developed and what they covered. PubMed was developed to make it easy for health care providers to find some relevant paper that their library had and that was considered "mainstream." It does NOT cover all literature, it is not intended as a means of making a thorough literature search, and it specifically excludes from coverage most non-MD literature. For example, very few chiropractic, optometric, osteopathic or homeopathic journals are indexed. One of the criteria for selecting which journals go into PubMed is that they cite mostly other journals that are indexed in PubMed. This decision makes PubMed indexed journals a self referential set that excludes other points of view – that is, they reflect only mainstream health plan and pharmaceutical company medicine.

Another general problem in the medical literature is that it is "sanitized" of historically relevant information needed to put today's attitudes regarding heavy metal poisoning in context. For example, not one textbook or medical journal paper on Acrodynia – epidemic childhood mercury poisoning – mentions that many parents used mercury containing teething powders on their children because it was recommended by their pharmacist or doctor. None of the papers on the seed grain poisoning disasters such as those in Iraq, Pakistan and Central America mentions that the grain containing mercury was provided by US foreign aid agencies and continued to be provided as mass poisoning disasters followed one after another. These and many other examples where expert advice and institutional oversight led to prolonged periods of mass poisoning are relevant. Many health care providers would humbly accept that current practices may miss some cases of heavy metal poisoning if they were aware of such history. Since the providers never see this kind of information they reasonably assume that medicine has not made such mistakes in the past and is unlikely to have recently started doing so. After all, do we not assume we have a much better level of scientific understanding regarding basic biochemistry and heavy metal toxicity than existed even 20 years ago?

Political correctness is not as big a problem as the simple reality that research is hard to do and many journal papers do not report useful information. Frankly, in a lot of the material I see in both mainstream and alternative medicine I have reason to believe the authors did not actually read or evaluate the papers they cited. In some cases they may have been looking for an abstract to use as an excuse for the story they want to tell rather than to try to gain any insight into what the truth is. For example, I could cite a particular paper[8] that states it shows how hair element testing distinguishes

[8] Trace Element Concentrations in Hair from Autistic Children, Wecker, Miller, Cochran, Dugger and Johnson, *Journal of Mental Deficiency Research*, volume **29**, pages 15-22, 1985.

between autistic and normal children with high accuracy. However, on analyzing the data presented in this paper it is clear there were methodological problems. First, the data on how much of each element is in the hair of normal and autistic children does not match what anyone else finds. Second, the authors use a discrimination methodology that is essentially guaranteed to separate the normals and autistics – first they use their limited data to generate equations with a lot of free parameters, then they apply these equations to their data. The equations of course work fine on the data they were generated from. There is no real reason to believe they'd work on any *other* data. I don't cite papers that do not stand up to rigorous evaluation even if they appear to support my position, nor did I use them in drawing the conclusions presented in this book. I cite a few information sources that are relevant, that after careful study I believe are accurate, and that may be worth your while reading. When I cite textbooks as authority, I use actual quotes and give you the citation so you can verify that I did not quote them out of context. I'm trying to provide you with accurate information you can use. I am not trying to use any available citation or literary device to support my arguments or to convince you to see things my way.

By using this approach of identifying sources of accurate information without regard to where they are indexed or made available, I am able to provide you with an authoritative book on the subject of hair test interpretation. Neither hair testing nor heavy metal poisoning from non-occupational sources are covered well in PubMed or in medical textbooks. This book is the result of finding all the relevant information wherever it was and verifying it was accurate by cross checking it against real hair tests as well as the situations of real people I am familiar with through case consulting, internet discussion groups or other means.

When you need accurate information beyond that provided in this book, I recommend reading textbooks and journal papers. Abstracts and brief quotations won't do. Old texts are economical and may conveniently be obtained through internet services such as ebay.com® and amazon.com®. They may also be found at most academic libraries and many major public libraries. Older textbooks are particularly informative regarding how the toxins make people act and feel (what health care professionals would call the clinical presentation of the toxic patient and the signs and symptoms observable by history taking and physical examination). Older textbooks and review papers are more concerned with the kinds of things a regular person can figure out, while the newer ones are more concerned with the kinds of things it takes a hospital and a research laboratory to understand.

Like an older text, this book is focused on what someone can figure out for themselves or what a doctor in private practice can figure out in his clinic. Unlike the old medical text, this book has the advantage of integrating recent biomedical advances with information on how people act, feel, and what happens to their bodies so you know what happens, why, and what to do about it.

Using this book you can get a hair test and figure out whether you have a heavy metal problem instead of ALS (Lou Gherig's disease, pages 44-46), a mentally retarded child (pages 216-217), interstitial cystitis (pages 132-134), colitis (pages 160-161), irritable bowel syndrome (pages 151-153), or allergies (pages 145-147). You can determine what kinds of interventions are suitable for any toxic element you have accumulated. In some cases you will even find a description of a real person who had a similar situation. This is the kind of information you need if you have a condition where your health plan doctor can't offer much more than drugs, drugs and more drugs to control your symptoms, then further drugs to control the side effects of the drugs you already are taking. This book contains the kind of information you can use to get yourself, your family member or your patients well.

Why Hair Tests can be Confusing

The fact that mercury causes deranged mineral transport appears to not be widely known in the medical community, leading some doctors (both alternative and mainstream) to view hair element tests as unreliable. For example, Dr. Charles Gantt MD Ph.D. has this to say on page 126 of his book *ADD and ADHD, Complimentary Medicine Solutions* (MindMender publications Syracuse, NY ISBN 0967573505).

"Even though the nutrient minerals are reported in these panels, I do not generally make conclusive clinical decisions based on them, because I have found the nutrient elements to be subject to wide variations and distortions."

This quotation discusses a *Great Smokies Diagnostic Laboratories* hair test shown as an example in Dr. Gantt's book. Someone familiar with the concept of mercury deranged mineral transport would recognize the signs of deranged mineral transport in the test results. This means that the child in question has mercury poisoning, and the levels of the toxic and essential elements are not a good guide to the actual amounts.

The most confusing issue in hair testing is a failure to understand that essential and toxic mineral levels usually do not reflect body inventory when mineral transport is deranged. Deranged mineral transport is relatively common in the population that ends up getting hair tests.

Even when mineral transport is orderly rather than deranged, some confusion flows from the fact that hair levels of certain elements (e. g. zinc, calcium) move contrary to body stores, while hair levels of other elements (e. g. sodium, potassium, iron) simply do not reflect body stores at all. Proper test interpretation requires understanding the specific behavior of each individual element.

Some further confusion is generated by the lack of a standard definition of "reference range" between different laboratories, and variations between alternative laboratory "reference ranges" and mainstream laboratory "normal ranges." For example, the red zone in a *Great Smokies* test corresponds to the yellow zone on a *Doctor's Data* or *Great Plains Laboratory* test.

Thus on a perfectly normal hair test performed by *Doctor's Data Laboratories* from a completely healthy person one might find a few toxic elements in the yellow as well as a few essential elements high or low in the yellow and one just making it into the high or low red. If you just throw this test in front of a doctor who has never seen such a thing before and demand he explain it, he might get confused and think the results all ought to be in the green. He might think the yellows and especially the red indicate real problems. The nice, colorful test report plotting the results of 39 elements contains a lot of information and it takes a while to learn how to interpret it. Once learned, the visual and colorful report is very easy to interpret and far more informative than a list of numbers flagged with "high" and "low."

Unlike blood counts and metabolic panels, few doctors were drilled and quizzed and tested incessantly in medical school about how to interpret hair tests. Countless hours are spent studying the physiology and pathology underlying how the results on standard tests can come out. For example, the MCV, or mean corpuscular volume, can be elevated in some conditions. This literally means the red blood cells are oversized. Big. Fat. However doctors know that this does NOT mean you are eating too much and you need to go on a diet so your blood cells lose weight!. Similarly a low MCV does not mean you need to eat more and fatten up your blood! However, when first viewing a hair test, both laymen and health care providers not trained in hair test interpretation may draw equally irrelevant conclusions. For example, elevated potassium does not mean you need to stop eating so many fruits and vegetables, and elevated calcium does not mean you should stop drinking milk. In fact these two results when they occur together mean your thyroid may be underactive, and high calcium by itself usually means that you need MORE calcium, not less.

Without extensive background information both health care professionals and laymen can easily draw the wrong conclusions from hair test results. This book is an attempt to provide enough background that the information in a hair test can be understood and put to use.

How Mineral Transport Works

Minerals, amino acids, and other substances do not just stir around in your body on their own, mixing around and diffusing wherever they want to go. Your body has cell walls and other barriers that keep these water soluble things on one side or the other, and many sophisticated mechanisms to move them back and forth across these barriers to where you need them.

The counting rules will be discussed in the next section, but first let's talk about how hair grows, how minerals move around, and how much mineral levels vary in different parts of the body.

Hair is not a secreted substance. It is dead hair cells, which grow in the hair follicle, are pushed out of it as the hair elongates, dehydrate, and leave their keratinous corpses as the building material of the hair shaft. Hair is the fastest growing body tissue under most circumstances.

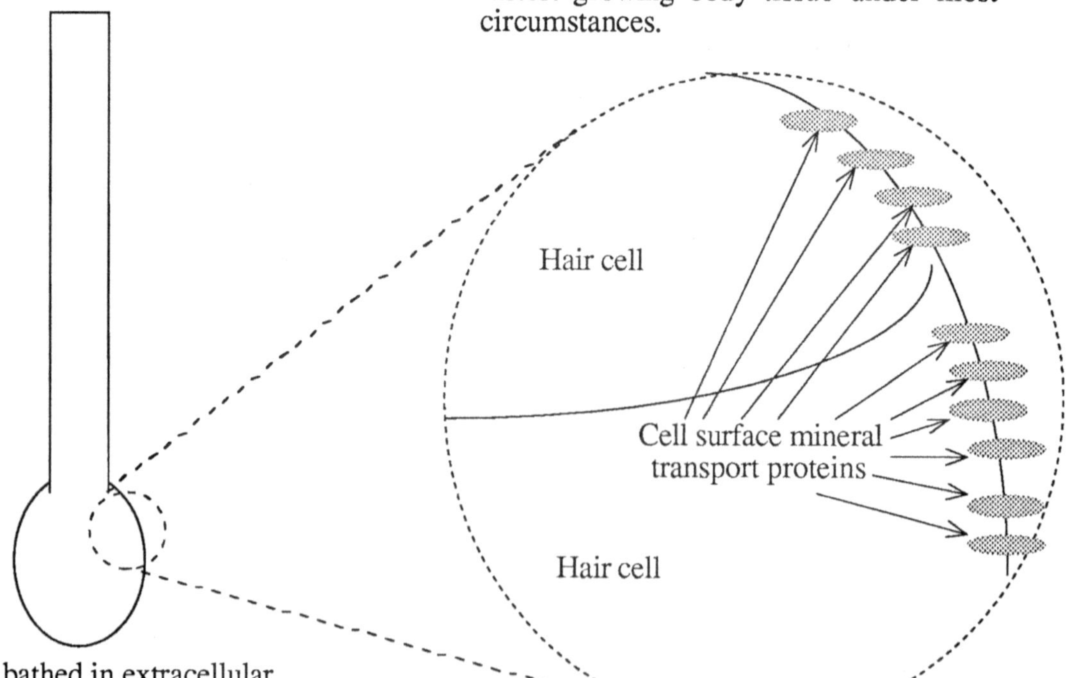

Hair root bathed in extracellular fluid which is of similar composition to blood plasma.

Minerals are actively absorbed from the intestines at various specific places along their length, actively secreted into bile by the liver, actively reabsorbed from the urine during kidney function, and actively pumped from the blood plasma or extracellular fluid into the various cells including the hair cells.

Examples of the mineral content of various body substances[9] on a common basis, mcg/g[10].

Mineral	Plasma / extracellular fluid/serum	Cytoplasm of "typical cell"[11]	Red blood cells	Urine[12]	Hair	Range of Variation[13]
Sodium (Na)	3,300	230	——	2,800	66	49 X
Potassium (K)	160	5500	79.0	1,600	40	140 X
Calcium (Ca)	96	0.0004[14]	16.0	140	640	1,600,000 X
Magnesium (Mg)	29	1400	620	96	75	48 X
Phosphorus (P)	120	2300	600	680	320	19 X
Zinc (Zn)	1.0	——	11	0.33	160	470 X
Copper (Cu)	1.0	——	0.680	0.023	15	660 X
Chromium (Cr)	0.32	——	0.029	0.055	0.28	11 X
Cobalt (Co)	0.63	——	——	0.0022	0.021	290 X
Manganese (Mn)	1.7	——	0.014	0.0020	0.31	840 X
Molybdenum (Mo)	0.00078	——	0.00100	0.056	0.040	72 X
Selenium (Se)	160	——	0.27	0.13	1.3	1,200 X

---

[9] For an adult.

[10] The values used in the table come from the following sources:
Na K Ca Mg P in extracellular fluid/plasma: *Guyton's Textbook of Medical Physiology.*
Zn and Cu in serum: Labcorp normal ranges, geometric mean, which agree well with values listed in *Tietz Textbook of Clinical Chemistry.*
Na K Ca Mg and P in cytoplasm: *Guyton's Textbook of Medical Physiology.*
Cr Co Mn Mo Se in serum: *Tietz Textbook of Clinical Chemistry.*
Red blood cells, all elements: *Doctor's Data* red blood cell elements test, geometric mean, cross checked with Great Smokies red blood cell elements reference ranges.
Urine, all elements: *Doctor's Data* 24 hour urine essential elements test, geometric mean, cross checked with *Tietz Textbook of Clinical Chemistry* and with Great Smokies urine test ranges.
Hair: *Doctor's Data* hair element profile, geometric mean, cross checked with *Tietz Textbook of Clinical Chemistry* and with *Great Smokies Diagnostic Laboratories, Inc.* and *Trace Elements Inc.* hair test ranges.
NB: The geometric means of the stated reference limits on the tests used corresponds to the 50th percentile, the arithmetic mean does not.

[11] According to *Guyton's Textbook of Medical Physiology*, ninth edition.

[12] Mcg/mg creatinine, which is numerically close to mg/l or mcg/g urine for those with normal fluid regulation.

[13] This is the ratio between the concentration in the tissue with the highest amount and the concentration in the tissue with the lowest amount. If these elements were just diffusing around the body the ratios would be about the same for elements with similar chemistry, for example sodium and potassium should have the same ratio, while calcium and magnesium should have the same ratio as each other.

[14] Note that there are very high concentrations in certain organelles, but calcium is low in cytoplasm.

The concentration variation in all these elements between different body fluids is significant. It means that these elements aren't just hanging around. Your body isn't letting these minerals come in, stir around your body, and go out in a passive manner. Your body isn't letting your mineral levels be controlled by how much is in your diet. Your body works very hard to regulate mineral levels in the different tissues, maintaining them within specific ranges needed for proper metabolic function in each type of cell. Your body does this because it has to – without well controlled levels of these essential minerals where they need to be, your health suffers. Because of this, deranged mineral transport directly impacts your health. Mercury causes the most widespread derangement of mineral transport, though other toxic elements may affect a few specific minerals (for example arsenic and antimony affecting magnesium, thallium affecting sodium and potassium, many toxins causing iron deficiency anemia).

The plasma and extracellular fluid have similar compositions.

Each cell expresses some subset of the available cell surface mineral transport proteins.

Passive diffusion of ions through the intact, healthy cell membrane (and the membranes of the intracellular organelles) is slow compared to active transport. If active transport fails for a few minutes, ischemic damage occurs due to unopposed passive diffusion and consequent osmotic disturbances.

There are three types of transport:

Primary active transport, in which ATP is used to move the minerals against a concentration gradient,

Secondary active transport, in which one type of ion (mineral) flows from high concentration to low concentration in order to move another ion against its concentration gradient from low to high concentration, and

Facilitated diffusion, which is typically in the form of ligand or electrically activated gated channels. These channels are normally closed.

Examples of things transported by the three different mechanisms:

Primary active transport – sodium, potassium, calcium, protons.

Secondary active transport – calcium, magnesium, potassium, protons, glucose, amino acids.

Facilitated diffusion – calcium, sodium, potassium, chloride.

Deranged Mineral Transport and Hair Test Interpretation

Mercury interferes with the process by which your body moves minerals around and regulates their concentrations. Some of the toxic effects it has are due to this, e. g. it interferes with calcium channels and can thus effect heart function.

While some other toxic metals are known to affect mineral transport, e. g. thallium disables the protein Na/K ATP-ase which uses ATP to pump sodium and potassium through cell membranes, and platinum in the form of the chemotherapy drug cisplatin interferes with proteins in the kidneys that recover magnesium from urine, only mercury affects a wide variety of mineral transport proteins in various different ways. One effect of this is that mercury levels may be normal in blood and urine and hair even when high concentrations of it are being held in some other part of the body. Another effect of this is that the hair element profiles of people with too much mercury in their systems for their own good often take on a characteristic appearance indicating deranged mineral transport. This can be identified using a simple technique I refer to as "the counting rules."

The counting rules are used to determine whether it is likely that mineral transport is deranged. Mercury causes deranged mineral transport in susceptible people. None of the other heavy metal toxins appear to do this. People who are toxic with some heavy metal OTHER than mercury but don't have a mercury problem will have normal (orderly) mineral transport along with, in most cases, a high level of that toxin. One exception to this is that long past exposures to lead sometimes give normal hair lead levels even in a very toxic person.

The most important thing to know about mercury induced mineral transport derangements is that it usually makes the hair, blood and urine mercury levels go DOWN. Thus the toxic people, the ones with lots of mercury in their bodies, often have lower mercury test results than healthy people. This is what makes tests hard to interpret – without the special technical knowledge of how to identify deranged mineral transport, laymen and unsophisticated health care providers aren't able to use laboratory tests to check if someone has mercury poisoning.

While mercury causes deranged mineral transport, another possible cause of apparent derangement of mineral transport is pregnancy and lactation, which may sometimes lead to an "all low" presentation (which meets counting rule #1) for normal healthy women.

Genetic and chromosomal defects can affect mineral transport into the hair but often do not. Whether they do or not can sometimes be found in the mainstream medical literature (that is by searching PubMed), but often needs to be sought out through more modern sources such as the internet, contacting health care providers who specialize in alternative care for such cases, or having someone with a health care license[15] contact the various laboratories and ask if they have any information and experience relating to how hair tests from people with the particular condition in question turn out.

Some people who do not have a susceptibility to mineral transport derangement get exposed to very high levels of mercury and do become toxic from it, in which case their hair test shows orderly mineral transport and very high mercury levels.

Thus in interpreting a hair test you first check whether mercury is affecting mineral transport, and then if it is not, just look for which toxic elements are very elevated and accepting these as significant. Make sure that the age and sex on the test report match the person the test is for and that the hair wasn't grown during pregnancy or lactation.

If mineral transport is normal and there are no red range toxic elements, the test is most likely not informative. This does NOT necessarily mean the person does not have a heavy metal problem, but it does mean that the metal is no longer available to get into the hair. For example, see the case on pages 198-200.

With normal mineral transport, no very elevated toxic elements, and one vastly elevated essential element there is some possibility that a metabolic defect relating to the handling of that particular element is causing problems. These kinds of issues can be either genetic (inborn) or acquired. They are not particularly rare and most of them are not yet recognized and understood by mainstream medicine[16]. Once contamination of the hair sample and some unsuspected but huge exposure are ruled out, it is worth asking if the person's relatives want to get hair tests too and see if it is something that runs in the family.

A lot of people get excited about the bars that go into the yellow stripe. These are seldom significant. *Doctor's Data* – the lab all the examples in this book are from – has carefully studied large numbers of healthy people in order to derive accurate age and sex related reference ranges. The green stripe in the toxic elements section means 68% of the people fall in that range, and the yellow stripe means that 27% fall in it. It really isn't that unlikely that people will have a few yellow band results. If they have eight or more this is quite unlikely and you could worry about them having some sort of "mixed metal intoxication." If they have six or seven in the yellow band and these are all in the higher part of the yellow band, maybe possibly they might be significant. It isn't too likely but if nothing else is turning up then mixed metal problems might be pursued. Generally if you see a few elements in the yellow band it doesn't mean much. In fact, the probability of ending up with every single toxic element in the green is about one percent – you really do expect to see a few lines going into the yellow region on most normal healthy people's hair tests.

The exception I make to this rule is lead. People are exposed to a lot more lead nowadays than is good for them, and

[15] Federal and state laws prohibit the laboratories from talking to ordinary people, they are only permitted to talk to people with health care licenses. While this is inconvenient for you and makes it much harder to gather necessary information, please respect this because the people at the laboratories can get in a lot of trouble if they help you. Licenses like DC, OD, ARNP, ND etc. are acceptable and often such doctors are more willing to do this or to authorize you to call "from their office" than more traditional MD's and DO's.

[16] This is quite apparent from having seen lots of hair test results, and three interesting cases of it are presented in this book – inherited variations in the transport of copper, manganese and organic mercury.

even the very conservative mainstream medical literature has many papers showing that acceptable levels of lead cause decreased IQ, developmental problems in children, and other issues[17]. You would not expect yellow zone lead to cause profound problems, but it certainly isn't good for you and going to some effort to remove it may well be worthwhile. Also, since sometimes body lead is a lot higher than the hair levels indicate it is worth doing something about the lead if there isn't another obvious candidate for whatever is causing the problem that first led to the hair test being ordered.

Statistics don't give certainty, don't answer questions you don't write the proper equation to ask, and pieces of paper that come from laboratories are not perfect reflections of what is happening inside living human beings. You need to use some common sense in applying the counting rules. Someone with a very normal looking test and a toxic element result, let us say bismuth for an example, that goes almost but not quite to the red zone who also has all kinds of symptoms that just scream "BISMUTH!" may well have bismuth poisoning. Someone with a very similar looking test and no problems is fine. Hair tests (and many other laboratory tests) have far more meaning when considered in the context of what is going on with the person they are from.

Any intervention that actually improves a person's health (biochemical status) can make the hair test look healthier too. The most common intervention is nutritional supplementation, however bed rest as in a hospital or long vacation, lifestyle changes that make them feel better, diets that make them feel better, or prescription medications that make them feel better also generally improve the appearance of the hair test if it would have deranged mineral transport in the absence of these interventions. Hair tests must be interpreted thoughtfully in the context of what else the person is doing or did just before the hair tested was grown.

---

[17] Recent Developments in Low-Level Lead Exposure and Intellectual Impairment in Children, Koller, Brown, Spurgeon and Levy, *Environmental Health Perspectives*, volume **112**, number 9, pages 987-94, 2004. Downloadable on the web for free at http://ehp.niehs.nih.gov/members/2004/6941/6941.pdf. Note especially figure one on page 991 that shows how earlier studies were "cooked" using illegitimate statistical techniques to reduce the apparent impact of low level lead exposure on intelligence. Even with statistical game playing the results suggested that the current "safe" threshold for pediatric lead exposure corresponds to an average IQ loss of seven points. A more accurate estimate based on consideration of the data actually taken is twelve points.

Recognizing Deranged Mineral Transport

Mercury causes a derangement of the body's ability to transport and handle essential and other elements. This deranged mineral transport can be identified by using a hair test to check whether the distribution of essential and other nontoxic minerals in the hair is orderly or not.

Traditional medical laboratory tests are (supposed to be) standardized by testing a bunch of normal, healthy people and determining "normal ranges" where 1 person in 40, on average, is above them and 1 in 40 below[18], for a total of 2 in 40 or 1 in 20 who do not fall "within normal limits." By these standard criteria, 95% of normal healthy people's results fall within the normal range. Note that many "alternative" labs choose different reference ranges but do not make this clear on their reports. Make sure you are aware of what the ranges on any test you use really mean. *Doctor's Data* and *Great Plains Laboratories* provide actual percentiles on their hair test telling what the color bands mean[19].

With this statistical definition, if you do many tests as part of a panel, some of them are expected to randomly be out of range even if the person the test is for is completely healthy. It is possible to use statistical methods to turn the whole panel into "one test" that is either "within normal limits" or "abnormal." I have done this and generated the following rules to interpret the *Doctor's Data Inc.* (*Doctor's Data*) "essential and other elements" portion of their Hair Element Profile, which includes 23 elements. *Great Plains Laboratories* offers the same hair test as *Doctor's Data.* Corresponding rules can be derived for other tests, however the counting rules approach taken here becomes MUCH less sensitive as the number of essential and other elements falls, and it is also necessary to know what the reference ranges mean statistically which some labs are not forthcoming about. Thus if you are going to use the test for this purpose I'd suggest sticking to a *Doctor's Data* or *Great Plains* test or using a very similar test.

Also note that the *Doctor's Data/Great Plains* test has very well normed reference ranges broken down by age and sex and is thus fine for children. Some laboratories only have normal ranges for human beings, some break it down further. I believe *Doctor's Data* (and *Great Plains* have the most well broken down reference ranges for hair testing.

Another concern is if you have a test in hand such as a *Doctor's Data* toxic element profile which does not have enough essential elements reported on it to check whether mineral transport is

[18] For a gaussian distributed value this corresponds to $\pm$ 1.96 standard deviations.

[19] Also note that the labs discuss taking hair from a particular location on the head. They chose to standardize based on samples collected this way. However it is clear from seeing an adequately large number of tests that head hair from any location is just fine, as is pubic hair.

orderly or not. Without being sure the person has orderly mineral transport misdiagnosis is easy if they are presumed to be toxic with the elements showing high since these elevations may in fact be an artifact of deranged mineral transport rather than an accurate reflection of body inventory.

If you DO end up with a Hair Toxic Element Exposure Profile test from *Doctor's Data*, have your doctor call them immediately and ask them to run the Hair Element Profile. If they are notified within 30 days they do not need another sample. They will charge you for the second test unless it was clearly their error.

I have provided figures showing which parts of the *Doctor's Data/Great Plains Laboratories* tests to count for the different rules, and which parts to ignore while you are applying the counting rules to determine whether mineral transport is normal or deranged. It is necessary to check for deranged mineral transport before considering elevations, especially mild ones, in the toxic elements section to be significant.

This is not to say that these labs endorse this "counting rules" concept – they do not. In fact the counting rules approach is relatively new and many laboratories are not even aware of them yet. This book applies the counting rules to *Doctor's Data* tests because those tests happen to be readily adaptable to this purpose and I was able to obtain a lot of actual test results to use as example cases. This use of one particular laboratory's tests for all the examples doesn't mean there is something wrong with other labs' tests. The various laboratories developed their tests for specific purposes for which they are well suited. I am making a new use of hair testing through the counting rules interpretation.

Interpreting a *Doctor's Data* or *Great Plains* Hair Element Profile

1. Count the number of results in the essential and other elements section that are above "average," that is, above the 50th percentile line in the middle of the page. The bars that extent to the right from the middle of the page are above the 50th percentile. If this number is five or less the test result indicates deranged mineral transport.

Exceptions to this rule: normal healthy lactating women will sometimes exhibit an "all low" profile, satisfying rule 1, even though they are not toxic. This is most likely to occur if they eat a vegetarian diet, especially a vegan[20] diet. I also do not believe the test can be considered completely accurate for pregnant women. In these cases either a hair sample needs to be collected that was grown at a time the woman wasn't pregnant or lactating[21], or great caution must be used when interpreting the results. Preferably some means other than a hair test must be found to explore mineral transport or otherwise address the problem of figuring out what is wrong if the hair test suggests a pregnant or lactating woman has a mineral transport problem that meets counting rule #1, especially if she is vegetarian. People otherwise on a vegetarian diet do come up normal on the counting rules if they don't have a mercury problem.

2. Count the number of results in the essential and other elements section that are below "average," that is below the 50th percentile line in the middle of the page. The bars extending to the left are below average. If this number is five or less the test result indicates deranged mineral transport[22].

The following pages show a color picture of the *Doctor's Data* Hair Element Profile test report, blank test reports with speckling showing where to count for each rule, and a test report's essential element section marked up to show exactly which bars to count for each rule.

[20] Some vegetarians do eat eggs and dairy products. Vegans do not eat any animal derived products at all.

[21] This may be done by clipping the ends of hair long enough to have grown before conception. Hair typically grows at a rate of about 1/4 to 1/2" per month.

[22] Rules 1 and 2 are mirror images of each other and could be stated as one rule by saying "between 6 and 17 bars should go to the right (or to the left) or the test is abnormal." It is a lot easier to count it as described in rules 1 and 2, and also which rule is met affects how you interpret what you see in the toxic element section.

Please reference book cover for example of a color picture of the *Doctor's Data* Hair Element Profile test report.

This page intentionally blank

HAIR ELEMENTS

DOCTOR'S DATA INC.

Counting Rule #1 - count bars extending into speckled area

POTENTIALLY TOXIC ELEMENTS

TOXIC ELEMENTS	RESULT µg/g	REFERENCE RANGE	PERCENTILE 68th 95th
Aluminum		< 7.0	
Antimony		< 0.05	
Arsenic		< 0.06	
Beryllium		< 0.02	
Bismuth		< 0.1	
Cadmium		< 0.1	
Lead		< 1.0	
Mercury		< 1.1	
Platinum		< 0.005	
Thallium		< 0.01	
Thorium		< 0.005	
Uranium		< 0.06	
Nickel		< 0.4	
Silver		< 0.15	
Tin		< 0.3	
Titanium		< 1.0	
Total Toxic Representation			

ESSENTIAL AND OTHER ELEMENTS

ELEMENTS	RESULT µg/g	REFERENCE RANGE	PERCENTILE 2.5th 16th 50th 84th 97.5th
Calcium		300 - 1200	
Magnesium		35 - 120	
Sodium		12 - 90	
Potassium		8.0 - 38	
Copper		12 - 35	
Zinc		140 - 220	
Manganese		0.15 - 0.65	
Chromium		0.2 - 0.4	
Vanadium		0.018 - 0.065	
Molybdenum		0.028 - 0.056	
Boron		0.3 - 2.0	
Iodine		0.25 - 1.3	
Lithium		0.007 - 0.023	
Phosphorus		160 - 250	
Selenium		0.95 - 1.7	
Strontium		0.5 - 7.6	
Sulfur		44500 - 52000	
Barium		0.26 - 3.0	
Cobalt		0.013 - 0.05	
Iron		5.4 - 14	
Germanium		0.045 - 0.065	
Rubidium		0.007 - 0.096	
Zirconium		0.02 - 0.42	

SPECIMEN DATA

COMMENTS:

Date Collected: Sample Size:
Date Received: Sample Type:
Date Completed: Hair Color:
Treatment:
Methodology: ICP-MS Shampoo:

V06.99

RATIOS

ELEMENTS	RATIOS	EXPECTED RANGE
Ca/Mg		4 - 30
Ca/P		1 - 12
Na/K		0.5 - 10
Zn/Cu		4 - 20
Zn/Cd		> 800

This form is provided to Dr. Cutler to be used as an educational tool and does not constitute endorsement by Doctor's Data.

HAIR ELEMENTS

DOCTOR'S DATA INC.

Counting Rule #2 - count bars extending into speckled area

POTENTIALLY TOXIC ELEMENTS

TOXIC ELEMENTS	RESULT µg/g	REFERENCE RANGE	PERCENTILE 68th	95th
Aluminum		< 7.0		
Antimony		< 0.05		
Arsenic		< 0.06		
Beryllium		< 0.02		
Bismuth		< 0.1		
Cadmium		< 0.1		
Lead		< 1.0		
Mercury		< 1.1		
Platinum		< 0.005		
Thallium		< 0.01		
Thorium		< 0.005		
Uranium		< 0.06		
Nickel		< 0.4		
Silver		< 0.15		
Tin		< 0.3		
Titanium		< 1.0		
Total Toxic Representation				

ESSENTIAL AND OTHER ELEMENTS

ELEMENTS	RESULT µg/g	REFERENCE RANGE	PERCENTILE 2.5th	16th	50th	84th	97.5th
Calcium		300 - 1200					
Magnesium		35 - 120					
Sodium		12 - 90					
Potassium		8.0 - 38					
Copper		12 - 35					
Zinc		140 - 220					
Manganese		0.15 - 0.65					
Chromium		0.2 - 0.4					
Vanadium		0.018 - 0.065					
Molybdenum		0.028 - 0.056					
Boron		0.3 - 2.0					
Iodine		0.25 - 1.3					
Lithium		0.007 - 0.023					
Phosphorus		160 - 250					
Selenium		0.95 - 1.7					
Strontium		0.5 - 7.6					
Sulfur		44500 - 52000					
Barium		0.26 - 3.0					
Cobalt		0.013 - 0.05					
Iron		5.4 - 14					
Germanium		0.045 - 0.065					
Rubidium		0.007 - 0.096					
Zirconium		0.02 - 0.42					

SPECIMEN DATA

COMMENTS:

Date Collected: Sample Size:

Date Received: Sample Type:

Date Completed: Hair Color:

Treatment:

Methodology: ICP-MS Shampoo:

V06.99

RATIOS

ELEMENTS	RATIOS	EXPECTED RANGE
Ca/Mg		4 - 30
Ca/P		1 - 12
Na/K		0.5 - 10
Zn/Cu		4 - 20
Zn/Cd		> 800

This form is provided to Dr. Cutler to be used as an educational tool and does not constitute endorsement by Doctor's Data.

This form is provided to Dr. Cutler to be used as an educational tool and does not constitute endorsement by Doctor's Data.

HAIR ELEMENTS

DOCTOR'S DATA INC

Counting Rule #3 - count bars extending into speckled area

POTENTIALLY TOXIC ELEMENTS

TOXIC ELEMENTS	RESULT µg/g	REFERENCE RANGE	PERCENTILE 68th 95th
Aluminum		< 7.0	
Antimony		< 0.05	
Arsenic		< 0.06	
Beryllium		< 0.02	
Bismuth		< 0.1	
Cadmium		< 0.1	
Lead		< 1.0	
Mercury		< 1.1	
Platinum		< 0.005	
Thallium		< 0.01	
Thorium		< 0.005	
Uranium		< 0.06	
Nickel		< 0.4	
Silver		< 0.15	
Tin		< 0.3	
Titanium		< 1.0	
Total Toxic Representation			

ESSENTIAL AND OTHER ELEMENTS

ELEMENTS	RESULT µg/g	REFERENCE RANGE	PERCENTILE 2.5th 16th 50th 84th 97.5th
Calcium		300 - 1200	
Magnesium		35 - 120	
Sodium		12 - 90	
Potassium		8.0 - 38	
Copper		12 - 35	
Zinc		140 - 220	
Manganese		0.15 - 0.65	
Chromium		0.2 - 0.4	
Vanadium		0.018 - 0.065	
Molybdenum		0.028 - 0.056	
Boron		0.3 - 2.0	
Iodine		0.25 - 1.3	
Lithium		0.007 - 0.023	
Phosphorus		160 - 250	
Selenium		0.95 - 1.7	
Strontium		0.5 - 7.6	
Sulfur		44500 - 52000	
Barium		0.26 - 3.0	
Cobalt		0.013 - 0.05	
Iron		5.4 - 14	
Germanium		0.045 - 0.065	
Rubidium		0.007 - 0.096	
Zirconium		0.02 - 0.42	

SPECIMEN DATA

COMMENTS:

Date Collected:	Sample Size:
Date Received:	Sample Type:
Date Completed:	Hair Color:
	Treatment:
Methodology: ICP-MS	Shampoo:

V06.99

RATIOS

ELEMENTS	RATIOS	EXPECTED RANGE
Ca/Mg		4 - 30
Ca/P		1 - 12
Na/K		0.5 - 10
Zn/Cu		4 - 20
Zn/Cd		> 800

This form is provided to Dr. Cutler to be used as an educational tool and does not constitute endorsement by Doctor's Data.

HAIR ELEMENTS

DOCTOR'S DATA INC.

Counting Rule #4 - count bars extending into speckled area

POTENTIALLY TOXIC ELEMENTS

TOXIC ELEMENTS	RESULT µg/g	REFERENCE RANGE	PERCENTILE 68th 95th
Aluminum		< 7.0	
Antimony		< 0.05	
Arsenic		< 0.06	
Beryllium		< 0.02	
Bismuth		< 0.1	
Cadmium		< 0.1	
Lead		< 1.0	
Mercury		< 1.1	
Platinum		< 0.005	
Thallium		< 0.01	
Thorium		< 0.005	
Uranium		< 0.06	
Nickel		< 0.4	
Silver		< 0.15	
Tin		< 0.3	
Titanium		< 1.0	
Total Toxic Representation			

ESSENTIAL AND OTHER ELEMENTS

ELEMENTS	RESULT µg/g	REFERENCE RANGE	PERCENTILE 2.5th 16th 50th 84th 97.5th
Calcium		300 - 1200	
Magnesium		35 - 120	
Sodium		12 - 90	
Potassium		8.0 - 38	
Copper		12 - 35	
Zinc		140 - 220	
Manganese		0.15 - 0.65	
Chromium		0.2 - 0.4	
Vanadium		0.018 - 0.065	
Molybdenum		0.028 - 0.056	
Boron		0.3 - 2.0	
Iodine		0.25 - 1.3	
Lithium		0.007 - 0.023	
Phosphorus		160 - 250	
Selenium		0.95 - 1.7	
Strontium		0.5 - 7.6	
Sulfur		44500 - 52000	
Barium		0.26 - 3.0	
Cobalt		0.013 - 0.05	
Iron		5.4 - 14	
Germanium		0.045 - 0.065	
Rubidium		0.007 - 0.096	
Zirconium		0.02 - 0.42	

SPECIMEN DATA

COMMENTS:

Date Collected:

Date Received:

Date Completed:

Methodology: ICP-MS

Sample Size:

Sample Type:

Hair Color:

Treatment:

Shampoo:

V06.99

RATIOS

ELEMENTS	RATIOS	EXPECTED RANGE
Ca/Mg		4 - 30
Ca/P		1 - 12
Na/K		0.5 - 10
Zn/Cu		4 - 20
Zn/Cd		> 800

This form is provided to Dr. Cutler to be used as an educational tool and does not constitute endorsement by Doctor's Data.

Example of which bars count for rule #1
the ends of the seventeen bars that count are circled

ESSENTIAL AND OTHER ELEMENTS

ELEMENTS	RESULT µg/g	REFERENCE RANGE
Calcium	238	125- 370
Magnesium	20	12- 30
Sodium	6.0	12- 90
Potassium	16	12- 40
Copper	21	8.0- 16
Zinc	120	100- 190
Manganese	0.81	0.2- 0.55
Chromium	0.5	0.26- 0.5
Vanadium	0.055	0.03- 0.1
Molybdenum	0.11	0.05- 0.13
Boron	9.3	0.6- 4.0
Iodine	3.1	0.25- 1.3
Lithium	0.049	0.007- 0.023
Phosphorus	189	160- 250
Selenium	1.6	0.95- 1.7
Strontium	0.4	0.16- 1.0
Sulfur	50600	45500- 53000
Barium	0.27	0.16- 0.8
Cobalt	0.021	0.013- 0.035
Iron	22	8.0- 19
Germanium	0.059	0.045- 0.065
Rubidium	0.029	0.016- 0.18
Zirconium	5.0	0.04- 1.0

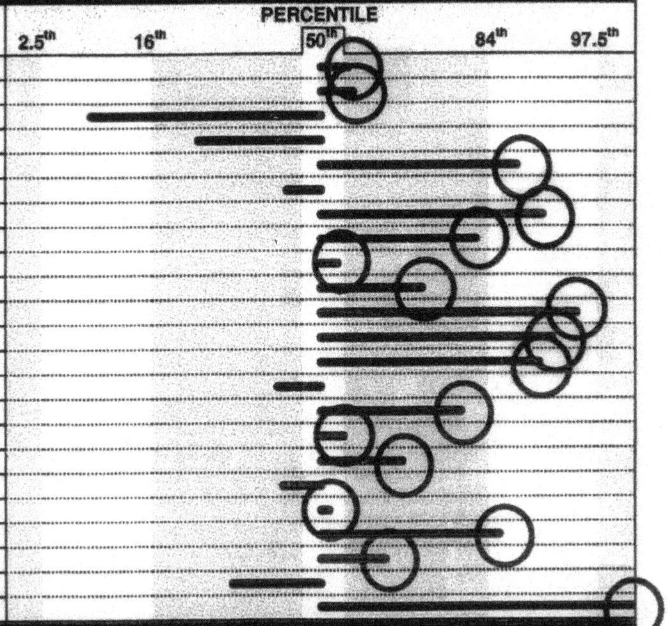

SPECIMEN DATA | RATIOS

Example of which bars count for rule #2
the ends of the six bars that count are circled

ESSENTIAL AND OTHER ELEMENTS

ELEMENTS	RESULT µg/g	REFERENCE RANGE
Calcium	238	125- 370
Magnesium	20	12- 30
Sodium	6.0	12- 90
Potassium	16	12- 40
Copper	21	8.0- 16
Zinc	120	100- 190
Manganese	0.81	0.2- 0.55
Chromium	0.5	0.26- 0.5
Vanadium	0.055	0.03- 0.1
Molybdenum	0.11	0.05- 0.13
Boron	9.3	0.6- 4.0
Iodine	3.1	0.25- 1.3
Lithium	0.049	0.007- 0.023
Phosphorus	189	160- 250
Selenium	1.6	0.95- 1.7
Strontium	0.4	0.16- 1.0
Sulfur	50600	45500- 53000
Barium	0.27	0.16- 0.8
Cobalt	0.021	0.013- 0.035
Iron	22	8.0- 19
Germanium	0.059	0.045- 0.065
Rubidium	0.029	0.016- 0.18
Zirconium	5.0	0.04- 1.0

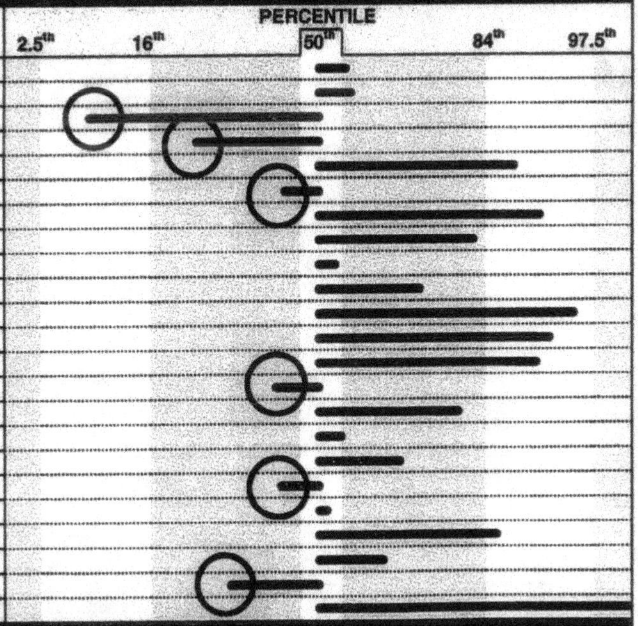

SPECIMEN DATA | RATIOS

This form is provided to Andrew Cutler to be used as an educational tool and does not constitute endorsement by *Doctor's Data.*

Example of which bars count for rule #3
the ends of the one bar that counts is circled

ESSENTIAL AND OTHER ELEMENTS

ELEMENTS	RESULT µg/g	REFERENCE RANGE
Calcium	238	125- 370
Magnesium	20	12- 30
Sodium	6.0	12- 90
Potassium	16	12- 40
Copper	21	8.0- 16
Zinc	120	100- 190
Manganese	0.81	0.2- 0.55
Chromium	0.5	0.26- 0.5
Vanadium	0.055	0.03- 0.1
Molybdenum	0.11	0.05- 0.13
Boron	9.3	0.6- 4.0
Iodine	3.1	0.25- 1.3
Lithium	0.049	0.007- 0.023
Phosphorus	189	160- 250
Selenium	1.6	0.95- 1.7
Strontium	0.4	0.16- 1.0
Sulfur	50600	45500- 53000
Barium	0.27	0.16- 0.8
Cobalt	0.021	0.013- 0.035
Iron	22	8.0- 19
Germanium	0.059	0.045- 0.065
Rubidium	0.029	0.016- 0.18
Zirconium	5.0	0.04- 1.0

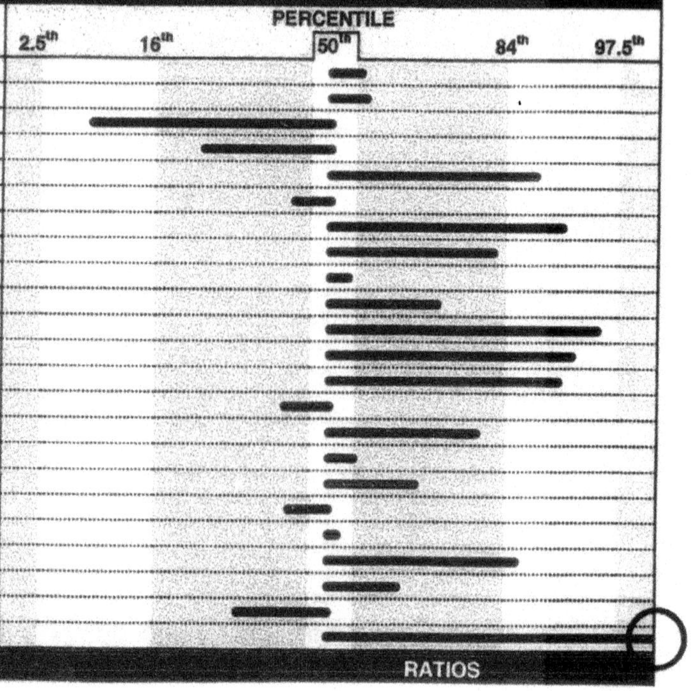

SPECIMEN DATA — RATIOS

Example of which bars count for rule #4
the ends of the fifteen bars that count are circled

ESSENTIAL AND OTHER ELEMENTS

ELEMENTS	RESULT µg/g	REFERENCE RANGE
Calcium	238	125- 370
Magnesium	20	12- 30
Sodium	6.0	12- 90
Potassium	16	12- 40
Copper	21	8.0- 16
Zinc	120	100- 190
Manganese	0.81	0.2- 0.55
Chromium	0.5	0.26- 0.5
Vanadium	0.055	0.03- 0.1
Molybdenum	0.11	0.05- 0.13
Boron	9.3	0.6- 4.0
Iodine	3.1	0.25- 1.3
Lithium	0.049	0.007- 0.023
Phosphorus	189	160- 250
Selenium	1.6	0.95- 1.7
Strontium	0.4	0.16- 1.0
Sulfur	50600	45500- 53000
Barium	0.27	0.16- 0.8
Cobalt	0.021	0.013- 0.035
Iron	22	8.0- 19
Germanium	0.059	0.045- 0.065
Rubidium	0.029	0.016- 0.18
Zirconium	5.0	0.04- 1.0

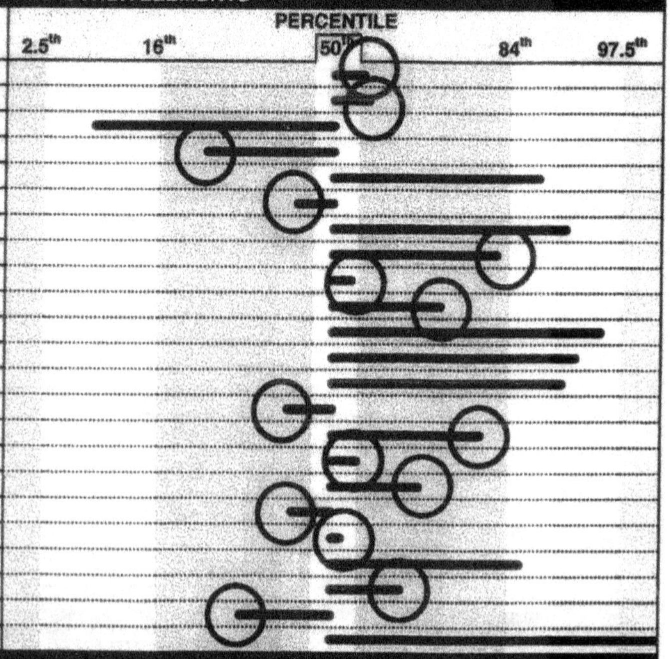

SPECIMEN DATA — RATIOS

This form is provided to Andrew Cutler to be used as an educational tool and does not constitute endorsement by *Doctor's Data.*

3. Count the number of results in the essential and other elements section that are all the way to either edge of the page, in the red zone, very elevated or very low. If this number is four or more mineral transport is deranged.

4. Count the number of results in the essential and other elements section that are in the middle band of the page (the green and white zone) with a short bar, which are neither very elevated nor very low. If this number is eleven or less mineral transport is deranged.

5. If any two of the above counts comes up one short of the rule, mineral transport is deranged (e. g. 3 in the red zone plus 12 in the middle band).

If any ONE of the above criteria is met, the test result is abnormal and shows derangement of mineral transport.

Vitamins, diet, medication and lifestyle can affect whether a hair test shows deranged mineral transport or not. Interventions that make people feel better often do so because they actually improve the physiological problems that were causing the people to feel bad. Hair tests done on people who are taking relevant medications, nutritional supplements, or performing other helpful interventions, often look better than if performed before the people started doing things that made them feel better.

If any one (or more) of these criteria is met, there is less than 1 chance in 40 that it happened randomly, and at least 39 chances in 40 that something which is changing the way the body handles minerals.

If mineral transport is abnormal, then the levels of the toxic elements in the hair do not necessarily reflect the amount in the body. What we know at that point is that there is a mercury problem. Other elevated toxic elements may or may not truly be a problem. Typically if copper is elevated (in the red zone, above 60 for adults) it is a real problem rather than artifactual. Elevations of aluminum, nickel, tin and titanium are frequently artifacts of the mineral transport problem rather than indications of those elements being a problem in and of themselves.

If rule 1 is met then very elevated potentially toxic elements may well be significant, and it is also possible that toxic elements the person has real problems with are not showing up high in the hair.

If rule 2 is met then very elevated potentially toxic elements are unlikely to have any meaning unless the toxic element is wildly more elevated than any of the essential elements.

If rule 3 or 4 is met then very elevated potentially toxic elements may or may not have any significance at all, and it is also possible that toxic elements the person has real problems with are not showing up high in the hair.

If rule 5 is met then the relevant combined caveats above apply but to a more limited extent.

If mineral transport is normal, then the potentially toxic elements in the hair test accurately reflect body stores and are a good guide to toxicity, with the exception that lead is sometimes falsely low in hair if the exposure was a long time ago.

Also if mineral transport is normal and there are one or two high red essential and other elements, the person may be toxic with those. Copper is the most common essential element to accumulate to a toxic level. Calcium and zinc are notorious for going up in hair when they are low in the body. Hair iron bears no real relationship to body iron levels.

In actuality, someone who is slowly being poisoned by chronic exposure to mercury does not have their hair element levels stay perfectly normal until one day they go wildly out of whack. Change is gradual and there are situations where people may get a hair test before they are toxic enough for the counting rules to be met. So let me give you some thresholds for how to count tests and decide how normal or abnormal they are.

	Number to right or left (rules 1 & 2)	Number of reds (rule 3)	Number in middle band (rule 4)
Ordinary and average	9 to 14	0 or 1	≥15
Not uncommon	8	2	14
Unusual	7	3	13
Suspicious	6	n/a	12
Meets counting rules, abnormal	≤5	≥4	≤11

You can bump the level of suspicion up one notch if it meets two or more of the criteria (the way counting rule 5 does).

This table gives you a substitute for the kind of intuition a health care provider can develop after seeing hundreds and hundreds of tests as to what "looks toxic" and "looks normal."

There are three important factors to keep in mind when using the counting rules or ranking how suspicious or average a test looks:

First, they provide a probability, not a certainty. Rarely you will find a normal healthy person who meets them by chance, and rarely a toxic person who by chance who does not meet them.

Second, being a probability, there are some results that are suspicious. That is, for rules 1 and 2, 6 or 7 bars to the right or left is suspicious in terms of not being very likely, yet not being as unlikely as 1 in 40. Similarly, 12 in the middle band is also suspicious. 3 red band results is odd but not extremely suspicious, there is an 8% chance of this happening randomly.

To explain these probabilities by analogy, if you fan a deck of cards, hold it out face down and ask someone to pick the ace of spades and they do it, everyone around will believe it was a magic trick. Picking the ace of spades is about as likely as meeting the counting rules by chance. If you fan the cards, say "pick an ace" and they do it, it is suspicious. If you tell them "pick a face card" and they do it, it is unusual. If you tell them "pick a black card" and they do, it is not uncommon. If you say "pick a card, any card" doing so is quite ordinary and average.

Third, the precise question being asked is "does it seem likely that the hair follicles were exposed to an environment causing deranged mineral transport when this hair was grown?" This means was the scalp or bloodstream toxic. If the brain or possibly liver was toxic but the rest of the body was not, the hair test would show normal mineral transport for a person who is very poisoned. The hair test really can only be interpreted in light of the person's health background and suspected history of toxic exposure.

Keeping in mind the real human being that the hair was snipped from, what is going on with them, what their history is, what probabilities mean and what the limitations of the hair test are will let you draw meaningful conclusions.

Here are Some Normal Looking Tests

Here are five absolutely normal looking tests from people between 24 and 33 with no significant medical history and no significant health problems. The sixth test is from a woman in this age range who also has no significant medical history and no significant health problems. She has very elevated silver. Given the statistical definition of the reference ranges there are occasional normal healthy people who have some toxic element very elevated in their hair but who have no health problems due to it.

	Case #1	Case #2	Case #3	Case #4	Case #5	Case #6
Counting rule #1	13	15	11	12	14	14
Counting rule #2	10	8	12	11	9	9
Counting rule #3	1	0	1	1	1	0
Counting rule #4	14	16	15	16	17	16

Note that cases 2 and 3 have the calcium – magnesium versus sodium – potassium stress sign indicating mild to moderate stress. Stress is not in and of itself a health problem or something pathological. The body's response to stress and its ability to tolerate stress determine whether it is a problem or not. All of these people work in a moderately stressful situation and are able to tolerate it well.

Also note that in case 1 the calcium to phosphorus ratio is slightly out of range while in case 4 the calcium to magnesium ratio is slightly out of range. Given the statistical definition of the expected ranges for the mineral ratios, one would expect to find two of them slightly out of range out of six tests.

The proper use of hair tests is to identify what the root cause of people's health problems are, or what risk factors they have for future health problems. The abuse of hair tests is in poring over them again and again to find some pretext for intervention unrelated to existing or reasonably suspected health problems. These normal healthy test results from normal healthy people are included so that you can compare them to sick people's tests and have a reasonable way to decide what kinds of results to pay attention to and what kinds of results to ignore.

It would not be irrational for these people to take supplements or modify diet to try to optimize their already good health based on these test results, but it also would not be irrational to do nothing based on them. Most reasonable people would choose to do nothing, a few who are very interested in optimal health or disease prevention might choose to supplement based on the hair test results.

HAIR ELEMENTS

DOCTOR'S DATA INC.

SEX: Male

POTENTIALLY TOXIC ELEMENTS

TOXIC ELEMENTS	RESULT µg/g	REFERENCE RANGE	PERCENTILE 68th 95th
Aluminum	5.6	< 7.0	
Antimony	0.032	< 0.066	
Arsenic	0.056	< 0.080	
Beryllium	< 0.01	< 0.020	
Bismuth	0.071	< 0.060	
Cadmium	0.023	< 0.15	
Lead	0.34	< 2.0	
Mercury	0.67	< 1.1	
Platinum	< 0.003	< 0.005	
Thallium	< 0.001	< 0.010	
Thorium	< 0.001	< 0.005	
Uranium	0.005	< 0.060	
Nickel	0.09	< 0.40	
Silver	0.09	< 0.12	
Tin	0.17	< 0.30	
Titanium	0.88	< 1.0	
Total Toxic Representation			

ESSENTIAL AND OTHER ELEMENTS

ELEMENTS	RESULT µg/g	REFERENCE RANGE	PERCENTILE 2.5th 16th 50th 84th 97.5th
Calcium	180	200- 750	
Magnesium	17	25- 75	
Sodium	49	12- 90	
Potassium	44	9- 40	
Copper	11	10- 28	
Zinc	160	130- 200	
Manganese	0.09	0.15- 0.65	
Chromium	0.60	0.20- 0.40	
Vanadium	0.048	0.018- 0.065	
Molybdenum	0.037	0.025- 0.064	
Boron	1.6	0.40- 3.0	
Iodine	0.56	0.25- 1.3	
Lithium	0.005	0.007- 0.023	
Phosphorus	226	160- 250	
Selenium	16	0.95- 1.7	
Strontium	0.25	0.30- 3.5	
Sulfur	49700	44500- 52000	
Barium	0.20	0.16- 1.6	
Cobalt	0.012	0.013- 0.035	
Iron	9.3	5.4- 13	
Germanium	0.064	0.045- 0.065	
Rubidium	0.055	0.011- 0.12	
Zirconium	0.46	0.020- 0.44	

SPECIMEN DATA

COMMENTS:

Date Collected: Sample Size:

Date Received: Sample Type:

Date Completed: Hair Color:

Treatment:

Methodology: ICP-MS Shampoo:

V06.99

RATIOS

ELEMENTS	RATIOS	EXPECTED RANGE
Ca/Mg	10.6	4- 30
Ca/P	0.796	0.8- 8
Na/K	1.11	0.5- 10
Zn/Cu	14.5	4- 20
Zn/Cd	> 999	> 800

This form is provided to Andrew Cutler to be used as an educational tool and does not constitute endorsement by *Doctor's Data*.

HAIR ELEMENTS

DOCTOR'S DATA INC

SEX: Male

POTENTIALLY TOXIC ELEMENTS

TOXIC ELEMENTS	RESULT µg/g	REFERENCE RANGE	PERCENTILE 68th 95th
Aluminum	7.3	< 7.0	
Antimony	0.032	< 0.066	
Arsenic	0.020	< 0.080	
Beryllium	0.010	< 0.020	
Bismuth	0.019	< 0.060	
Cadmium	0.076	< 0.15	
Lead	0.81	< 2.0	
Mercury	0.09	< 1.1	
Platinum	< 0.003	< 0.005	
Thallium	< 0.001	< 0.010	
Thorium	< 0.001	< 0.005	
Uranium	0.011	< 0.060	
Nickel	0.17	< 0.40	
Silver	0.08	< 0.12	
Tin	0.38	< 0.30	
Titanium	0.71	< 1.0	
Total Toxic Representation			

ESSENTIAL AND OTHER ELEMENTS

ELEMENTS	RESULT µg/g	REFERENCE RANGE	PERCENTILE 2.5th 16th 50th 84th 97.5th
Calcium	1020	200- 750	
Magnesium	140	25- 75	
Sodium	26	12- 90	
Potassium	14	9- 40	
Copper	21	10- 28	
Zinc	200	130- 200	
Manganese	0.33	0.15- 0.65	
Chromium	0.54	0.20- 0.40	
Vanadium	0.049	0.018- 0.065	
Molybdenum	0.026	0.025- 0.064	
Boron	1.6	0.40- 3.0	
Iodine	1.7	0.25- 1.3	
Lithium	0.008	0.007- 0.023	
Phosphorus	185	160- 250	
Selenium	0.79	0.95- 1.7	
Strontium	4.2	0.30- 3.5	
Sulfur	49500	44500- 52000	
Barium	3.3	0.16- 1.6	
Cobalt	0.019	0.013- 0.035	
Iron	8.7	5.4- 13	
Germanium	0.063	0.045- 0.065	
Rubidium	0.021	0.011- 0.12	
Zirconium	0.57	0.020- 0.44	

SPECIMEN DATA

COMMENTS:

Date Collected:

Date Received:

Date Completed:

Methodology: ICP-MS

Sample Size:

Sample Type:

Hair Color:

Treatment:

Shampoo:

V06.99

RATIOS

ELEMENTS	RATIOS	EXPECTED RANGE
Ca/Mg	7.29	4- 30
Ca/P	5.51	0.8- 8
Na/K	1.86	0.5- 10
Zn/Cu	9.52	4- 20
Zn/Cd	> 999	> 800

This form is provided to Andrew Cutler to be used as an educational tool and does not constitute endorsement by *Doctor's Data.*

HAIR ELEMENTS

DOCTOR'S DATA INC

SEX: Female

POTENTIALLY TOXIC ELEMENTS

TOXIC ELEMENTS	RESULT µg/g	REFERENCE RANGE
Aluminum	10	< 7.0
Antimony	0.027	< 0.050
Arsenic	0.031	< 0.060
Beryllium	< 0.01	< 0.020
Bismuth	0.12	< 0.10
Cadmium	0.051	< 0.10
Lead	0.42	< 1.0
Mercury	0.90	< 1.1
Platinum	< 0.003	< 0.005
Thallium	< 0.001	< 0.010
Thorium	< 0.001	< 0.005
Uranium	0.003	< 0.060
Nickel	0.30	< 0.40
Silver	0.57	< 0.15
Tin	0.31	< 0.30
Titanium	1.7	< 1.0
Total Toxic Representation		

PERCENTILE: 68th, 95th

ESSENTIAL AND OTHER ELEMENTS

ELEMENTS	RESULT µg/g	REFERENCE RANGE
Calcium	607	300- 1200
Magnesium	53	35- 120
Sodium	120	12- 90
Potassium	52	8- 38
Copper	14	12- 35
Zinc	170	140- 220
Manganese	0.36	0.15- 0.65
Chromium	0.34	0.20- 0.40
Vanadium	0.048	0.018- 0.065
Molybdenum	0.023	0.028- 0.056
Boron	0.75	0.30- 2.0
Iodine	2.9	0.25- 1.3
Lithium	0.007	0.007- 0.023
Phosphorus	167	160- 250
Selenium	0.83	0.95- 1.7
Strontium	3.1	0.50- 7.6
Sulfur	45500	44500- 52000
Barium	3.9	0.26- 3.0
Cobalt	0.025	0.013- 0.050
Iron	7.5	5.4- 14
Germanium	0.040	0.045- 0.065
Rubidium	0.049	0.007- 0.096
Zirconium	1.9	0.020- 0.42

PERCENTILE: 2.5th, 16th, 50th, 84th, 97.5th

SPECIMEN DATA

COMMENTS:

Date Collected: | Sample Size:
Date Received: | Sample Type:
Date Completed: | Hair Color:
| Treatment:
Methodology: ICP-MS | Shampoo:

V06.99

RATIOS

ELEMENTS	RATIOS	EXPECTED RANGE
Ca/Mg	11.5	4- 30
Ca/P	3.63	1- 12
Na/K	2.31	0.5- 10
Zn/Cu	12.1	4- 20
Zn/Cd	> 999	> 800

This form is provided to Andrew Cutler to be used as an educational tool and does not constitute endorsement by *Doctor's Data.*

HAIR ELEMENTS

DOCTOR'S DATA INC.

SEX: Male

POTENTIALLY TOXIC ELEMENTS

TOXIC ELEMENTS	RESULT µg/g	REFERENCE RANGE	PERCENTILE 68th 95th
Aluminum	6.2	< 7.0	
Antimony	0.035	< 0.066	
Arsenic	0.059	< 0.080	
Beryllium	< 0.01	< 0.020	
Bismuth	0.043	< 0.060	
Cadmium	0.047	< 0.15	
Lead	0.14	< 2.0	
Mercury	0.50	< 1.1	
Platinum	< 0.003	< 0.005	
Thallium	< 0.001	< 0.010	
Thorium	< 0.001	< 0.005	
Uranium	0.009	< 0.060	
Nickel	0.07	< 0.40	
Silver	0.06	< 0.12	
Tin	0.25	< 0.30	
Titanium	0.74	< 1.0	
Total Toxic Representation			

ESSENTIAL AND OTHER ELEMENTS

ELEMENTS	RESULT µg/g	REFERENCE RANGE	PERCENTILE 2.5th 16th 50th 84th 97.5th
Calcium	432	200- 750	
Magnesium	110	25- 75	
Sodium	92	12- 90	
Potassium	19	9- 40	
Copper	9.5	10- 28	
Zinc	190	130- 200	
Manganese	0.12	0.15- 0.65	
Chromium	0.38	0.20- 0.40	
Vanadium	0.029	0.018- 0.065	
Molybdenum	0.024	0.025- 0.064	
Boron	6.5	0.40- 3.0	
Iodine	0.65	0.25- 1.3	
Lithium	0.009	0.007- 0.023	
Phosphorus	196	160- 250	
Selenium	0.99	0.95- 1.7	
Strontium	2.2	0.30- 3.5	
Sulfur	50600	44500- 52000	
Barium	0.90	0.16- 1.6	
Cobalt	0.009	0.013- 0.035	
Iron	9.4	5.4- 13	
Germanium	0.049	0.045- 0.065	
Rubidium	0.022	0.011- 0.12	
Zirconium	2.3	0.020- 0.44	

SPECIMEN DATA

COMMENTS:

Date Collected:

Date Received:

Date Completed:

Methodology: ICP-MS

Sample Size:

Sample Type:

Hair Color:

Treatment:

Shampoo:

V06.99

RATIOS

ELEMENTS	RATIOS	EXPECTED RANGE
Ca/Mg	3.93	4- 30
Ca/P	2.2	0.8- 8
Na/K	4.84	0.5- 10
Zn/Cu	20	4- 20
Zn/Cd	> 999	> 800

This form is provided to Andrew Cutler to be used as an educational tool and does not constitute endorsement by *Doctor's Data*.

HAIR ELEMENTS

DOCTOR'S DATA INC

SEX: Female

POTENTIALLY TOXIC ELEMENTS			
TOXIC ELEMENTS	RESULT µg/g	REFERENCE RANGE	PERCENTILE 68th 95th
Aluminum	4.0	< 7.0	
Antimony	< 0.01	< 0.050	
Arsenic	0.022	< 0.060	
Beryllium	< 0.01	< 0.020	
Bismuth	0.067	< 0.10	
Cadmium	0.045	< 0.10	
Lead	0.19	< 1.0	
Mercury	0.51	< 1.1	
Platinum	< 0.003	< 0.005	
Thallium	< 0.001	< 0.010	
Thorium	< 0.001	< 0.005	
Uranium	0.010	< 0.060	
Nickel	0.46	< 0.40	
Silver	0.11	< 0.15	
Tin	0.16	< 0.30	
Titanium	1.0	< 1.0	
Total Toxic Representation			

ESSENTIAL AND OTHER ELEMENTS			
ELEMENTS	RESULT µg/g	REFERENCE RANGE	PERCENTILE 2.5th 16th 50th 84th 97.5th
Calcium	2010	300- 1200	
Magnesium	280	35- 120	
Sodium	47	12- 90	
Potassium	9	8- 38	
Copper	30	12- 35	
Zinc	170	140- 220	
Manganese	0.33	0.15- 0.65	
Chromium	0.33	0.20- 0.40	
Vanadium	0.033	0.018- 0.065	
Molybdenum	0.048	0.028- 0.056	
Boron	0.83	0.30- 2.0	
Iodine	3.8	0.25- 1.3	
Lithium	0.004	0.007- 0.023	
Phosphorus	208	160- 250	
Selenium	1.1	0.95- 1.7	
Strontium	9.8	0.50- 7.6	
Sulfur	45500	44500- 52000	
Barium	6.6	0.26- 3.0	
Cobalt	0.021	0.013- 0.050	
Iron	8.1	5.4- 14	
Germanium	0.047	0.045- 0.065	
Rubidium	0.017	0.007- 0.096	
Zirconium	0.15	0.020- 0.42	

SPECIMEN DATA

COMMENTS:

Date Collected:	Sample Size:
Date Received:	Sample Type:
Date Completed:	Hair Color:
	Treatment:
Methodology: ICP-MS	Shampoo:

V06.99

RATIOS

ELEMENTS	RATIOS	EXPECTED RANGE
Ca/Mg	7.18	4- 30
Ca/P	9.66	1- 12
Na/K	5.22	0.5- 10
Zn/Cu	5.67	4- 20
Zn/Cd	> 999	> 800

This form is provided to Andrew Cutler to be used as an educational tool and does not constitute endorsement by *Doctor's Data*.

HAIR ELEMENTS

SEX: Female

POTENTIALLY TOXIC ELEMENTS

TOXIC ELEMENTS	RESULT µg/g	REFERENCE RANGE
Aluminum	13	< 7.0
Antimony	0.077	< 0.050
Arsenic	0.030	< 0.060
Beryllium	< 0.01	< 0.020
Bismuth	0.040	< 0.10
Cadmium	0.055	< 0.10
Lead	0.85	< 1.0
Mercury	0.95	< 1.1
Platinum	< 0.003	< 0.005
Thallium	< 0.001	< 0.010
Thorium	< 0.001	< 0.005
Uranium	0.005	< 0.060
Nickel	0.12	< 0.40
Silver	0.35	< 0.15
Tin	1.5	< 0.30
Titanium	0.66	< 1.0
Total Toxic Representation		

PERCENTILE: 68th, 95th

ESSENTIAL AND OTHER ELEMENTS

ELEMENTS	RESULT µg/g	REFERENCE RANGE
Calcium	664	300- 1200
Magnesium	48	35- 120
Sodium	78	12- 90
Potassium	72	8- 38
Copper	11	12- 35
Zinc	180	140- 220
Manganese	0.34	0.15- 0.65
Chromium	0.47	0.20- 0.40
Vanadium	0.033	0.018- 0.065
Molybdenum	0.043	0.028- 0.056
Boron	2.1	0.30- 2.0
Iodine	1.9	0.25- 1.3
Lithium	0.012	0.007- 0.023
Phosphorus	177	160- 250
Selenium	0.93	0.95- 1.7
Strontium	2.3	0.50- 7.6
Sulfur	43600	44500- 52000
Barium	2.2	0.26- 3.0
Cobalt	0.025	0.013- 0.050
Iron	8.1	5.4- 14
Germanium	0.052	0.045- 0.065
Rubidium	0.083	0.007- 0.096
Zirconium	0.51	0.020- 0.42

PERCENTILE: 2.5th, 16th, 50th, 84th, 97.5th

SPECIMEN DATA

COMMENTS:

Date Collected:

Date Received:

Date Completed:

Methodology: ICP-MS

Sample Size:

Sample Type:

Hair Color:

Treatment:

Shampoo:

V06.99

RATIOS

ELEMENTS	RATIOS	EXPECTED RANGE
Ca/Mg	13.8	4- 30
Ca/P	3.75	1- 12
Na/K	1.08	0.5- 10
Zn/Cu	16.4	4- 20
Zn/Cd	> 999	> 800

This form is provided to Andrew Cutler to be used as an educational tool and does not constitute endorsement by *Doctor's Data*.

Medical Textbooks say Many Terrible Conditions can be from Heavy Metals and Hair Tests show it

What Textbooks say about Parkinson's Disease

Some recent editions of *Harrison's Textbook of Internal Medicine* discuss the physician's ethical obligation to rule out all possible similar conditions when faced with a patient the doctor believes has an incurable disease with a poor prognosis, and use the example that Wilson's disease – a rare genetic defect that causes copper to build up to toxic levels – should be ruled out in all cases of suspected Parkinson's disease since, even though it is rare, Wilson's disease is treatable and can sometimes be confused with Parkinson's disease for which the treatments are not very good. Yet few physicians actually do this, much less generalizing their ethical obligation to require them to screen for heavy metal toxicity among patients with neurological and psychiatric problems.

In the specific case of Parkinson's disease, Merritt's *Textbook of Neurology*, 9th edition, 1995, says: "Exposure to manganese ... followed by Parkinsonism months to 40 years after toxic exposure."

Textbook of Clinical Neurology, first edition., Goetz, 1999 says in the section discussing mercury poisoning: "At times, these patients present a typical picture of Parkinsonism."

Clinical Toxicology, Ford, first edition, 2001 says: "Parkinsonian symptoms are frequently associated with chronic elementary mercury toxicity."

Clinical Management of Poisoning and Drug Overdose, Third edition, 1998, Haddad, Shannon and Winchester (editors), says: "Chronic mercurialism may be misdiagnosed as Parkinsonism."

As you can see, there is more than adequate reason given in standard medical texts to check any Parkinson's disease patient for heavy metal toxicity, as mercury, manganese and copper toxicity can all be confused with it. The facing page contains the hair element profile of a gentleman with Parkinson's disease.

As a young boy of 8 years this gentleman had a bottle of mercury to play with. As an adolescent he was high strung and got into trouble sometimes. He always felt separate from the rest of the world in some odd and unexplainable way and this feeling continued throughout his life though it has recently been diminishing with chelation.

This gentleman received 4 amalgam fillings in his 20's and developed "traumatic arthritis" within weeks of their being placed.

He had fine tremors from a young age, and when no doctor could diagnose the cause he was referred for counseling. This was ineffective. Wellbutrin (an antidepressant) was also tried because the psychologist and medical doctors decided he had obsessive-compulsive disorder and depression. He went along with them and gave the Wellbutrin a try even though he didn't believe they were correct in their diagnosis. The Wellbutrin was ineffective.

During his adult life this gentleman had ample exposure to mercury through using mercury manometers (U tubes filled with mercury which are used to measure air pressure differences in flow systems). While barometers, manometers, blood pressure meters (sphygmomanometers), etc. can be substantial sources of occupational exposure to mercury few physicians recognize them as such.

The onset of Parkinson's disease was typical, with forgetfulness, clumsiness, a short temper, cramped and illegible writing, lessening fine motor control and tremors starting in the hands that would sometimes spread to the wrists.

This gentleman was familiar with the fact that the anti-Parkinson's medications often make the patient more miserable than the disease does since his father had passed away from Parkinson's disease. He was also familiar with mercury as the unsuspected cause of neurological problems since he was detoxing his autistic and mercury toxic sons. So he got himself a hair test. Referring to the facing page, we see that the counting rules suggest a high probability of mercury problems.

Counting rule #1: 15 bars go right. This does not meet rule #1.

Counting rule #2: 8 bars go left. This does not meet counting rule #2.

Counting rule #3: 3 bars go into the red bands. This does not meet counting rule #3.

Counting rule #4: 12 bars are contained entirely in the middle green + white stripe. This does not meet counting rule #4.

Counting rule #5: 3 bars enter the red zone and 12 bars are entirely within the middle band. This meets counting rule #5, that the test counts one short of positive on any other two rules.

Also note that mercury is abnormally elevated and does make it into the red zone. Everything about this test points to mercury poisoning.

After amalgam filling removal and nine months of chelation, but without using any drugs to treat his Parkinson's disease, his tremors had all but subsided, his handwriting improved, he lost the annoying nervous tic of accidentally biting his tongue, his balance had improved, his sense of separateness from everyone else had decreased and he remained strong, active, kayaking, running on the beach for exercise and he remodeled his house.

His wife reports that his temper and personality greatly improved on chelation.

He also tried digestive enzymes and found that his frequent acid reflux (GERD) stopped completely.

HAIR ELEMENTS

DOCTOR'S DATA INC

SEX: Male
AGE: 51

POTENTIALLY TOXIC ELEMENTS

TOXIC ELEMENTS	RESULT µg/g	REFERENCE RANGE
Aluminum	9.3	< 7.0
Antimony	0.038	< 0.066
Arsenic	0.1	< 0.08
Beryllium	< 0.01	< 0.02
Bismuth	0.081	< 0.06
Cadmium	0.034	< 0.15
Lead	1.4	< 2.0
Mercury	3.6	< 1.1
Platinum	< 0.003	< 0.005
Thallium	< 0.001	< 0.01
Thorium	< 0.001	< 0.005
Uranium	0.003	< 0.06
Nickel	0.34	< 0.4
Silver	0.14	< 0.12
Tin	0.15	< 0.3
Titanium	1.2	< 1.0
Total Toxic Representation		

Percentile: 68th, 95th

ESSENTIAL AND OTHER ELEMENTS

ELEMENTS	RESULT µg/g	REFERENCE RANGE
Calcium	286	200- 750
Magnesium	60	25- 75
Sodium	700	12- 90
Potassium	170	9.0- 40
Copper	23	10- 28
Zinc	110	130- 200
Manganese	0.23	0.15- 0.65
Chromium	0.35	0.2- 0.4
Vanadium	1.2	0.018- 0.065
Molybdenum	0.068	0.025- 0.064
Boron	2.5	0.4- 3.0
Iodine	1.7	0.25- 1.3
Lithium	0.013	0.007- 0.023
Phosphorus	152	160- 250
Selenium	1.0	0.95- 1.7
Strontium	1.1	0.3- 3.5
Sulfur	43300	44500- 52000
Barium	0.17	0.16- 1.6
Cobalt	0.022	0.013- 0.035
Iron	14	5.4- 13
Germanium	0.043	0.045- 0.065
Rubidium	0.2	0.011- 0.12
Zirconium	0.11	0.02- 0.44

Percentile: 2.5th, 16th, 50th, 84th, 97.5th

SPECIMEN DATA

COMMENTS:

Date Collected:
Date Received:
Date Completed:
Methodology: ICP-MS

Sample Size:
Sample Type:
Hair Color:
Treatment:
Shampoo:

V06.99

RATIOS

ELEMENTS	RATIOS	EXPECTED RANGE
Ca/Mg	4.77	4- 30
Ca/P	1.88	0.8- 8
Na/K	4.12	0.5- 10
Zn/Cu	4.78	4- 20
Zn/Cd	> 999	> 800

This form is provided to Andrew Cutler to be used as an educational tool and does not constitute endorsement by *Doctor's Data.*

What Textbooks say about Problems with Children's Development

Standard medical textbooks point out the need to check for heavy metal problems in children with developmental disorders. For example, *Clinical Toxicology*, Ford, first edition, 2001: "Chronic lead poisoning should be considered in the differential diagnosis of any child with neurocognitive deficits, attentional impairment, or developmental delay."

Textbooks also point out that despite what is in essence an advertising and public relations campaign by the makers and users of childhood vaccines there really is no reason to consider them proven safe.

Child Neurology, sixth edition, 2000, John Menkes, editor. From Chapter 16, Disorders of Mental Development, section on Autism, p. 1172, by Marcel Kinsbourne: "The increase in the number of children diagnosed with autism is so spectacular, that it is unlikely to be merely because health care workers are paying greater attention to this condition. Despite apparent links between measles, mumps, and rubella (MMR) immunizations and autism, opposed by energetic disclaimers, the definitive epidemiologic study that would distinguish causation, a rare inordinate susceptibility, from chance association has not been performed."

Recent research demonstrates that heavy metal problems are quite common in children – so it is important to check for them early on. The US Government's *Centers for Disease Control* periodically conducts studies to determine what toxic exposure members of the general population have. The 2003 study found that 10% of children and 10% of women of childbearing age had mercury levels above safe limits. They also found that 2.2% of children had lead above safe levels. This is down from 4% of children who had too much lead in similar studies 20-30 years ago[23].

Despite this, no states require children to be screened for mercury and most do not require children to be screened for lead. In those states which do require children to be screened for lead, it is common that many children either are not screened or that if they show up with elevated lead levels, the doctors don't actually do anything about it other than keep repeating the test until they get a low number.

For example, records indicate that in Ohio only about one third of the children under the age of six enrolled in Medicaid who are legally required to have lead tests performed actually were checked for lead, even though the Medicaid program paid for all the tests in advance[24].

With admitted uncertainty regarding the possible causes of autism even in conservative mainstream textbooks, it makes sense to check for all possible causes before deciding that no medical or biochemical treatment is going to help and limiting interventions to behavioral approaches such as applied behavioral analysis (ABA) and special education.

There is also much uncertainty regarding the causes of ADHD, learning disabilities, pervasive developmental disorders, Asperger's syndrome and other developmental problems, all of which appear to have dramatically increased in incidence over the last 20 years.

Even the conservative textbook literature says lead must be considered in developmental disorders. The journal literature is full of material suggesting other heavy metals may cause autism and

---

[23] The study is not yet published as of the time this book was written. It is expected to appear in *Morbidity and Mortality Weekly Reports*, MMWR. The prior study for 1999 is reported there and can be downloaded on the internet at http://www.cdc.gov/mmwr/PDF/wk/mm5008.pdf.

[24] see http://www.ewg.org/reports/ohiolead/

other developmental disorders in some children. Lead is such a well known cause of developmental problems that this information also appears in consumer literature: for example, Savon/Albertson's pharmacies use a brochure entitled *ADD & ADHD, Pharmacy Answers* (© 2003) that says "Attention deficit hyperactivity disorder (ADHD) is the most common behavioral disorder diagnosed in children and teens." After describing ADHD it says "Certain environmental factors, such as lead exposure ... may also contribute to the risk of developing ADHD." Later the brochure elaborates on this topic as follows: "Metal Exposure: Lead and other heavy metal exposures have been linked to ADHD."

A hair test (and consideration of the individual child's history and situation) seems like a reasonable way to screen children with developmental disorders for heavy metals. As you can see from the tests below, when people do get hair tests on their impaired children heavy metal problems often turn up! While genetically based autism and related conditions are often presumed to be completely unresponsive to medical treatment, heavy metal problems are easily treated and children with them show dramatic improvement in their neurodevelopmental status on detox.

The incidence of classical (Kanner's) autism, generally believed to be of genetic origin in the medical community, was well established at 1 child in 40,000 fifty to seventy five years ago. The incidence of autism today is about 1 in 150, which is 267 in 40,000. Since the incidence of a genetic condition cannot change from generation to generation under the circumstances prevailing in modern society, the only reasonable conclusion to draw is that 266 out of 267 of today's autistic children – 99.6% – acquired their autism from some environmental cause and most likely can be improved by treatment. It is also worth noting that even among those children with a genetic cause, there are often treatments available if the exact cause is identified by proper testing and evaluation.

A pretty obvious case where screening for heavy metal exposure would be appropriate is an autistic little girl who has no known risk factors, no family history suggestive of any genetic issues, and whom the pediatrician thinks is a textbook case. This little girl's hair test is on the next page. It counts as follows:

Counting rule #1: six bars go right. This does not meet rule #1.

Counting rule #2: fifteen bars go left. This does not meet counting rule #2.

Counting rule #3: three bars go into the red bands. This does not meet counting rule #3.

Counting rule #4: sixteen bars are contained entirely in the middle green + white stripe. This does not meet counting rule #4.

Counting rule #5: 6 bars go right left and 3 bars enter the red zone. This meets counting rule #5, that the test counts one short of positive on any other two rules.

Since the test meets a counting rule, rule #5, it suggests a high probability of mercury induced derangement of mineral transport.

Looking at rules #1 and #2, you may note that 6 + 15 = 21, not the 23 total bars there are. The bars for selenium and zirconium are right smack dab in the middle and I didn't count them as going either way. It would also be consistent to count them as half either direction.

The antimony, arsenic and tin being higher than the other toxics is also consistent with mercury being a problem.

This hair test suggests that the girl's autism is a consequence of mercury intoxication. Chelation led to improvement in her autistic symptoms, confirming this.

HAIR ELEMENTS

DOCTOR'S DATA INC

SEX: Female
AGE: 3

POTENTIALLY TOXIC ELEMENTS

TOXIC ELEMENTS	RESULT µg/g	REFERENCE RANGE	PERCENTILE (68th, 95th)
Aluminum	3.1	< 8.0	
Antimony	0.14	< 0.066	
Arsenic	0.14	< 0.080	
Beryllium	< 0.01	< 0.020	
Bismuth	0.077	< 0.13	
Cadmium	< 0.009	< 0.10	
Lead	0.20	< 1.0	
Mercury	0.38	< 0.40	
Platinum	< 0.003	< 0.005	
Thallium	< 0.001	< 0.010	
Thorium	< 0.001	< 0.005	
Uranium	0.005	< 0.060	
Nickel	0.05	< 0.40	
Silver	0.01	< 0.20	
Tin	0.50	< 0.30	
Titanium	0.78	< 1.0	
Total Toxic Representation			

ESSENTIAL AND OTHER ELEMENTS

ELEMENTS	RESULT µg/g	REFERENCE RANGE	PERCENTILE (2.5th, 16th, 50th, 84th, 97.5th)
Calcium	328	140- 500	
Magnesium	19	15- 45	
Sodium	21	12- 90	
Potassium	19	9- 60	
Copper	19	10- 22	
Zinc	140	100- 190	
Manganese	0.09	0.20- 0.55	
Chromium	0.54	0.26- 0.50	
Vanadium	0.062	0.030- 0.10	
Molybdenum	0.21	0.048- 0.13	
Boron	0.86	0.35- 3.0	
Iodine	0.20	0.25- 1.3	
Lithium	< 0.004	0.007- 0.023	
Phosphorus	171	160- 250	
Selenium	1.2	0.95- 1.7	
Strontium	0.27	0.19- 2.0	
Sulfur	47300	45500- 53000	
Barium	0.12	0.16- 1.2	
Cobalt	0.010	0.013- 0.035	
Iron	8.8	8.0- 19	
Germanium	0.054	0.045- 0.065	
Rubidium	0.019	0.012- 0.16	
Zirconium	0.23	0.030- 1.0	

SPECIMEN DATA

COMMENTS:

Date Collected:
Date Received:
Date Completed:
Methodology: ICP-MS

Sample Size:
Sample Type:
Hair Color:
Treatment:
Shampoo:

V06.99

RATIOS

ELEMENTS	RATIOS	EXPECTED RANGE
Ca/Mg	17.3	4- 30
Ca/P	1.92	1- 12
Na/K	1.11	0.5- 10
Zn/Cu	7.37	4- 20
Zn/Cd	> 999	> 800

This form is provided to Andrew Cutler to be used as an educational tool and does not constitute endorsement by *Doctor's Data.*

A more interesting case involves an autistic boy and his learning disabled sister. The girl is six years older than her brother. The girl's learning problems became apparent during grade school. The school conducted a thorough evaluation of her and developed an individualized educational program (IEP) to try to keep her at grade level with her peers even though they were not really able to pinpoint anything more definite than "short term memory problems."

At about the same time as the sister was being evaluated, her younger brother began to exhibit problems. He had shown normal development for the first 16-18 months, then regressed into autism. At age 6 this boy had no expressive language. He had never been able to say a single word. It was the speech therapist's impression that he did try to say some words but was unable to pronounce them well enough that even his mother could figure it out.

Two siblings with developmental problems certainly does suggest a genetic cause. However, hair testing was performed on both of them and the results on the following pages count as follows:

The brother, page 41:

Counting rule #1: thirteen bars go right. This does not meet rule #1.

Counting rule #2: ten bars go left. This does not meet counting rule #2.

Counting rule #3: four bars go into the red bands. This does meet counting rule #3.

Counting rule #4: fourteen bars are contained entirely in the middle green + white stripe. This does not meet counting rule #4.

Counting rule #5: is irrelevant since another rule was satisfied.

Also note the relative elevation of sodium and potassium compared to calcium and magnesium. This suggests that his adrenals may be slightly under performing.

The sister, page 42:

Counting rule #1: eleven bars go right. This does not meet rule #1.

Counting rule #2: eleven bars go left. This does not meet counting rule #2.

Counting rule #3: one bar goes into the red bands. This does not meet counting rule #3.

Counting rule #4: fifteen bars are contained entirely in the middle green + white stripe. This does not meet counting rule #4.

Counting rule #5: is not met.

This is a very normal and ordinary hair test in terms of the essential and other elements section.

Now looking at the toxic elements portion, where the bars do represent body burden of the relevant elements since mineral transport is orderly, we can see what the sister's problem is. Vast quantities of antimony, lead and tin.

The relatively high ratio of potassium to calcium suggests it would be a good idea for her to have her thyroid checked.

HAIR ELEMENTS

DOCTOR'S DATA INC

SEX: Male
AGE: 7

POTENTIALLY TOXIC ELEMENTS

TOXIC ELEMENTS	RESULT µg/g	REFERENCE RANGE
Aluminum	15	< 8.0
Antimony	0.066	< 0.066
Arsenic	0.044	< 0.08
Beryllium	< 0.01	< 0.02
Bismuth	0.075	< 0.12
Cadmium	0.021	< 0.15
Lead	0.4	< 1.0
Mercury	< 0.03	< 0.4
Platinum	< 0.003	< 0.005
Thallium	0.003	< 0.01
Thorium	< 0.001	< 0.005
Uranium	0.014	< 0.06
Nickel	0.09	< 0.4
Silver	0.06	< 0.13
Tin	0.34	< 0.3
Titanium	1.0	< 1.0
Total Toxic Representation		

PERCENTILE: 68th, 95th

ESSENTIAL AND OTHER ELEMENTS

ELEMENTS	RESULT µg/g	REFERENCE RANGE
Calcium	227	160- 500
Magnesium	34	12- 50
Sodium	720	12- 90
Potassium	220	10- 40
Copper	8.2	9.0- 30
Zinc	180	110- 190
Manganese	0.25	0.18- 0.6
Chromium	0.22	0.23- 0.5
Vanadium	0.041	0.025- 0.1
Molybdenum	0.051	0.04- 0.089
Boron	2.2	0.5- 3.5
Iodine	0.25	0.25- 1.3
Lithium	0.023	0.007- 0.023
Phosphorus	268	160- 250
Selenium	1.6	0.95- 1.7
Strontium	0.42	0.21- 2.1
Sulfur	52300	45500- 53000
Barium	0.34	0.19- 1.6
Cobalt	0.14	0.013- 0.035
Iron	13	6.0- 17
Germanium	0.025	0.045- 0.065
Rubidium	0.21	0.008- 0.08
Zirconium	0.8	0.06- 0.7

PERCENTILE: 2.5th, 16th, 50th, 84th, 97.5th

SPECIMEN DATA

COMMENTS:

Date Collected:
Date Received:
Date Completed:

Methodology: ICP-MS

Sample Size:
Sample Type:
Hair Color:
Treatment:
Shampoo:

V06.99

RATIOS

ELEMENTS	RATIOS	EXPECTED RANGE
Ca/Mg	6.68	4- 30
Ca/P	0.847	0.8- 8
Na/K	3.27	0.5- 10
Zn/Cu	22	4- 20
Zn/Cd	> 999	> 800

This form is provided to Andrew Cutler to be used as an educational tool and does not constitute endorsement by *Doctor's Data.*

HAIR ELEMENTS

DOCTOR'S DATA INC.

SEX: Female
AGE: 13

POTENTIALLY TOXIC ELEMENTS

TOXIC ELEMENTS	RESULT µg/g	REFERENCE RANGE
Aluminum	14	< 8.0
Antimony	2.1	< 0.066
Arsenic	0.047	< 0.06
Beryllium	< 0.01	< 0.02
Bismuth	0.022	< 0.1
Cadmium	0.2	< 0.1
Lead	13	< 1.0
Mercury	< 0.03	< 0.4
Platinum	< 0.003	< 0.005
Thallium	< 0.001	< 0.01
Thorium	< 0.001	< 0.005
Uranium	0.048	< 0.06
Nickel	1.1	< 0.4
Silver	0.38	< 0.16
Tin	2.4	< 0.3
Titanium	0.92	< 1.0
Total Toxic Representation		

PERCENTILE: 68th, 95th

ESSENTIAL AND OTHER ELEMENTS

ELEMENTS	RESULT µg/g	REFERENCE RANGE
Calcium	396	350- 1000
Magnesium	19	35- 120
Sodium	140	12- 90
Potassium	96	7.0- 36
Copper	25	12- 35
Zinc	130	130- 220
Manganese	0.14	0.15- 0.65
Chromium	0.29	0.2- 0.4
Vanadium	0.02	0.02- 0.075
Molybdenum	0.044	0.029- 0.058
Boron	5.0	0.3- 1.5
Iodine	0.84	0.25- 1.3
Lithium	0.006	0.007- 0.023
Phosphorus	235	160- 250
Selenium	1.2	0.95- 1.7
Strontium	1.1	0.86- 6.2
Sulfur	50500	44000- 52000
Barium	0.41	0.34- 2.5
Cobalt	0.013	0.013- 0.05
Iron	7.5	5.8- 15
Germanium	0.068	0.045- 0.065
Rubidium	0.073	0.006- 0.06
Zirconium	0.21	0.025- 0.5

PERCENTILE: 2.5th, 16th, 50th, 84th, 97.5th

SPECIMEN DATA

COMMENTS:

Date Collected:
Date Received:
Date Completed:

Sample Size:
Sample Type:
Hair Color:
Treatment:
Shampoo:

Methodology: ICP-MS

V06 99

RATIOS

ELEMENTS	RATIOS	EXPECTED RANGE
Ca/Mg	20.8	4- 30
Ca/P	1.69	1- 12
Na/K	1.46	0.5- 10
Zn/Cu	5.2	4- 20
Zn/Cd	650	> 800

This form is provided to Andrew Cutler to be used as an educational tool and does not constitute endorsement by *Doctor's Data*.

It turns out the house the parents were living in when they had the daughter was being remodeled during pregnancy. The home was old. Old paint contains lead, and since lead may no longer legally be used in paint, antimony oxide is now used as a replacement for it. Water pipe solder is either lead and tin (old) or antimony and tin (current). It is likely the sister absorbed toxic amounts of lead, antimony and tin during pregnancy and infancy.

The family moved out of this house when the sister was 3 – long before the brother was born.

The brother received far more vaccinations than the sister and they all contained mercury as a preservative (in the form of thimerosal). Since his body holds onto mercury unusually well this apparently was enough, when combined with whatever unknown exposures he may have been receiving from other sources, to make him autistic.

Based on these test results the parents chose to detoxify their children through chelation. They gave their son a combination of DMSA and alpha lipoic acid, 25 mg of each every 3 hours for three days in a row, skip the rest of the week and repeat.

Despite having no expressive language at all at age 6, by age 8, after 61 weekend rounds of chelation, he was articulate, normally expressive for his age, and took part in his second grade class' stage production, singing and participating so well that he was indistinguishable from the other children much to the amazement of his parents, teachers, and other children's parents who had known him in Kindergarten and first grade.

After 61 rounds of chelation this autistic boy is far from cured, but has undergone dramatic improvement and continues to improve with further chelation. If he had continued on in the state he was in at age 6, he would have been institutionalized for life as an adult because he couldn't even make his needs known, much less take care of himself. Now, at age 8, even if no further progress occurs, he will likely grow up to be able to live independently, hold a job and take care of himself, a dramatic improvement in his quality of life. The progress reasonably expected on further chelation should enhance his quality of life much more in terms of future health, happiness, social relationships, educational opportunities and employment options.

Less intervention has been performed for the sister. She has not been chelated. She did have her mercury amalgam dental fillings replaced with mercury free materials and in the subsequent year has made much more academic progress than expected.

Developmental Disorders: Lead or Mercury?

You may be curious as to why the textbooks talk about lead but two of my three examples are mercury. In fact, if you went out collecting hair tests on learning disabled an autistic children you would find mercury accounts for at least 3/4 of the cases today. So why don't the textbooks mention mercury?

The reason is very interesting and illustrates how self referencing logical systems like PubMed indexed medical journals can head off into never never land for decades on end.

There is an enormous amount of research on the effects of fish mercury (methyl mercury) on child development. There have recently been two very large studies of this, one in the Seychelles islands, one in the Faeroe islands. For both economic and practical reasons hair mercury levels are used. The data for hair mercury is correlated to intelligence and other measures of development to determine what the safe level of mercury is.

This is considered a valid methodology because hair and blood mercury levels have been shown to correlate well, which means if you put hair and blood mercury measurements on a graph and draw a straight line through it a lot of the points are near the line.

The problem with this methodology is that when people finally get enough mercury in them to be poisoned, their hair mercury goes down. Thus we have one paper that shows that the predicted hair mercury level based on the amount of mercury injected with childhood vaccinations (of the sort used until this year when most became thimerosal free) is high enough to suggest the children should have neurological problems as a consequence[25], but another that shows that the actual hair mercury level in autistic children is unusually low[26]. Despite having low hair mercury, autistic children appear to have more mercury in them as they excrete a lot more than their normal counterparts when given chelating agents[27].

What is really happening is that the most neurologically impaired children have lower hair mercury than everyone else and the correlation between hair mercury levels and development is much weaker than the actual correlation between mercury body burden or mercury's toxic effects and development. Thus this reasonable but incorrect assumption that hair (and blood) mercury are good indicators of body burden has led both to a serious underestimation of the dangers of mercury for growing children, and also to a false sense of security among doctors who order mercury tests at parents' urgings, see low results, and assume this means that the child has low mercury levels and can't be poisoned when in fact it means the child is poisoned by a high body burden of mercury which is being retained rather than made available for testing in blood and hair due to deranged mineral transport.

So when textbooks say things like this:

Rakel: Integrative Medicine, first Edition, 2003, in discussing organic mercury: "Fetal exposure: high (hair 10-20 ppm) – infant: seizures, mental retardation, and disturbed gait, vision and hearing; maternal hair (3-20 ppm) – school age: impaired language, attention, memory, and visual recognition correlated with prenatal level of exposure; effects occur even without maternal symptoms."

A reasonable person now has to suspect mercury as a potential cause for all these conditions in children with low, even unusually low, hair mercury levels.

What Textbooks say about ALS (Lou Gherig's disease)

Standard medical textbooks mention that both mercury and lead poisoning can easily be confused with Lou Gherig's disease[28].

Cecil Textbook of Medicine, 20th edition, 1996, p. 68, in the section on mercury: "The spinal cord may also be involved, with manifestations having some similarity to ALS."

Textbook of Clinical Neurology, Goetz, second edition, 2003, page 851 in the section on mercury: "Motor neuron disease resembling

[25] Predicted Mercury Concentrations in Hair From Infant Immunizations: Cause for Concern, Redwood, Bernard and Brown, *NeuroToxicology*, volume **22**, pages 691-697, 2001.

[26] Reduced levels of mercury in first baby haircuts of autistic children, by Holmes, Blaxill and Haley, *International Journal of Toxicology*, volume **22**, issue 4, pages 277-85, July-August (2003).

[27] A Case-Control Study of Mercury Burden in Children with Autistic Spectrum Disorders, Bradstreet, Geier, Kartzinel, Adams and Geier, *Journal of American Physicians and Surgeons*, volume **8**, number 3, pages 76-9, summer 2003.

[28] the medical term in the US is amylotropic lateral sclerosis or ALS, in the UK the terms motor neuron disease is used.

amylotrophic lateral sclerosis is another notable and prominent clinical pattern."

Merritt's Textbook of Neurology, 9th edition, 1995, page 668, in the section on mercury: "Because of the prominent motor manifestations, confusion with ALS can occur," and in the section on lead: "Rarely, upper motor neuron signs may occur with lower motor neuron disorder and mimic amylotrophic lateral sclerosis."

Handbook of Medical Psychiatry, Moore and Jefferson, second edition, 2004, page 338: "Rarely, an amylotrophic lateral sclerosis-like syndrome may be seen in lead or mercury poisoning."

With this background it seems reasonable that anyone who appears to have ALS would be screened for heavy metal toxicity since with proper treatment heavy metal poisoning has good outcomes, while even with the best treatments available, the outcome for ALS is always death and disease progression is usually rapid.

The woman whose hair test is on the facing page was healthy and active until age 70, when she felt a "snap" in her left leg (this was not a fracture, but rather a subjective sensation and this is how the woman always described it) and began deteriorating.

She went through a long, frustrating diagnostic process. Unfortunately this diagnosis process did not include any attempt to rule out heavy metal toxicity and she did not have a hair test done until after 3 years of illness and shortly before her death. Her niece convinced her she needed a hair test, which her attending doctors refused to order thus adding much delay. By the time the results were available she was in the final stages of the disease, had wasted away to 75 pounds, and no intervention based on a diagnosis of mercury toxicity was possible before she passed away.

She grew up on a farm, was a member of a religious order, and was a teacher. She had a mouthful of mercury amalgam fillings. Her test on page 46 counts this way:

Counting rule #1: Four bars go right. This satisfies rule #1.

Counting rule #2: 19 bars go left. This does not meet counting rule #2.

Counting rule #3: Six bars go into the red bands. This satisfies counting rule #3.

Counting rule #4: Nine bars are contained entirely in the middle green + white stripe. This satisfies counting rule #4.

Counting rule #5: not relevant since other counting rules are met.

This test in fact meets all possible criteria for deranged mineral transport – in theory a maximum of three counting rules can be met, and this test meets three counting rules.

The all low presentation of the essential elements means that the low values of the toxic elements aren't relevant and do not necessarily indicate that she had a low body burden of them.

The relative elevation of zinc suggests zinc wasting and low body zinc levels (common with deranged mineral transport) and that large amounts of supplemental zinc would have helped.

The all low presentation of the hair test suggests that measures to increase absorption of nutrients would have helped. Simple use of supplementary acid and digestive enzymes with meals is one way to do this, another is additional supplementation with vitamins, minerals and amino acids, and a third – which probably would have been justified and quite helpful in the case of someone near death as this woman was at the time the hair test was obtained – would have been to supply the nutrients, vitamins and minerals intravenously until she could absorb them adequately when taking them by mouth.

This case highlights the need to check for heavy metals early enough to do something about them if found, and also the health care provider's ethical obligation to help people check for things like heavy metals even if the health care provider believes they are unlikely to be relevant.

HAIR ELEMENTS

DOCTOR'S DATA INC.

SEX: Female
AGE: 73

POTENTIALLY TOXIC ELEMENTS

TOXIC ELEMENTS	RESULT µg/g	REFERENCE RANGE	PERCENTILE 68th 95th
Aluminum	1.9	< 7.0	
Antimony	0.016	< 0.05	
Arsenic	0.027	< 0.06	
Beryllium	< 0.01	< 0.02	
Bismuth	0.014	< 0.1	
Cadmium	0.026	< 0.1	
Lead	0.2	< 1.0	
Mercury	0.72	< 1.1	
Platinum	< 0.003	< 0.005	
Thallium	< 0.001	< 0.01	
Thorium	< 0.001	< 0.005	
Uranium	0.01	< 0.06	
Nickel	0.04	< 0.4	
Silver	0.01	< 0.15	
Tin	0.13	< 0.3	
Titanium	0.86	< 1.0	
Total Toxic Representation			

ESSENTIAL AND OTHER ELEMENTS

ELEMENTS	RESULT µg/g	REFERENCE RANGE	PERCENTILE 2.5th 16th 50th 84th 97.5th
Calcium	201	300- 1200	
Magnesium	45	35- 120	
Sodium	18	12- 90	
Potassium	3.0	8.0- 38	
Copper	11	12 35	
Zinc	250	140- 220	
Manganese	0.03	0.15- 0.65	
Chromium	0.21	0.2- 0.4	
Vanadium	0.013	0.018- 0.065	
Molybdenum	0.012	0.028- 0.056	
Boron	0.1	0.3- 2.0	
Iodine	6.0	0.25- 1.3	
Lithium	0.005	0.007- 0.023	
Phosphorus	201	160 250	
Selenium	1.0	0.95- 1.7	
Strontium	0.22	0.5- 7.6	
Sulfur	51100	44500- 52000	
Barium	0.16	0.26- 3.0	
Cobalt	0.005	0.013- 0.05	
Iron	5.8	5.4- 14	
Germanium	0.026	0.045- 0.065	
Rubidium	0.009	0.007- 0.096	
Zirconium	0.12	0.02- 0.42	

SPECIMEN DATA

COMMENTS:

Date Collected:
Date Received:
Date Completed:

Methodology: ICP-MS

Sample Size:
Sample Type:
Hair Color:
Treatment:
Shampoo:

V06.99

RATIOS

ELEMENTS	RATIOS	EXPECTED RANGE
Ca/Mg	4.47	4- 30
Ca/P	1	1- 12
Na/K	6	0.5- 10
Zn/Cu	22.7	4- 20
Zn/Cd	> 999	> 800

This form is provided to Andrew Cutler to be used as an educational tool and does not constitute endorsement by *Doctor's Data.*

What Textbooks say about Schizophrenia

Textbook of Clinical Neurology, by Goetz, second edition, 2003, says in the section on mercury poisoning: "Psychotic episodes with delirium, hallucinations, and motor hyperactivity have been reported."

Lead, mercury, carbon disulfide are listed as possible causes of acute psychosis in table 11-2 on page 107 of the *Textbook of Primary Care Medicine*, Noble, third edition, 2001.

Several other standard medical texts say that mercury exposure should be ruled out in anyone showing up at the hospital with an unexpected psychotic episode. In addition to this, there is journal literature directly linking mercury amalgam dental fillings to schizophrenia[29].

The sixteen year old boy whose hair test is presented on page 49 started out as a happy and friendly child. He was athletic, he enjoyed sports a lot and he was quite involved in them. He would sometimes play more than one sport in a season – great fun for him but his parents still remember how tiring it was to bring him to all the practices and games! School often did not come easy to him. He would spend long hours doing homework (without the need for parental pressure) and get by with average grades despite this great effort. He was able to relate to people of all ages, was well liked and got along with everyone. He passed the lifeguard test and planned to be a lifeguard the next summer. One of his goals was to play college football.

Around his sixteenth birthday he started to complain to his parents that the kids at school were talking about him. The parents didn't recognize this as suggesting any mental disturbance and simply hoped things would work themselves out. Then suddenly one day the boy was locking doors and saying that people were trying to break in to get him. He told his parents that the people on TV were talking to him. He was also hearing voices. He was very upset by this. Realizing at this point that something was seriously wrong, the parents took him to the emergency psychiatric ward, where he was kept as a patient for three weeks until he was stabilized on medication.

He was initially diagnosed as having manic depression[30] and treated aggressively with Risperdal and lithium. He gained weight, paced constantly, couldn't focus on anything, showed no emotions and had a very different personality. The parents thought he was overmedicated and, with physician supervision, started reducing the Rispderal dose. The boy subsequently developed heavy breathing, stopped talking, and acted like a zombie. He was brought back to the doctors who decided he was having another psychotic episode. By this time the parents had educated themselves, declined to let the doctors keep their son as an inpatient, and convinced the doctors to treat him as a schizophrenic rather than bipolar since his symptoms fit that more closely. The lithium was dropped, and Abilify was substituted for the Risperdal in hopes of lowering the side effects, and he was released to his parents' care.

The boy still had profound personality changes, still paced a lot, and did a lot of smiling for no apparent reason. He could not stay focused on anything and really couldn't do his schoolwork. He would show up and dress for sports but didn't play in competition. He didn't talk much and couldn't really carry on a conversation.

---

[29] Psychometric Evidence that Dental Amalgam Mercury May be an Etiological Factor in Schizophrenia, Siblerud, Motl and Kienholz, *Journal of Orthomolecular Medicine*, volume 14, number 4, pages 201-9, 1999.

[30] The medical term is bipolar disorder.

His new psychiatrist asked the parents not to dismiss the diagnosis of bipolar disorder too quickly, but held out no real hope of anything other than a lifetime of being medicated to the gills and nonfunctional regardless of whether they eventually decided the boy was schizophrenic or bipolar. From the parents' perspective the first year was Hell. The (mainstream medical) doctors offered little help, seemed to have a narrow view of the possible differential diagnosis, and mostly seemed to make their son worse with their prescribed treatments. These doctors told the parents that their son would be like this for the rest of his life and the parents needed to prepare for that.

Based on their own research regarding possible etiologic agents for their son's condition, the parents decided to get a hair test done to check for heavy metal problems since some of these (mercury, manganese) are well known to cause mental disorders and are mentioned in standard textbooks as problems that should be ruled out when someone unexpectedly has a psychotic episode. The boy's hair test is shown on the facing page and counts as follows:

Counting rule #1: 17 bars go right. This does not meet rule #1.

Counting rule #2: 6 bars go left. This does not meet counting rule #2.

Counting rule #3: no bars go into the red bands. This does not meet counting rule #3.

Counting rule #4: 21 bars are contained entirely in the middle green + white stripe. This does not meet counting rule #4.

Counting rule #5: is not met.

The count is only one short on rule number 2, which is suspicious that mineral transport may be deranged. Giving chelation a try seemed like a reasonable thing to do on this basis as a possible alternative to a lifetime of dysfunction and heavy medication.

The uranium level is elevated enough to possibly be significant (over 0.5, well into the red). Uranium causes fatigue and achiness. It does not as a rule cause psychiatric symptoms. It is sometimes elevated as a result of mercury induced derangement of mineral transport.

The father was immediately convinced that his son needed chelation, but the mother resisted it at first because the doctors told her it was dangerous, couldn't possibly help, that a lot of people got hurt by it, that DMSA was not an FDA approved drug (it is), and managed to frighten the mother into believing that she would hurt her son in some manner if she tried to chelate him.

The father had read enough on chelation and mercury – mostly on the internet – to be able to discuss these issues intelligently with the doctors. He would attempt to have reasoned discussions with the doctors while his wife listened in. A typical encounter would go like this: first the doctors would say that the son couldn't possibly have a mercury problem because he didn't have high levels in his blood or urine. The when the father pointed out that even mainstream medical textbooks like *Harrison's Textbook of Internal Medicine* and *Cecil Textbook of Medicine* said that blood levels only showed current exposure the doctors would reluctantly admit he was right, that they weren't familiar with the subject, then would refuse to prescribe DMSA because they weren't familiar with it!

The mother became convinced that chelation was worth trying when her husband showed her on the FDA website that DMSA was an approved medication for heavy metal chelation, and when she saw her husband argue rings around doctor after doctor, none of whom offered any reasonable discussion as to why heavy metal toxicity should not be considered and treated. Eventually the

HAIR ELEMENTS

DOCTOR'S DATA INC.

SEX: Male
AGE: 16

POTENTIALLY TOXIC ELEMENTS

TOXIC ELEMENTS	RESULT µg/g	REFERENCE RANGE
Aluminum	6.6	< 8.0
Antimony	0.052	< 0.066
Arsenic	0.04	< 0.08
Beryllium	< 0.01	< 0.02
Bismuth	0.024	< 0.06
Cadmium	0.04	< 0.15
Lead	0.29	< 1.5
Mercury	0.56	< 0.4
Platinum	< 0.003	< 0.005
Thallium	< 0.001	< 0.01
Thorium	0.004	< 0.005
Uranium	0.53	< 0.06
Nickel	0.25	< 0.4
Silver	0.07	< 0.13
Tin	0.19	< 0.3
Titanium	1.1	< 1.0
Total Toxic Representation		

ESSENTIAL AND OTHER ELEMENTS

ELEMENTS	RESULT µg/g	REFERENCE RANGE
Calcium	1080	200- 700
Magnesium	67	18- 70
Sodium	38	12- 90
Potassium	33	10- 40
Copper	28	10- 30
Zinc	160	130- 200
Manganese	0.31	0.15- 0.65
Chromium	0.31	0.2- 0.4
Vanadium	0.05	0.02- 0.075
Molybdenum	0.063	0.031- 0.062
Boron	2.4	0.34- 3.0
Iodine	0.61	0.25- 1.3
Lithium	0.009	0.007- 0.023
Phosphorus	210	160- 250
Selenium	0.83	0.95- 1.7
Strontium	3.2	0.3- 3.2
Sulfur	47300	44000- 52000
Barium	0.53	0.17- 1.5
Cobalt	0.016	0.013- 0.035
Iron	12	5.8- 13
Germanium	0.058	0.045- 0.065
Rubidium	0.029	0.009- 0.09
Zirconium	0.51	0.047- 0.7

SPECIMEN DATA

COMMENTS:
Date Collected:
Date Received:
Date Completed:
Methodology: ICP-MS

Sample Size:
Sample Type:
Hair Color:
Treatment:
Shampoo:

V06.99

RATIOS

ELEMENTS	RATIOS	EXPECTED RANGE
Ca/Mg	16.1	4- 30
Ca/P	5.14	0.8- 8
Na/K	1.15	0.5- 10
Zn/Cu	5.71	4- 20
Zn/Cd	> 999	> 800

This form is provided to Andrew Cutler to be used as an educational tool and does not constitute endorsement by *Doctor's Data.*

parents were able to find a doctor who was willing to prescribe DMSA and provide appropriate care for their son.

After 6 months and 23 chelation rounds (from 6 PM Friday to 6 AM Monday, every 3 hours, 50 mg each alpha lipoic acid and DMSA), the parents have been able to lower their son's dose of Abilify substantially. The school called and said the son was suddenly talking to people again! School officials continued to call every so often reporting further improvement. The wrestling coach said he could see the boy growing up before his eyes – by the end of the season he was acting somewhat normal, was able to compete in wrestling, and took first place in one of the wrestling tournaments.

While their son still has a ways to go, the parents could see after 23 chelation weekends that their son would become normal again in the foreseeable future. In fact, they were able to see progress after the third weekend. In addition to behavioral improvements, chronic acne on the boy's back suddenly disappeared and hasn't come back. The parents say the difference in their son's condition after six months of chelation is like night and day.

The parents report that chelation can cause increased symptoms as a side effect. While the symptoms come and go, they don't all come at once, and when they go, they are usually gone for good.

The boy's psychiatrist continues to insist that chelation can't possibly be doing any good and now offers the explanation that the boy must never have had bipolar disorder or schizophrenia and is just getting better on his own.

This is what the father has to say about his experiences over the last year and a half: "This whole situation with my son really opened my eyes. These mainstream doctors will talk against chelation like they are experts. But, I will sit and talk the facts with them. They will then tell me that they really do not know much about it. You would think that they see the improvements in my son and look into it. With what happen to my son my eyes are now open to the toxic metal problem. I am amazed at how many people I see that they or their children could possibly have this problem and are not being treated properly. My sister has found out that her problems could be due to this toxic metal poisoning. She had her fillings taken out and will be chelating. Too bad a doctor did not look into this possibility for her years ago. I believe chelation saved my son a life of misery. Look at how many other people who are suffering that could be helped as well. I try my best to let other people know. I just hope some of them will give it a try and look into it as a possible solution to their problems."

Another interesting case is a 56 year old woman who has been schizophrenic since age 15. She is currently well controlled while on the antipsychotic medication Stellazine as long as she doesn't work. If she tries to hold a job, she has a nervous breakdown. She also has fatigue, anxiety, depression, dizziness, irritable bowel, wheezing and weight gain. The weight gain and dizziness are side effects of the Stellazine.

Her hair test is presented on the facing page. It shows that she has the deranged mineral transport characteristic of mercury toxicity. It counts as follows:

Counting rule #1: 12 bars go right. This does not meet rule #1.

Counting rule #2: 11 bars go left. This does not meet counting rule #2.

Counting rule #3: 8 bars go into the red bands. This meets counting rule #3.

Counting rule #4: 7 bars are contained entirely in the middle green + white stripe. This meets counting rule #4.

Counting rule #5: other counting rules are met so it doesn't matter.

HAIR ELEMENTS

DOCTOR'S DATA INC

SEX: Female
AGE: 56

POTENTIALLY TOXIC ELEMENTS

TOXIC ELEMENTS	RESULT µg/g	REFERENCE RANGE	PERCENTILE 68th 95th
Aluminum	4.8	< 7.0	
Antimony	0.12	< 0.050	
Arsenic	0.26	< 0.060	
Beryllium	< 0.01	< 0.020	
Bismuth	0.093	< 0.10	
Cadmium	0.096	< 0.10	
Lead	2.1	< 1.0	
Mercury	0.19	< 1.1	
Platinum	< 0.003	< 0.005	
Thallium	< 0.001	< 0.010	
Thorium	< 0.001	< 0.005	
Uranium	0.001	< 0.060	
Nickel	0.22	< 0.40	
Silver	2.4	< 0.15	
Tin	0.55	< 0.30	
Titanium	0.39	< 1.0	
Total Toxic Representation			

ESSENTIAL AND OTHER ELEMENTS

ELEMENTS	RESULT µg/g	REFERENCE RANGE	PERCENTILE 2.5th 16th 50th 84th 97.5th
Calcium	106	300- 1200	
Magnesium	8	35- 120	
Sodium	270	12- 90	
Potassium	100	8- 38	
Copper	14	12- 35	
Zinc	120	140- 220	
Manganese	0.09	0.15- 0.65	
Chromium	0.57	0.20- 0.40	
Vanadium	0.075	0.018- 0.065	
Molybdenum	0.087	0.028- 0.056	
Boron	2.8	0.30- 2.0	
Iodine	2.4	0.25- 1.3	
Lithium	< 0.004	0.007- 0.023	
Phosphorus	145	160- 250	
Selenium	6.0	0.95- 1.7	
Strontium	0.62	0.50- 7.6	
Sulfur	52000	44500- 52000	
Barium	0.14	0.26- 3.0	
Cobalt	0.025	0.013- 0.050	
Iron	8.7	5.4- 14	
Germanium	0.025	0.045- 0.065	
Rubidium	0.094	0.007- 0.096	
Zirconium	0.084	0.020- 0.42	

SPECIMEN DATA

COMMENTS:

Date Collected:
Date Received:
Date Completed:
Methodology: ICP-MS

Sample Size:
Sample Type:
Hair Color:
Treatment:
Shampoo:

V06.99

RATIOS

ELEMENTS	RATIOS	EXPECTED RANGE
Ca/Mg	13.3	4- 30
Ca/P	0.731	1- 12
Na/K	2.7	0.5- 10
Zn/Cu	8.57	4- 20
Zn/Cd	> 999	> 800

This form is provided to Andrew Cutler to be used as an educational tool and does not constitute endorsement by *Doctor's Data.*

Elevated hair selenium might in theory be due to unreported use of a selenium containing shampoo, and this woman does live in an area with low lithium. However even if the selenium and lithium were considered false abnormals this woman still more than meets two counting rules. This test is so wild that the probability of it occurring by chance is incredibly small[31]. Most likely this woman has suffered from mercury poisoning for 41 years because it was misdiagnosed as schizophrenia.

The very elevated arsenic and silver may or may not have any significance since the essential element results are so scattered. This woman does not report being chemically sensitive even though chemical sensitivity is very common in people who have a lot of arsenic and mercury in their systems.

She does have a dramatic sign of inadequate adrenal function in the very low calcium and magnesium compared to the very high sodium and potassium. This would be expected to make her quite stress intolerant, and in fact she has nervous breakdowns if she tries to hold a job.

Her cholesterol is high – over the real upper normal limit of 275, not just the moralistic limit of 200 – which is also consistent with adrenal problems since the adrenal steroid hormones are made from cholesterol and the body's natural response to needing more adrenal hormones is to provide more cholesterol to make them from.

The sodium and potassium being up suggests her adrenals are shot but her thyroid continues to function properly – and proper thyroid function was confirmed by blood tests.

Other laboratory testing showed that her cysteine and sulfate were elevated, which is characteristic of heavy metal intoxication. She also has a large number of IgG mediated food allergies which suggests she has a leaky gut[32] – another common consequence of mercury toxicity – and helps explain the irritable bowel problems.

This woman had, on her own initiative, chosen to have her amalgam fillings replaced with mercury free materials about 3 years prior to the date of the hair test.

What Textbooks say about Cancer

Because the dominant belief in medicine is that toxicity problems are an occupational hazard, most research on metals causing cancer (and other diseases) has been performed on people exposed in the workplace. Regardless of where toxic metals come from, if they get into someone's body they behave however they are going to behave. The real issue is whether the metals are there, and one way to determine that is a hair test.

One would expect toxicology books to mention heavy metals and cancer if the association is well known, and in fact they do. In *Clinical Toxicology*, by Ford, first edition (2001), page 718:

"Inorganic arsenic is a recognized human carcinogen, primarily of the respiratory tract and skin (International Agency for Research on Cancer group 1 and U.S. Environmental Protection Agency group A)."

Textbooks on cancer would be expected to have significant coverage of any links between heavy metals and cancer, and they do. Perusing *Clinical Oncology* by Abeloff, second edition

[31] 10^{-5} to be exact.

[32] a diagnosis primarily considered by alternative doctors in which the cells in the walls of the intestine do not adhere to each other as closely as they should, letting toxins and undigested protein fragments get into the bloodstream.

(2000), we find much useful information.

Page 289: "Medicinal use of inorganic arsenic was associated with skin cancers in the early 20th century. More recently, excess skin cancer has been observed in populations exposed to arsenic-contaminated drinking water."

and

"Reports of skin and lung cancers among vineyard workers with exposure to arsenic fungicides and pesticides appeared during the late 1950s."

and

Page 290: "Elevated risk of prostate and lung cancer among workers exposed to cadmium has been reported."

and

"Experimental investigations indicate that the hexavalent salts of chromium are highly carcinogenic, whereas trivalent chromium is not carcinogenic."

With mainstream medicine's focus on occupational exposure which usually happens by breathing dust at work we would expect to find this topic covered in relevant texts if an association between heavy metals and cancer was believed to be important. In fact, the *Textbook of Respiratory Medicine* by Murray and Nadel, third edition (2000) has a lot to say on this topic:

(page 1400) "Occupational exposure to inorganic arsenic in copper smelting causes a 2- to 14-fold increase in respiratory cancer risk."

(P 1401) "Hexavalent chromium compounds are established human carcinogens."

and

"The increased lung cancer risk prevalent in nickel refinery workers was initially attributed to exposure to metallic nickel and nickel carbonyl, which are carcinogenic in animals. However, subsequent epidemiologic evidence has suggested that exposure to combinations of nickel sulfides and oxides encountered in the refining industry increases lung cancer risk."

(page 1402): "Two cohort mortality studies have reported a significant excess of lung cancer associated with exposure to beryllium."

(page 1403): "Follow-up from a cohort study of cadmium smelter workers reported a relative risk of 1.49 not explainable by smoking." (The risk is for lung cancer).

The role of heavy metals in causing cancer is so well known that basic medical texts covering internal medicine and pathology also discuss it.

In table 193-2 of *Cecil Textbook of Medicine*, 21st edition (2000) arsenic, chromium and nickel are listed as being carcinogenic.

Table 8-3 on page 274 of *Robbins Pathologic Basis of Disease*, by Cotran, Kumar and Collins, sixth edition (1999) lists arsenic and arsenic compounds as causing lung and skin cancer as well as hemangiosarcoma[33], beryllium and its compounds as causing lung cancer, cadmium and its compounds as causing prostate cancer, chromium and its compounds as causing lung cancer, as well as nickel and its compounds as causing lung and nasal cancer.

Checking the *Textbook of Natural Medicine* by Pizzorno and Murray, second edition (1999) we find on page 168:

"One study which evaluated bladder cancer mortality over a 5-year period in 26 counties in the US, found that bladder cancer was significantly higher in counties with documented arsenic exposure."

---

[33] a malignant tumor derived from blood vessel cells.

In addition to excessive levels of toxic elements being carcinogenic, inadequate levels of certain nutrient elements may also increase the risk of cancer. For example, in *Sleisenger & Fordtran's Gastrointestinal and Liver Disease* by Feldman, seventh edition (2002), the authors state that low dietary selenium can cause colon cancer. Low levels of selenium (and other relevant nutrient elements) may be detected by a hair test as long as mineral transport is normal and orderly.

There is an extremely large amount of journal literature on the topic of what causes cancer. The most useful study is Environmental and Heritable Factors in the Causation of Cancer: Analyses of Cohorts of Twins from Sweden, Denmark and Finland, by Lichtenstein *et al*., New England Journal of Medicine, volume 343, number 2, pages 78-85 (2000).

"Inherited Genetic factors make a minor contribution to susceptibility of most types of neoplasms. This finding indicates that the environment has the principal role in causing sporadic cancer."

and

"We conclude that the overwhelming contributor to the causation of cancer in the population of twins that we studied was the environment."

This study covered 44,788 pairs of twins. Due to this enormous number of subjects statistical issues are minimized and the percentage likelihoods of various cancers being genetic or environmental can be established with good accuracy.

In this twin study, even the cancers with the strongest genetic component (prostate, colorectal and breast cancer) environmental influences were conclusively shown to be stronger than inheritance – in technical terms, at the 95% confidence level the upper limit on the number of cancers that were genetic in origin was 50% or less. On average, 75% of the cancers were found to be of environmental origin, 25% due to genetic risk factors.

Most medical research has been directed at identifying the genetic and occupational causes of cancer. A lot of effort has gone into studying a limited number of chemicals. Those few chemicals where there was medically recognized widespread population exposure and a clear cancer problem due to them have also been studied. However, the link between cancer and most environmental chemicals that people are exposed to at home, through play, hobbies or by medical procedures or government regulation has for the most part not been studied. Obviously, if little research is done in an area it is unlikely that any cause and effect relationships will be found even if they are obvious and strong. Since most heavy metals have similar physiological effects and cause toxicity through similar mechanisms it is reasonable to suspect that all of them are carcinogenic to some extent since it turns out all the ones that have been extensively studied are.

If you go from the research literature to the clinics of health care providers who actually do screen for heavy metal problems in cancer patients and look over the tests it is quite apparent that mercury and lead show up in a lot of these people in addition to arsenic, nickel and chromium.

Heavy metal exposure is one risk factor that can be easily screened for with a hair test, and for which there are effective means of removing the cancer causing material from the body if it is found.

The hair test on the facing page is from someone with recurrent thyroid cancer[34]

[34] One paper of particular interest is Trace Elements and Thyroid Cancer, by Zuichik, Tsib and Vtyurin in *Analyst*, volume 120, pages 817-821 (1995). The authors found vastly elevated levels of mercury in cancerous thyroid nodules removed at surgery, but not in benign nodules.

HAIR ELEMENTS

DOCTOR'S DATA INC.

SEX: Male
AGE: 40

POTENTIALLY TOXIC ELEMENTS

TOXIC ELEMENTS	RESULT µg/g	REFERENCE RANGE	PERCENTILE (68th, 95th)
Aluminum	6.9	< 7.0	
Antimony	0.016	< 0.066	
Arsenic	0.039	< 0.08	
Beryllium	< 0.01	< 0.02	
Bismuth	0.008	< 0.06	
Cadmium	0.098	< 0.15	
Lead	0.91	< 2.0	
Mercury	0.62	< 1.1	
Platinum	< 0.003	< 0.005	
Thallium	< 0.001	< 0.01	
Thorium	< 0.001	< 0.005	
Uranium	0.004	< 0.06	
Nickel	0.37	< 0.4	
Silver	0.1	< 0.12	
Tin	0.14	< 0.3	
Titanium	0.62	< 1.0	
Total Toxic Representation			

ESSENTIAL AND OTHER ELEMENTS

ELEMENTS	RESULT µg/g	REFERENCE RANGE	PERCENTILE (2.5th, 16th, 50th, 84th, 97.5th)
Calcium	979	200- 750	
Magnesium	21	25- 75	
Sodium	6.0	12- 90	
Potassium	5.0	9.0- 40	
Copper	16	10- 28	
Zinc	170	130- 200	
Manganese	0.2	0.15- 0.65	
Chromium	0.18	0.2- 0.4	
Vanadium	0.033	0.018- 0.065	
Molybdenum	0.026	0.025- 0.064	
Boron	0.61	0.4- 3.0	
Iodine	0.28	0.25- 1.3	
Lithium	0.004	0.007- 0.023	
Phosphorus	155	160- 250	
Selenium	1.0	0.95- 1.7	
Strontium	2.3	0.3- 3.5	
Sulfur	41900	44500- 52000	
Barium	0.47	0.16- 1.6	
Cobalt	0.054	0.013- 0.035	
Iron	9.2	5.4- 13	
Germanium	0.018	0.045- 0.065	
Rubidium	0.004	0.011- 0.12	
Zirconium	0.044	0.02- 0.44	

SPECIMEN DATA

COMMENTS:

Date Collected:
Date Received:
Date Completed
Methodology: ICP-MS

Sample Size:
Sample Type:
Hair Color:
Treatment:
Shampoo:

V06.99

RATIOS

ELEMENTS	RATIOS	EXPECTED RANGE
Ca/Mg	46.6	4- 30
Ca/P	6.32	0.8- 8
Na/K	1.2	0.5- 10
Zn/Cu	10.6	4- 20
Zn/Cd	> 999	> 800

This form is provided to Andrew Cutler to be used as an educational tool and does not constitute endorsement by *Doctor's Data.*

that began in his teenage years and recurred at age 38. He is alive at the time of this writing, but the cancer is expected to kill him. He needs to take large amounts of thyroid hormone to control the cancer (by suppressing thyroid activity) and this makes him quite miserable. He found that Thyrolar is tolerable but Levoxyl is not[35]. He is a small, thin, very nice person. His mother died of melanoma and his grandfather had eye cancer. Due to the debilitating effects of the very high amounts of thyroid hormone he takes he is not able to work and is on welfare. He takes many nutritional supplements and has learned to give himself nutritional solutions intravenously at home pursuant to doctor's orders. His hair test is on the prior page and counts as follows:

Counting rule #1: six bars go right. This does not meet rule #1.

Counting rule #2: sixteen bars go left. This does not meet counting rule #2.

Counting rule #3: two bars go into the red bands. This does not meet counting rule #3.

Counting rule #4: twelve bars are contained entirely in the middle green + white stripe. This does not meet counting rule #4.

Counting rule #5: six bars go right and twelve bars are entirely within the middle band. This meets counting rule #5, that two other rules are one short of being met.

This test result suggests a high probability of mercury induced derangement of mineral transport.

Also note the relatively great separation of the potassium and calcium results (without magnesium following calcium). This suggests a physiologically inappropriate level of thyroid hormones which this gentleman indeed does have – his health care provider is prescribing so as to purposefully create such a situation in order to control his cancer.

[35] Thyrolar provides a physiologically realistic ratio of 1 part of the thyroid hormone T3 to 4 parts of the hormone T4. Armour thyroid, Westhroid and Biothroid are similar. Levoxyl and Synthroid provide only the T4 form of thyroid hormone. For a discussion of the relevance of T3 and T4 in thyroid treatment see pages 114-118 of *Amalgam Illness: Diagnosis and Treatment* (information about the book can be obtained at www.noamalgam.com).

The hair test on the next page is from a gentleman who died at age 59 from a recurrence of brain cancer. He had earlier had a brain tumor which was surgically removed and he received radiation therapy. Subsequently he developed a second similar brain tumor and was sent to a nursing home for care until his shortly expected demise. A family member sought alternative care for him (which did not include heavy metal detoxification) and was only able to try a limited number of interventions before his death. This gentleman's hair test, shown on the next page, counts as follows:

Counting rule #1: fifteen bars go right. This does not meet rule #1.

Counting rule #2: eight bars go left. This does not meet counting rule #2.

Counting rule #3: three bars go into the red bands. This does not meet counting rule #3.

Counting rule #4: twelve bars are contained entirely in the middle green + white stripe. This does not meet counting rule #4.

Counting rule #5: 3 bars enter the red zone and 12 bars are entirely within the middle band. This meets counting rule #5, that the test counts one short of positive on any other two rules.

This suggests a high probability of mercury induced derangement of mineral transport. Also note that there is frankly elevated hair mercury. while this is not common when mineral transport is deranged it certainly does confirm that excessive mercury is present and is interfering with physiological functions.

Note that due to the derangement of mineral transport, the high vanadium level may not accurately reflect body stores. It would certainly have been worth following up on given its vast elevation but there is a reasonable likelihood that blood and urine vanadium levels, and body stores, would be normal. Also note that the high sodium and potassium in relation to the magnesium and calcium suggests inadequate adrenal production of hormones like cortisol and DHEA with compensating excessive production of adrenaline, another issue on which follow up testing would have been relevant. This gentleman passed away as a result of his cancer before such follow up was performed.

HAIR ELEMENTS

DOCTOR'S DATA INC

SEX: Male
AGE: 59

POTENTIALLY TOXIC ELEMENTS

TOXIC ELEMENTS	RESULT µg/g	REFERENCE RANGE
Aluminum	2.8	< 7.0
Antimony	0.017	< 0.066
Arsenic	0.04	< 0.08
Beryllium	< 0.01	< 0.02
Bismuth	0.011	< 0.06
Cadmium	0.046	< 0.15
Lead	0.19	< 2.0
Mercury	0.66	< 1.1
Platinum	0.003	< 0.005
Thallium	< 0.001	< 0.01
Thorium	< 0.001	< 0.005
Uranium	0.018	< 0.06
Nickel	0.08	< 0.4
Silver	0.03	< 0.12
Tin	0.62	< 0.3
Titanium	0.82	< 1.0
Total Toxic Representation		

PERCENTILE: 68th, 95th

ESSENTIAL AND OTHER ELEMENTS

ELEMENTS	RESULT µg/g	REFERENCE RANGE
Calcium	711	200- 750
Magnesium	35	25- 75
Sodium	2.0	12- 90
Potassium	3.0	9.0- 40
Copper	22	10- 28
Zinc	260	130- 200
Manganese	0.17	0.15- 0.65
Chromium	0.21	0.2- 0.4
Vanadium	0.02	0.018- 0.065
Molybdenum	0.012	0.025- 0.064
Boron	0.29	0.4- 3.0
Iodine	2.3	0.25- 1.3
Lithium	0.004	0.007- 0.023
Phosphorus	203	160- 250
Selenium	2.0	0.95- 1.7
Strontium	1.2	0.3- 3.5
Sulfur	54700	44500- 52000
Barium	0.26	0.16- 1.6
Cobalt	0.007	0.013- 0.035
Iron	11	5.4- 13
Germanium	0.028	0.045- 0.065
Rubidium	0.005	0.011- 0.12
Zirconium	0.025	0.02- 0.44

PERCENTILE: 2.5th, 16th, 50th, 84th, 97.5th

SPECIMEN DATA

COMMENTS:

Date Collected:
Date Received:
Date Completed:

Sample Size:
Sample Type:
Hair Color:
Treatment:
Shampoo:

Methodology: ICP-MS

V06.99

RATIOS

ELEMENTS	RATIOS	EXPECTED RANGE
Ca/Mg	20.3	4- 30
Ca/P	3.5	0.8- 8
Na/K	0.667	0.5- 10
Zn/Cu	11.8	4- 20
Zn/Cd	> 999	> 800

This form is provided to Andrew Cutler to be used as an educational tool and does not constitute endorsement by *Doctor's Data*.

Reasons to check for Heavy Metal Poisoning

There are six basic reasons to check for heavy metal poisoning:

- Problems suggestive of heavy metals
- A health condition with a poor expected outcome
- A history of or a high risk of cancer
- Health problems the doctor can't find a cause (or a cure) for
- A relative has a heavy metal problem without a large exposure
- One person has several apparently unrelated health problems

Problems suggestive of heavy metals include combinations of fatigue, malaise, anxiety, depression, emotional distress and allergies that are not otherwise explained.

Health conditions with poor expected outcomes are things like Parkinson's disease, borderline personality disorder, autism, multiple sclerosis, lupus erythrematosus. After all, heavy metal intoxication is easily treatable so it is prudent to make sure the person doesn't really have it.

History of cancer is a reason to exclude heavy metal intoxication since most cancers are environmental in origin, not genetically determined[36].

Health problems that the doctor can't find a cause (or if they assert a cause, can't find a cure) for are things like fibromyalgia, chronic fatigue, irritable bowel syndrome, ADHD, chemical sensitivity, or severe allergies.

Most cases of heavy metal poisoning today are due to people with a genetic predisposition to accumulate some metal. The relatives of these people have a good chance of sharing this predisposition and need to be checked. Also note that I have seen many family sets of hair tests, and the husbands and wives almost always both have heavy metal problems or neither do. Most of the heavy metal toxicities are known to affect mood and personality which apparently leads to people with heavy metal problems getting together.

One person having the misfortune to have several unrelated health problems is quite unlikely. When one person DOES have a lot of health problems, even if they are mundane ones, it makes sense to check and be sure they aren't all due to heavy metal toxicity.

---

[36] Environmental and heritable factors in the causation of cancer--analyses of cohorts of twins from Sweden, Denmark, and Finland. By Lichtenstein *et al.* in the *New England Journal of Medicine*, volume **343**, number 2, pages 78-85, July 13th, 2000.

Why bother to do a Hair Test?

Figures 1 and 2 show the various possible causes of serious childhood developmental disorders and also of adult disorders that are poorly handled by conventional medicine. The areas taken up by the different causes correspond approximately to the prevalence of that particular cause among the pool of people with the given diagnosis. That is to say, on the autism and PDD figure about 75% of the area is covered by mercury and the other 25% by other things since about 75% of autistics have mercury poisoning.

These figures are helpful in understanding why you want to do a hair test and also in understanding why you want to double or triple check certain diagnoses. If you lay the open book on the floor and throw a dart at it, you aren't very likely to hit one of the small boxes. This is statistically the same as the likelihood that a given person has the cause given in that box for their problems. If you were to throw a dart at the book and the dart hit a teensy little box, you'd go look to make sure the dart was really in there, just like you look carefully at the dartboard when it looks like you threw a bulls-eye. This is the same as saying that if a doctor says a child has genetically determined autism that the parents ought to get further testing done – the medical equivalent of looking real close – before they believe that.

These figures are also helpful in choosing doctors, since it is unlikely the doctors who say everyone has the same cause for their problem really know what they are doing, and if they also insist it is a relatively uncommon cause it is not likely they are going to make very many people better. On the other hand, if they treat something moderately common the doctor may well benefit a fair number of the people who walk through their door by treating them all the same regardless of what they really have. If you happen to have one of the less common conditions that doctor isn't going to help you even though he does help a lot of other patients.

Figures 3 and 4 compare what fraction of the causes of these conditions standard mainstream medical tests check for versus the fraction checked for by hair testing. Since hair testing covers most of the common causes and is inexpensive, while regular mainstream testing covers few causes and is expensive, it makes sense to try hair testing early on.

Figure 1. Probabilities of different causes for autism, PDD, etc. The areas of the shaded regions correspond to the relative likelihoods.

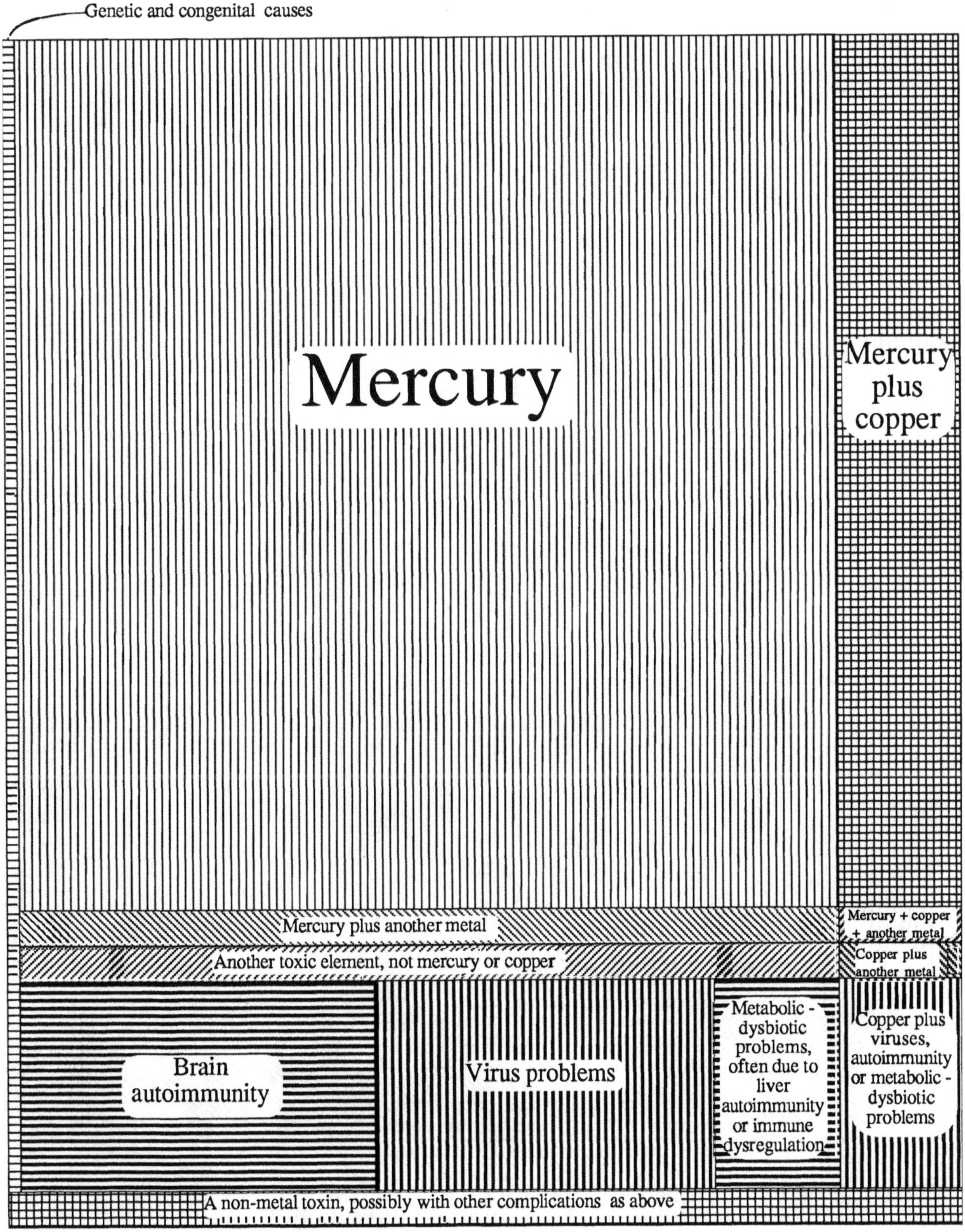

Figure 2. Probabilities of different causes for chronic fatigue, fibromyalgia, chemical sensitivity, psychiatric problems. The areas of the shaded regions correspond to the relative likelihoods.

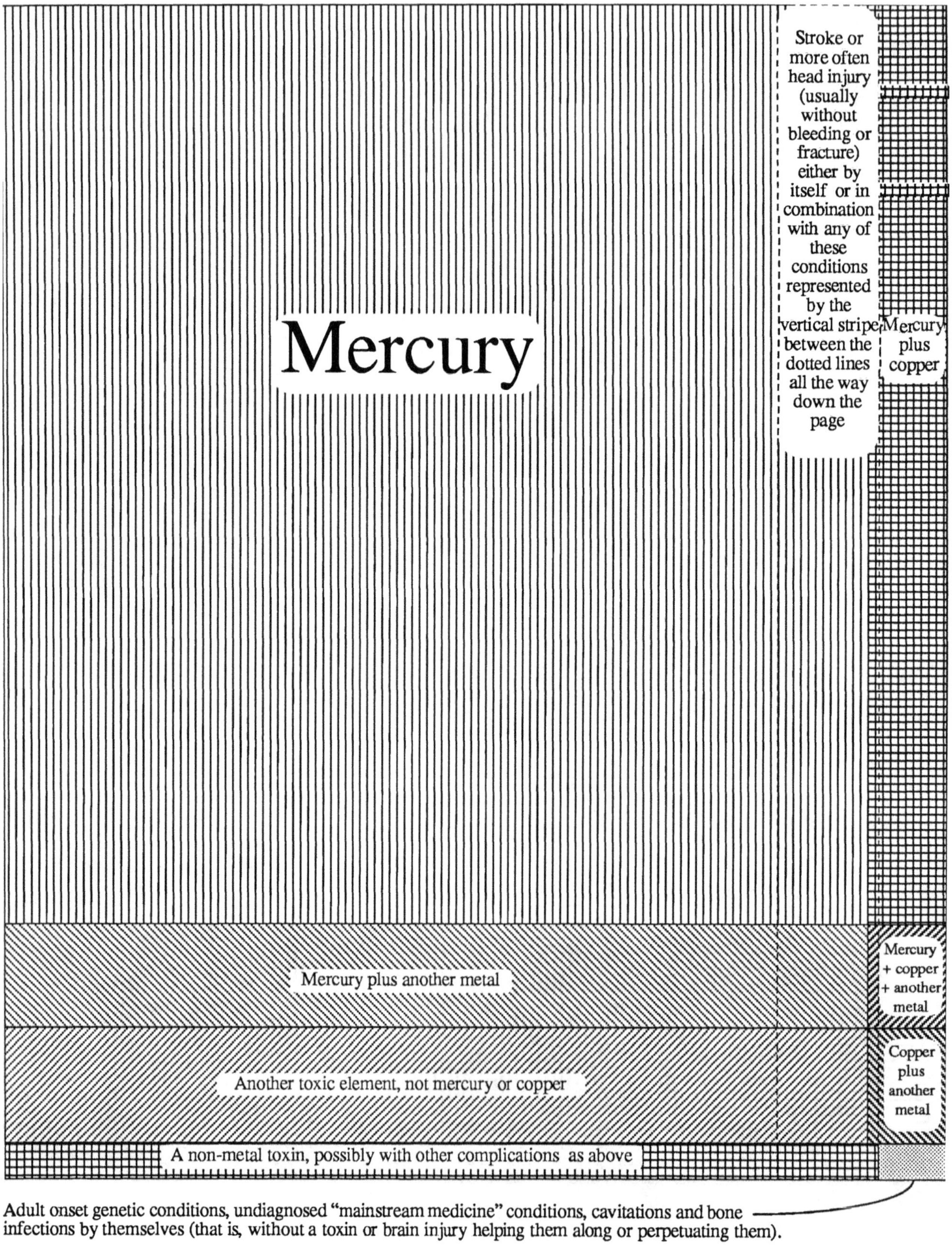

Figure 3. Where different kinds of medical testing look for the cause of a child's problems

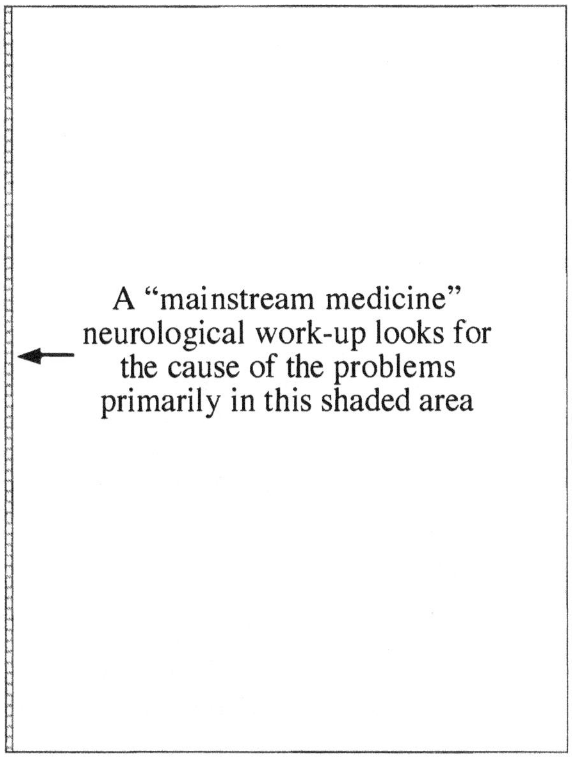

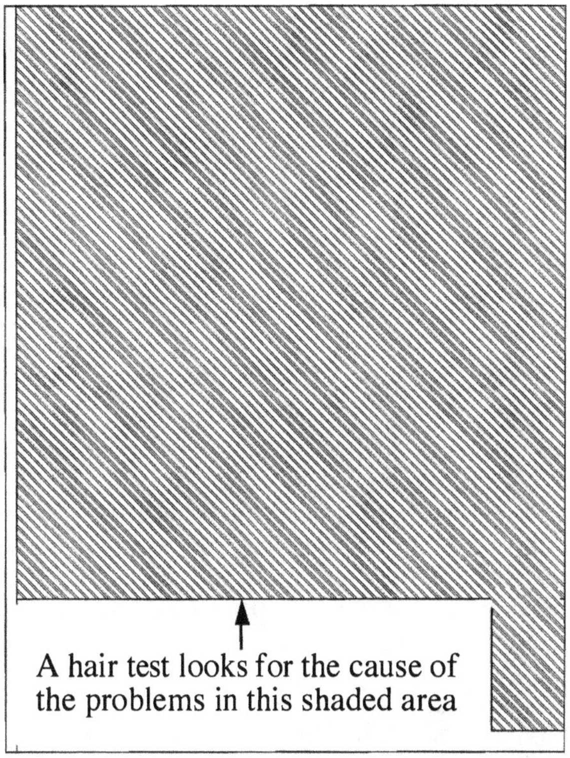

Figure 4. Where different kinds of medical testing look for the cause of adult problems

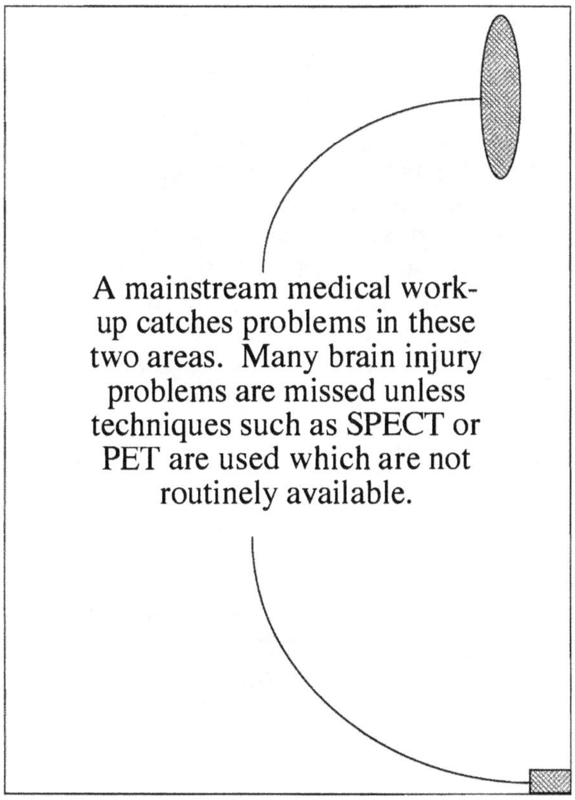

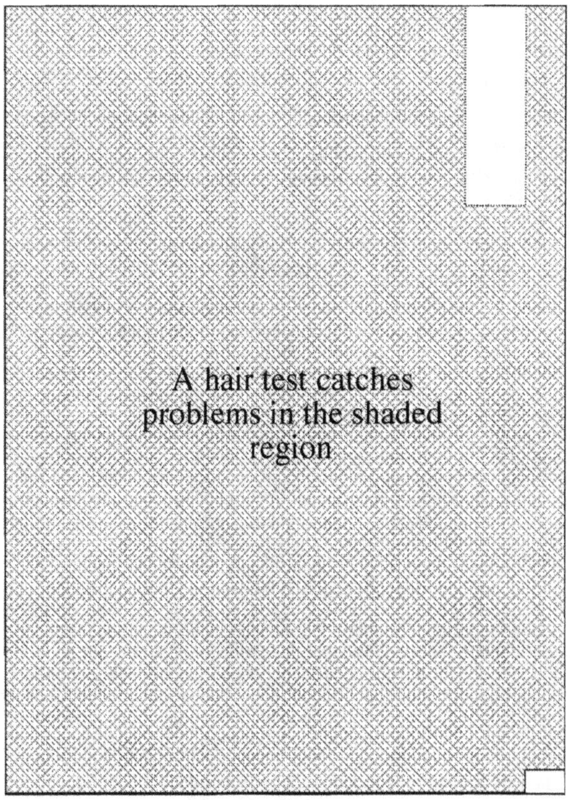

Why Worry about Heavy Metals?

The federal Centers for Disease Control (CDC) did a study[37] based on blood and urine samples taken from 3,800 people in 1999. CDC found that 10% of young children and women of child bearing age had mercury levels high enough that the children or unborn babies may reasonably be expected to suffer health damage.

Previously the National Research Council (NRC) had estimated that 60,000 children were born annually with too much mercury in their system - a HUGE number - but this CDC study shows that was a substantial underestimate.

To make matters worse, until 2004[38] these children received vast quantities of highly toxic organic mercury in their vaccinations in addition to the already high levels present in their bodies!

Other CDC studies showed that in the 1990's, 4.4% of children had too much lead in them. Since almost none of them received treatment, it is still there, making them less intelligent and functional than they should be. The good news is that in 2001, only 2.2% of young children had too much lead. The bad news is that 2.2% of the population is an incredibly high number. We are talking about roughly a million little children with brain damaging chemicals in their system that nobody is doing anything about, and maybe 2 million older children in the same situation.

Over the last 30 years the "threshold" for lead in children's blood has been reduced from 45 to 10. Current studies[39] show that a blood level of 10 corresponds to an average loss of 7 IQ points. Of course, half the people lose more than this, half less, since it is an average. And of course adults who were children back when 45 was an acceptable blood level might have lost 32 IQ points due to lead levels physicians didn't consider worth doing anything about. To put this in proper perspective, 100 is an average IQ. People with an IQ of 68 usually can't live on their own as adults, they need to be taken care of in a group home or institution.

---

[37] The study is not yet published as of the time this book was written. It is expected to appear in *Morbidity and Mortality Weekly Reports*, MMWR. The prior study for 1999 is reported there and can be downloaded on the internet at http://www.cdc.gov/mmwr/PDF/wk/mm5008.pdf.

[38] While thimerosal containing vaccines were taken off the market earlier, many doctors continued to administer their stock of thimerosal containing vaccinations until those expired – a year or more after their manufacturing date. For practical purposes vaccines are made and bought a year in advance so children were still receiving mercury laden vaccines until 2004.

[39] Recent Developments in Low-Level Lead Exposure and Intellectual Impairment in Children, Koller, Brown, Spurgeon and Levy, Environmental Health Perspectives, volume **112**, number 9, pages 987-94, 2004. Downloadable on the web for free at http://ehp.niehs.nih.gov/members/2004/6941/6941.pdf.

There is a lot of discussion of the hazards of mercury in fish. In fact, women of childbearing age should not eat more than one can of tuna a week *if that is their only source of mercury exposure* due to the risks it poses to their unborn children. It is politically incorrect to talk about the facts that fish mercury is a much smaller source of exposure for the mother and her unborn child than dental amalgam fillings, and is much smaller for the child than infant vaccinations were until last year. Unfortunately, once mercury was removed from all existing childhood vaccines, flu shots were added to the list of vaccinations children routinely receive and these still do contain mercury.

Also curiously absent from the news is much discussion of the effect of the arsenic treated wood that suddenly replaced creosote treated wood in the early 1990's. This new kind of wood has dramatically increased children's and adults' exposure to arsenic. There is also little discussion of antimony from flame retardant pajamas and bedding. Both arsenic and antimony are well known to be toxic, and are being observed at high levels in surprising numbers of people by those few doctors who check. "Mainstream medicine" says don't bother to check for antimony or arsenic unless people work in factories or occupations using it. Unfortunately this attitude comes from older experience gathered before these poisons were placed in every home as wood preservatives and flame retardants.

The primary problems with heavy metals are that people's metabolism of them varies greatly, and that individual exposure can also vary greatly. Most people are not exposed to unusual amounts, and clear the metals at an "average" rate. These people are fine. It is the people who clear the metals slowly or who are exposed to more than usual that have trouble. This presents health care providers with a very confusing situation where everyone *seems* to be exposed to about the same amount of heavy metals and most people are fine. Thus many doctors "don't get it," because it is often impossible to figure out that someone has an unusual exposure level, or had a large past exposure, and it is also difficult to figure out how a given individual's metabolism for heavy metals works.

For a detailed discussion of individual variations in the metabolism of mercury, see pages 16-25 of *Amalgam Illness: Diagnosis and Treatment*[40].

We have a situation today where we know large numbers of people accumulate too much toxic metal and suffer ill effects from that. Unfortunately we have a system full of health care providers who weren't taught how to tell if you are one of them, but were taught not to order any tests to check. Since metal toxicity doesn't fit the general "medical model" of disease, which is that either you are OK or you are obviously sick (that is, your arm is either broken or not, you either have the flu or you don't) health care providers may have trouble recognizing that there is a problem until it is severe.

The most important reason to worry about heavy metals is quite straightforward. The kinds of problems heavy metals cause are generally considered to be untreatable by mainstream medicine, and to have poor outcomes. Heavy metal poisoning IS treatable, treatment usually has a good outcome, and neither testing nor treatment are particularly expensive or unpleasant. So whenever someone has a condition that is chronic and has a negative impact on their life it makes a lot of sense to check for heavy metals.

---

[40] See www.noamalgam.com for more information or to obtain the book.

Diagnostic Difficulties in Modern Medical Practice

Medicine is taught by apprenticeship, sometimes at the expense of scholarship. The new doctors spend years listening to and following older doctors around. After watching how they practice these new doctors then practice under the close supervision of older doctors. Great stress is placed on learning how to do it the way the senior doctor does it. Critical thinking skills, innovative problem solving and approaching clinical problems with an open mind are not stressed. Learning how to practice medicine from books is seldom on the agenda.

This is done because medicine is an art. While medicine uses scientific principles and techniques, the goal is not to turn patients into statistics in some grand experiment, it is to make all of them as much better as they can be. It is important for patient safety and comfort that doctors intuitively recognize dangerous or debilitating conditions early on, and select the proper treatment for every patient, including those who don't have the same response to treatment that most people do. Also, while newly graduated scientists and engineers can be sent into the job market with lots of theoretical knowledge but little practical skill and then spend the first year or two learning the way things are really done on the job, it would be unethical to let doctors care for patients before they have a very practical and realistic understanding of what drugs really do, how patients really respond to different therapies, etc. Medicine is a performing art where aspiring doctors are expected to be understudies until there is great certainty that every performance will be perfect when they finally go on stage in the clinic or hospital and are responsible for the health of real people.

Unfortunately, this means that once an error creeps into medical practice, it is almost impossible to correct even if all the textbooks are rewritten to address it. This appears to be what is what has happened in the area of toxicology as it is relevant to heavy metals in non-occupational settings.

Many years ago, severe poisoning in the workplace was quite common. There wasn't much that doctors could do about it, either, until the poor worker got so sick the doctor had to order them to quit their job so as to reduce exposure. Given the huge exposures in industry, the lack of much that could be done other than ceasing further exposure, and the relatively greater expense of sophisticated tests in earlier times, the general approach was only to diagnose heavy metal poisoning when someone had an occupational exposure, vastly elevated blood or urine levels of the metal, and was so sick that the loss of their job was a secondary consideration. This general view has passed down from senior doctor to aspiring young students and residents for a few generations now.

In the meantime industry has become adequately regulated so that they don't poison many people. Thus today's doctors are taught that heavy metal poisoning does not occur without industrial exposure and high blood or urine levels. They are taught this by doctors who in fact have seen few if any such cases themselves and thus don't recognize the signs and symptoms of the various heavy metal poisonings clinically anyway.

Research over the last 50 years has shown that levels of the toxic metals far lower than those used in industry have significant harmful effects on people and in fact are enough to poison some severely. Despite the fact that this appears in textbooks and the journal literature, doctors still rely on the lack of high levels of industrial exposure to decide someone does not have heavy metal poisoning.

Typically heavy metal poisoned people are labeled with incorrect psychiatric diagnoses (depression being the most common, somatization disorder the next most common) because they really do have lots of symptoms they complain about. The doctor doesn't recognize the symptoms as those related to some heavy metal because this was not part of apprenticeship training.

With the stress on apprenticeship rather than scholarship, it never occurs to the doctor to read *old* medical books which were written by doctors who actually saw, diagnosed, and treated a lot of people with heavy metal poisoning. Instead the doctor calls a toxicologist, who also learned from his mentor that heavy metal poisoning only happens with large industrial exposures.

To make matters worse, insurance companies have implemented a reimbursement schedule that punishes book research (or scholarship) and rewards knee jerk diagnosis and treatment. Thus there is financial pressure to call a colleague who has never diagnosed heavy metal intoxication under the relevant conditions rather than to make reference to useful literature written by people who have done so.

The information below is compiled from a variety of literature sources (that were all written by doctors who really did see, recognize and diagnose heavy metal poisoning) as illuminated by my experience in talking to people with it, consulting to their physicians, and seeing the results of many different kinds of lab tests and therapies with these people.

One of the frustrating parts of diagnosing heavy metal problems for physicians is that there are a large number of possible symptoms and no one person has them all. In fact, there is no one symptom everyone actually has for a given heavy metal. Thus the physician needs to be familiar with the overall syndrome each heavy metal causes and take some time to examine and interview the patient to make a good diagnosis. This again is not something insurance companies pay for. Insurance companies pay on the theory that each problem may be rapidly identified by recognizing one or a few cardinal symptoms and perhaps checking one thing on exam or ordering a test. This does not work for most cases of heavy metal intoxication.

Since intoxications are not like infections, cancer or broken bones in terms of being something you either have or don't have, less severe cases are harder to diagnose because there are fewer symptoms and most lab tests have not been driven to abnormal values yet. However this is the best time to diagnose them in terms of preventing unnecessary suffering. To catch intoxications at an early stage, where they are most readily treated and easily cured, the health care provider needs to be familiar enough with the various intoxications to suspect them when the signs and symptoms are still subtle. One major advantage of hair testing is that it is a good way to screen for most of the common toxins using a test that shows abnormalities before most other tests do.

For those who want more discussion of what the different heavy metal poisonings are like, I suggest finding 1960's or earlier editions of *Hunter's Diseases of Occupations*, any reasonable internal medicine or toxicology book, and checking *Mineral Tolerance of Domestic Animals* [41] for pre-PubMed journal papers

Homeopathy books are excellent sources for clinical information on what the different toxins do to people. Whether or not you believe in the theories on which homeopathic prescribing is based, most of the homeopathic literature is simply a clinical description of poisoned people that homeopathic doctors would spend days or weeks interviewing and observing so they could have a complete and thorough understanding of precisely what effects each toxin has.

---

[41] which may conveniently be read online at http://books.nap.edu/books/0309030226/html.

Sources of Exposure to Toxic Metals

Many Elements are Ubiquitous

Elements that are a component of the normal diet and don't have widely unappreciated sources of exposure are not listed below.

Essentially every element has some use in industry or research, even if quite limited. Academic laboratories and industrial R&D facilities, or R&D companies, are particularly likely sources of exposure to just about anything.

Most of the less common elements do have specific geographic locales where they have been concentrated by geological processes and get into the local soil and food.

Sources of Exposure may never be Determined

Please note: in many cases of poisoning with obvious symptoms and high exposure levels, the actual source of exposure is never identified. Health care providers who fail to order relevant testing due to a lack of a recognized exposure source are not acting responsibly.

Aluminum (Al) Sources

Aluminum can be found in innumerable items, both naturally occurring and man made. A few sources that are relevant to how aluminum gets into people are given here.

Aluminum salts are clarifying and flocculating agent used in water treatment plants. Most municipal water has some added aluminum, sometimes very large amounts of aluminum are added. In addition, many fluoridated water supplies have aluminum fluoride complexes added and these carry aluminum into the brain quite efficiently. These complexes decompose very slowly in solution so even though simple minded equilibrium calculations suggest that no fluoroaluminate complexes will be present in the water they in fact are there when it is consumed by people.

Many antacid preparations contain aluminum and it is absorbed from them.

Preparations to keep people with kidney problems from absorbing too much phosphorus often contain aluminum.

Many baking powders contain aluminum in a form that is well absorbed when eaten.

Processed cheese products (American cheese, Velveeta® and similar cheese products) are high in aluminum due to the way they are manufactured.

Acidic foods cooked in uncoated aluminum pans often absorb a lot of aluminum, e. g. tomato sauce. If cooking discolors the pan, some aluminum got into the food.

Some colloidal mineral products and some clays that are taken internally contain aluminum.

The active ingredient in many antiperspirants is aluminum – specifically including the "natural" antiperspirant crystals sold at heath food stores.

Styptic pencils contain aluminum and as they are applied to cuts there is significant absorption into the bloodstream.

Antimony (Sb) Sources

The largest use of antimony is in flame retardants in plastics and textiles. Bedding and children's pajamas must be treated and by law are not labeled as having been treated. Most curtains, upholstery and carpets are also treated.

Lead containing solder for water pipes was replaced by antimony containing solder for water pipes.

Antimony is used in many pottery glazes, mostly replacing lead for this application.

Antimony is alloyed with the lead in lead-acid batteries.

Antimony oxide has replaced lead compounds in many paints to provide sun resistance.

Antimony oxide is used as a catalyst to produce polyesters and polyethylene terepthalate (PET) resins, fibers and films. Many of these are used for food and beverage containers or clothing.

Antimony oxide is used as a phosphor in fluorescent light tubes.

Antimony oxide is added to many lubricants such as motor oil and Moly Dee®.

Arsenic (As) Sources

There are an amazing number of sources of arsenic exposure so this list should not be considered complete. Also many of these sources are of the sort where there could be old unrecognized contamination leading to an exposure nobody would ever expect so it is common for people to have a lot of arsenic without any good idea of where it came from.

Arsenic (CCA[42]) treated wood is the green tinged stuff that replaced creosote treated wood for foundations, fence posts, and is used in decks and to raise garden beds. The arsenic in this wood is water soluble and comes out easily. Workmen who handle it are supposed to wear gloves and wash immediately afterwards. Anyone growing a garden in a yard where such treated wood is used is getting a lot of arsenic in their food, and they are very likely getting poisoned if they used it to raise their garden bed. Children playing on treated decking also are at great risk of being poisoned.

Arsenic treated wood is slowly being replaced with copper treated wood that looks the same. Without chemical testing it is not possible to tell if wood in place is arsenic or copper treated and the great likelihood is that it has arsenic. Most people who installed the wood are not aware of what treatment it had.

[42] This acronym is for chromated cupric arsenate.

Lead arsenate was commonly used as an agricultural pesticide. Thus it contaminates large areas of current and former agricultural land, is taken up by food grown there, and gets into people. People who worked with this pesticide also frequently became poisoned.

Certain kinds of ant traps contain 10% arsenic trioxide in corn syrup. This tastes really good to children as well as to ants. Also it is sticky and the ants track it all over the place in the kitchen where it may easily contaminate food.

Arsenic compounds are used as feed additives for poultry and swine. Most commercial food animals receive it. It concentrates in their flesh, particularly their livers. People then absorb it from eating these animals. Only organic animals or those specifically labeled as not having received vermifuges were not treated with arsenic.

Arsenic compounds are sometimes used in pottery glazes. Arsenic can be absorbed by using such ceramic products for food handling.

While there is much medical literature complaining that the occasional herbal remedy contains arsenic, arsenic trioxide is also an FDA approved mainstream medical drug that is used for cancer treatment with the entirely predictable side effect that everyone who receives it becomes arsenic poisoned.

Studies have shown that a major source of arsenic in infants is cereal (grain) in the diet.

Barium (Ba) Sources

Barium compounds have a variety of commercial uses, the most common one for most people to come into contact with being as white pigments for paper, paint and ceramics, or in specialty glass. These barium compounds are not absorbed unless mobilized by acids.

The highest volume use of barium is as an additive to drilling muds in oil and gas well drilling. It is used in the form of the sulfate which is not absorbed unless mobilized by acid.

Barium sulfate is used as a contrast agent in medical imaging and is proclaimed to be entirely passive and not absorbed. No doubt this is true for most people. Whether it is true for everyone is less clear. Mixing this barium sulfate with significant amounts of acid will release absorbable barium compounds.

Beryllium (Be) Sources

Beryllium is used in certain very specialized aerospace, semiconductor and nuclear applications. Those who work with it in industrial applications tend to be quite well aware of its toxicity, those who work with it in research situations are sometimes cavalier about the hazards.

Beryllium is a component of the gemstone Beryl. Grinding, cutting and polishing this may lead to some slight exposure, having a rough or finished stone will not.

The most likely source of exposure for most people is the few percent beryllium that is added to many copper alloys used for electrical purposes in order to improve their properties. This beryllium can not be absorbed unless the metal is oxidized or corroded and the tarnish then comes in contact with someone.

Bismuth (Bi) Sources

The most common source of exposure to bismuth is through medicines for indigestion. For example, Pepto-bismol® and similar products, certain stomach acid blockers combined with bismuth salts, and many other products intended to control or treat problems in the digestive tract. Significant amounts of bismuth are absorbed into the bloodstream during use of these products.

Bismuth is an ingredient in many cosmetic preparations, including some that are labeled as natural and nontoxic.

Bismuth is used in some diaper rash remedies.

Bismuth is also a component of certain specialty alloys for scientific and industrial use. These are mostly low melting compositions.

Boron (B) Sources

The principal sources of boron that people may absorb are borax based personal care and laundry products, boron based flame retardant chemicals, and boric acid used for medicinal purposes.

Boron compounds are also used in glass and ceramics but are inert in this form except at certain points during their manufacture.

Cadmium (Cd) Sources

Cadmium is found in rechargeable batteries, certain types of plated steel, some silver solders used in jewelry making and industrial applications, certain pigments in oil painting, in the glaze on certain ceramics which may be used for cooking and eating, as a stabilizer in some plastics, and as a coloring agent in denture plates.

Cadmium has innumerable industrial uses.

Cigarette smoking leads to significant cadmium intake.

Cadmium is the coloring agent in some vivid yellow, red, and orange paint pigments.

Calcium (Ca) Sources

Calcium is ubiquitous in nature (it is a major component of most rocks and minerals). It is present in many foods and is a component of the normal diet in large amounts. A detailed discussion of calcium sources is not necessary for the purposes of this book.

Chromium (Cr) Sources

Exposure sources to hexavalent chromium include the manufacture of ferrochromium and stainless steel, welding or grinding stainless steel, tanning leather, certain types of wood finishing, handling cement, and using chromic acid or any dichromate as a radiator cleaner, glass cleaner, or for some other chemical or industrial use.

Note regarding stainless steels: 300 series alloys have chromium and nickel in them and will not hold a magnet. 400 series alloys have chromium but no nickel and will hold a magnet. There is a slightly "brighter" appearance to 300 versus 400 series alloys and one can quickly learn to tell them apart by eye if they are not tarnished. Both 300 and 400 series stainless steel alloys often have other elements added in modest amounts.

Cobalt (Co) Sources

Cobalt is the binding metal in hard metal machine tool bits, holding together powders of tungsten carbide and other carbides. Cobalt and the other metals are released in easily absorbable form when these are ground to shape or sharpen them.

Cobalt is also a component of certain metal alloys that are primarily used in aerospace applications.
Cobalt salts are used to make indicators that turn from pink to blue depending on the humidity.

Cobalt is used in blue pigments, ultramarine blue in oil painting, a deep dark blue in glass or ceramics. Getting cobalt containing pigments or glazes on oneself can easily lead to absorption, particularly if food or drink is consumed without washing beforehand.

Copper (Cu) Sources

Copper: Copper is a part of the normal diet and some intoxications are due to acquired or inborn problems with its metabolism. Other intoxications are caused by alternative health care providers since it is wildly overprescribed due to superstitions about its competition with zinc, and its role in health and disease. Sometimes people decide to take too much nutritional supplement copper on their own.

Copper sulfate is used to prevent algae from growing in swimming pools and this sometimes does lead to people swimming there accumulating quite a bit of copper by absorbing it from the water.

Copper is a very common metal used in wiring, piping, and mechanical parts. The corrosion from these can be absorbed through the skin or by transfer to food or drink.

The amount of copper released from copper containing dental alloys is usually insignificant compared to the amount in the diet.

Treated wood that is green colored may contain a copper-arsenic salt, in which the arsenic is so much wildly more toxic than the copper that copper can be considered unimportant, or just copper salts. Copper-arsenic preserved wood is being taken off the market as this book is being written and replaced with copper only treated wood. This copper treated wood is potentially a new source of exposure to enough copper that it might be problematic for some people.

Gadolinium (Gd) Sources

The primary source of gadolinium exposure is through its use as an injectable contrast agent in MRI imaging performed for medical purposes. People who routinely perform such imaging and do the injections (technicians, nurses, doctors) are potentially also exposed as are the few people who actually work in the factories making the imaging agent. Gadolinium is also used in certain scientific applications.

Germanium (Ge) Sources

Germanium is used in certain specialized electronic and infrared optical applications.

Germanium is also available as a nutritional supplement and this is the most likely source of exposure for someone to high levels of germanium.

Gold (Au) Sources

The primary source of absorbable gold is as gold compounds used in the treatment of arthritis. Metallic gold and its alloys is not absorbed to any significant extent. Gold salts are absorbed if ingested but few people run across them due to their expense.

Iodine (I) Sources

Seafood and seaweed contain substantial amounts of iodine.

Much table salt is iodized to prevent iodine deficiency, and is labeled as such.

Iodine and iodide dissolved in alcohol is sold as an antiseptic. Significant amounts of this are absorbed when it is used.

Potassium iodide tablets are available as nutritional supplements for the purpose of preventing the absorption of radioactive iodine after nuclear incidents – these tablets contain vast quantities of iodide compared to nutritional sources.

Iron (Fe) Sources

Iron is ubiquitous in nature and in technology.

Much food also contains iron. The high iron content of dark green leafy vegetables is well known. This iron is much less well absorbed than the iron in animal derived foods.

Many plant foods (fertilizers) contain iron in the ferrous form which is the most well absorbed form of iron.

Lead (Pb) Sources

There are innumerable sources of lead exposure. Since exposure and toxicity are so common I will provide a long list here, though it should not be considered exhaustive.

Lead is found in water pipe solder made before 1986. After 1986 it was supposedly not used, but anyone who actually checks will find that lead solder was commonly used after this date and continues to be used for potable water systems. This is easily understood by going to the hardware store where solder suitable for use in potable water systems is displayed along with expensive price tags, while solder labeled as NOT FOR USE ON POTABLE WATER SYSTEMS is displayed next to it at a much lower price.

Lead based pigments were used in many paints up until the 1970's. Sanding during repainting and remodeling projects is the principal route of exposure and is a very serious problem if proper dust control techniques and respirator use are not scrupulously followed, as they never are during "do it yourself" projects.

Lead is so handy that innumerable uses were found for it in home construction, such as lead flashing for roofs, or bathroom and shower floors, lead headed roof nails, lead sheating for cables, or the red layer on the bottom of old linoleum tiles.

Much lead is handled, melted, trimmed and sanded during ammunition reloading. Improper ventilation in indoor shooting galleries is a known risk for lead poisoning as well.

Crafts projects such as making stained glass or cast figures use lead.

Leaded crystal is very popular for storing and drinking certain wines and spirits. Significant amounts of lead can leach out of the glass into the beverage. Lead salts are sweet and the Romans considered wine boiled in lead cauldrons to be superior in taste.

Leaded glass is used for optical applications as well as for beverage containers.

Pewter plates and mugs also contain lead if they are not of recent manufacture.

Lead was used to solder food containing tin can seams through 1996.

There are many different kinds of batteries that contain lead. The red and white corrosion on the terminals of an automobile battery are lead salts that can easily be absorbed by working on the battery and then not scrupulously washing.

Weights for any number of things are made from lead. Over time the lead oxidizes and the soft white lead oxide powder is easily rubbed off onto anyone who handles the object.

While lead has been removed from automobile gasoline, it is still used in aviation gasoline (for piston engines, not for jet engines).

Lead is used in pottery glazes (this has been banned in the United States, but much lead containing pottery is still around and it is still being made in much of the world) and lead can leach into foods prepared, stored or served in lead glazed pottery.

Lead is used as a hair color treatment (Grecian Formula®) and has been used in a few cosmetic formulations at various times. It is also occasionally used in jewelry. In addition, galena mineral specimens are a popular item that few people realize is a lead ore that can easily be crumbled.

Lead arsenate was previously used as an agricultural pesticide, especially on apples and tobacco and wine grapes. Areas where it was used and stored may still be heavily contaminated.

Candies from various sources have repeatedly been found to have toxic amounts of lead in them due to manufacturing issues. Since lead salts taste sweet, many foods can have unnoticed lead contamination to very high levels.

Older public schools are often found to have very high lead levels in drinking water (which can be presumed to have been the case since shortly after they were built, sometimes leading to generations of potential exposure victims). Other old public buildings likely also have high lead levels in drinking water, but unlike public schools they are not required to check lead levels and report them so it seldom becomes publicly known.

Smoking causes elevation of hair lead.

Magnesium (Mg) Sources

Magnesium is ubiquitous in nature (it is a major component of most rocks and minerals). It is present in many foods and is a component of the normal diet in large amounts. A detailed discussion of magnesium sources is not necessary for the purposes of this book.

Manganese (Mn) Sources

Manganese has a very large number of industrial applications. Generally the soluble salts are easily absorbed. It is also used in iron alloys such as those for the shovels, blades and hoes of construction equipment. Welding and grinding on these may release manganese in absorbable form.

Manganese salts are used in dry cell batteries, an application which will bring essentially everyone in contact with it from time to time.

Potassium permanganate is a deep purple solid that makes a deep purple strongly colored water solution. This salt is particularly well absorbed if taken internally. While it is not supposed to be sold as a nutritional supplement, it often is and it is the only form of supplementary manganese that has a significant risk of acute toxicity if too much is consumed.

Manganese is also used as a nutritional supplement in animal feed and the person mixing it into the feed may be exposed to substantial quantities.

Manganese is also used in certain blue, purple and green pigments in oil painting.

Mercury (Hg) Sources

Mercury is ubiquitous due to its tremendous technological utility and peculiar biochemistry. A large number of sources are listed below, but it must always be kept in mind that due to mercury's widespread use and high vapor pressure many people are exposed to high levels of mercury without their knowledge and in cases of even severe, high level toxicity a source of exposure is often never identified.

The primary source of exposure to mercury for most people comes from "silver" amalgam dental fillings. See *Amalgam Illness: Diagnosis and Treatment*[43] for a more detailed discussion of this issue.

Childhood vaccinations contained mercury through 2003. While it was announced that mercury would be removed several years earlier and many public proclamations were made that mercury had already been removed, what actually happened is a deadline was established several years in the future for not manufacturing any more vaccines with mercury, and even after this deadline, vaccines manufactured with mercury remained on the shelf and in use until their expiration date. At present many vaccines still contain mercury – it was only removed from those intended for small children – and there is pressure to start using mercury containing flu vaccines for children.

Most allergy shots have contained mercury until recently. People who received years and years of allergy shots did receive toxic amounts of mercury due to this.

Mercurochrome and Merthiolate were popular antiseptics that contain mercury. Merthiolate under its trade names thimerosal or Thiomersal was also used in a wide variety of hand and surface cleaners used in hospitals and other health care institutions. Thimerosal was also used in innumerable reagent solutions for medical laboratory use. Thimerosal was used as a preservative in many medicinal solutions applied to the body. It was removed from eye care products in the 1990's but continues to be used in some ear and nose drops. A mixture of mercury and cadmium salts was used to provide the pink color in denture plates until recently.

Mercury compounds were often used as a preservative in paints to prevent mold growth.

Mercury compounds are also used for embalming. It is also used in taxidermy.

Cinnabar is a red stone used in jewelry, art, and some older forms of tattooing and make-up. Cinnabar is mercuric sulfide.

Mercury was commonly used in gold mining, and extensive environmental contamination occurred and is still present in many older gold mining areas.

Fish concentrate mercury from the environment, and fresh water fish from small highly contaminated waterways may be especially dangerous. These fish are generally caught for sport, not commercially, so there is no organized program of monitoring in place to ensure fish with vastly high mercury levels are not eaten.

Liquid mercury is especially dangerous since if dispersed it is hard to find and it continues to evaporate,

[43] see www.noamalgam.com for more information or to obtain the book.

exposing people to it for years on end. 10% or so of dental offices have excessive mercury vapor levels and mercury contamination. This is not regulated by OSHA or the states. When the dental office moves, there is no clean-up procedure and future tenants may be exposed unknowingly.

The utility of mercury in pressure gauge manifolds leads to it being used (and splashed all over the place as it is frequently blown out of the gauges when someone turns the wrong knob) in all kinds of engine repair shops, in gauges for ship's ballast, or in any application where pressure differences due to air flow must be measured.

Mercury is also used to wash out the glass tubes used to make neon signs.

Liquid mercury was a commonly used ballast on submarines, especially in world war 2. Mercury was also used to fill counterweights in any number of applications, from medical ones such as instrumentation that had to be held in the stomach to things like using mercury filled glass vials as weights to run pendulum (grandfather style) clocks which were commonly used in schools, factories and institutions 50-100 years ago.

Liquid mercury is fun to play with. Breaking open thermometers, tilt switches, fluorescent lamps or street light bulbs, etc.to get at it is a common childhood pastime. After playing with mercury and getting it into carpets, furniture upholstery, under floorboards, and other inaccessible locations children and most adults do not realize they have created a toxic hazard, nor do they know how to clean it up or reduce exposure.

Liquid mercury may be purposely sprinkled around homes as part of religious ceremonies or by playing children. It remains there causing high level toxic exposure for future tenants who are completely unaware of its presence.

High school and college laboratories and classrooms, older medical offices and older hospitals have a high likelihood of being contaminated by mercury due to the ease of spilling it out of instruments (blood pressure meters[44], barometers, thermometers) and the difficulty cleaning it up properly.

The mercury in injections, Merthiolate or thimerosal containing antiseptics, embalming fluid, paints and in fish is organic mercury. The mercury in amalgam fillings is metallic mercury. The mercury in mercurochrome, mercury salts and cinnabar is inorganic mercury.

Molybdenum (Mo) Sources

Molybdenum metal has a wide variety of industrial applications.

Molybdenum disulfide is a popular lubricant and if ingested can be absorbed somewhat.

Certain red pigments contain molybdenum and lead. The lead content is high enough that the amount of molybdenum absorbed is insignificant compared to the amount of lead if these do manage to get into people.

Molybdenum is a normal component of the diet and occurs in many foods.

Nickel (Ni) Sources

Nickel has a very wide variety of uses so it is impractical to list all possible sources of exposure. Stainless steel that magnets won't stick to has nickel in it, stainless that will hold a magnet does not have much nickel if any.

[44] The medical term is sphygmomanometer.

Many "superalloys" used in aerospace or other specialty applications are high in nickel.

Stainless steels of the 300 series contain nickel, 400 series alloys do not. Magnets will stick to 400 series alloys but not 300 series alloys. 300 series alloys are brighter and shinier appearing than 400 series alloys in the fresh, uncorroded state.

Welding, cutting or grinding nickel containing alloys is a source of readily absorbable nickel. Welding on stainless steel is typically done using high nickel welding rods.

The greenish powder on old bumpers is nickel oxide – most chrome is plated over nickel that is put on the steel part so the chrome will actually stick to it. With aging, the nickel oxidizes and this is a potential source of exposure.

Much jewelry has nickel in it.

Older surgical implants and certain modern large ones are made of nickel containing alloys such as stainless steel or Hastelloy. Also many alloys used for dental work contain nickel, including most of the gold used[45].

Nickel is a commonly used catalyst in the chemical industry. It is used to catalyze the hydrogenation of fats and oils so shortening and margarine have substantial amounts of nickel in them.

Smoking causes significant elevation of nickel in the hair.

Palladium (Pd) Sources

Palladium is used in some automobile catalytic converters, in certain alloys used for dental work, and for a variety of industrial applications. There is also a nutritional supplement containing it.

[45] High noble gold alloy typically contains platinum but no nickel or palladium. The laboratory that makes the metal pieces can provide composition data.

Phosphorus (P) Sources

Phosphorus is present in significant amounts in all foods since it is essential for all life processes. Many common fertilizers contain phosphorus in the form of phosphate. Soda pop contains phosphorus in the form of phosphate.

Platinum (Pt) Sources

Platinum is used in jewelry, industrial catalysts, catalytic converters and scientific apparatus. Platinum metal does not corrode under any reasonable circumstances and does not dissolve in most acids. Absorption is only likely when the platinum is present as a salt or something special is done to dissolve it.

Platinum alloys are used in certain surgical implants, and platinum salts are also used by injection as anticancer agents.

Potassium (K) Sources

Potassium is ubiquitous in nature. It is present in many foods and is a component of the normal diet in large amounts. Potassium is primarily present in plant materials. Some salt substitutes and low sodium salts contain potassium chloride. A detailed discussion of potassium sources is not necessary for the purposes of this book.

Rubidium (Rb) Sources

Rubidium associates with potassium in minerals and brines as well as in foods.

It is used for a limited variety of things such as making special kinds of glass. A few people take it as a nutritional supplement to treat cancer or as an antidepressant.

Silicon (Si) Sources

Silicon is ubiquitous in nature (it is a major component of most rocks and minerals) and is not very soluble or well absorbed. A detailed discussion of silicon sources is not necessary for the purposes of this book.

Silver (Ag) Sources

Silver is found in many items, but it is poorly absorbed from those items generally thought of as "silver," such as coins and jewelry. Only solid silver in long term contact with the blood stream, e. g. a surgical implant, can be significantly absorbed.

Many photographic chemicals contain silver and it can be absorbed by contact of chemical solutions with the skin. This takes a fair amount of contact, someone developing film as a hobby or profession may be at risk, someone who did it a few times for a class is not.

Silver salts are often used for medicinal purposes and this may lead to absorption. Silver nitrate sticks are used to sterilize wounds and stop blood flow, sometimes are used to treat throat or nose infections, colloidal silver is used internally, silver nitrate used to be used on burns or prevent infection, and there are a wide variety of other health care uses.

Silver compounds are used for certain chemical laboratory procedures.

Sodium (Na) Sources

Sodium is ubiquitous in nature. It is present in many foods and is a component of the normal diet in large amounts. Sodium is primarily present in animal flesh and table salt is a very concentrated source of it. A detailed discussion of sodium sources is not necessary for the purposes of this book.

Sulfur (S) Sources

Sulfur in various chemical forms is a signifcant part of the diet. Discussion of these various forms and their physiological function is outside the scope of this book.

Thallium (Tl) Sources

Thallium is used in pesticides and rat poison though this is no longer permitted in the United States.

Thallium used to be a component of ringworm remedies and depilatory creams.

Thallium forms a stainless alloy with silver and is used in jewelry, it forms a corrosion resistant alloy with lead.

Thallium has innumerable industrial and scientific applications. Among these are its use in optical glasses, in radiation detectors, in low temperature thermometers, in a very dense liquid used for sink-float separation of minerals, and in the synthesis of organic compounds and in luminescent tubes, dyes and pigments.

Thallium imparts a green color to some fireworks.

Radioactive thallium compounds are increasingly administered to people for use in medical imaging.

Thorium (Th) Sources

Thorium oxide is added to tungsten electrodes used in welding and in some light bulbs, and to carbon electrodes used in arc lamps. Thorium is also often added to gas mantles (in gas lanterns) to make them burn brighter.

Ceramic items, such as laboratory crucibles, are made more heat resistant by adding Thorium.

Certain optical glasses used in cameras and binoculars contain thorium.

Thorium oxide is used as a catalyst in the chemical industry.

Thorium is also used in nuclear applications, but primarily in India and not much in the rest of the world.

Thorium intake for average people is about 98% through food, 1% each through air and water.

Cereal grains appear to provide most of the thorium taken in with the diet.

Tin (Sn) Sources

Tin is used in a wide variety of applications. It is the traditional coating on "tin cans" and old fashioned copper cooking pots to protect them from corrosion and protect the food from contamination. It is used solder, in bronze and pewter, and in applications like cosmetics, preservatives and biocides for nonfood items, and as a stabilizer in plastics such as polyethyleneterpthalate which is widely used for beverage bottles.

The form of tin used in stabilizing plastics and as a biocide or preservative in agriculture is organic tin. Organic tin was formerly used in antifouling paints on ships but this has been discontinued for environmental reasons.

Titanium (Ti) Sources

Titanium dioxide is a very common white pigment used in everything from paint to sunscreen. It is very likely you can reach out and touch something containing titanium dioxide right now. Titanium in this form is poorly absorbed.

Titanium dioxide is also a very common welding flux for electric arc welding rods. In this application it can be turned into a fine dust which may lead to some titanium being absorbed.

Titanium metal is also used in certain aerospace and sporting applications. For example, high tech golf clubs. The metal is very corrosion resistant and little absorption is possible unless it is burned.

Titanium and its alloys are often used for surgical and dental implants. Significant amounts of titanium are absorbed from these.

Titanium also occurs in certain common rocks and minerals.

Titanium metal, anodized to a variety of attractive colors, is also often used as jewelry.

Tungsten (W) Sources

Fumes from burning up light bulb filaments in the air, fumes from certain kinds of welding[46], the use of hard metal machine tool bits such containing tungsten carbide and especially grinding these to sharpen them. Tungsten is also used to fireproof fabrics and as a mordant for textiles. In addition, tungsten is also

[46] TIG welding which is short for Tungsten Inert Gas welding.

commonly used in scientific apparatus for filaments and heaters and as the core in certain kinds of armor piercing ammunition. Tungsten salts are soluble (and some compounds are volatile) and have many scientific and industrial uses, one being in the manufacture of certain semiconductor components.

Uranium (U) Sources

Most often uranium comes from food and water in geographic areas where the rocks and soil are high in it. Granite and certain sandstones are high in uranium.

Uranium is used in certain industrial applications such as aircraft counterweights as well as the more well known use in nuclear weapons and certain military armor piercing ammunition.

Uranium salts are often the yellow color in glazes on ceramic pottery and cookware.

Vanadium (V) Sources

Vanadium is widely used and it is impractical to list everything containing it. It is used to improve the rusting resistance of springs and tool steel and for other metallurgical uses. Vanadium pentoxide is used in ceramics and as a chemical catalyst. Vanadium compounds are used as fabric dyes and inks. Vanadium is present in certain crude oils. Vanadium in the vanadyl form, for example vanadyl sulfate, is a commonly used nutritional supplement.

Zinc (Zn) Sources

Zinc is present in many foods and is a component of the normal diet. It is also used in many industrial applications and in many products people run into every day. Galvanized iron – iron or steel coated with zinc – is a very common thing and everyone encounters this on a daily basis. A detailed discussion of zinc sources is not necessary for the purposes of this book.

Zirconium (Zr) Sources

Zirconium is used in esoteric industrial applications, in nuclear technology, in certain scientific equipment, and in some alloys that are used for surgical implants.

Zirconium salts are the active ingredient in certain underarm antiperspirants and are well absorbed this way.

Zirconium oxide has a very high melting point and is used in many industrial applications that depend on this property, and its electrical conductivity lead to other applications as heating elements and fuel cells.

Zirconium is also used in cubic zirconia, fake diamonds used in jewelry. Another form of zirconia is used for oxygen sensors in car engines. Zirconium is not absorbed from these sources.

Alloys used in Surgical Implants

There are a wide variety of alloys used for surgical implants, staples, dental implants, and pins and plates for setting bones. Each alloy usually contains some of the following materials: titanium, zirconium, cobalt, nickel, tantalum, chromium, iron, vanadium and molybdenum. Each alloy has a different composition and not all elements are present in every alloy.

Mercury Poisoning

Mercury poisoning is the most common heavy metal problem. It is also difficult to diagnose because the exact problems it causes vary so much from person to person. There is also some confusion in the health care community as to how to interpret clinical laboratory test results. There is intense controversy because the major cause of mercury poisoning today is health care and many doctors understandably find this very upsetting. Given these factors I am including a separate fairly long section on mercury. Additional descriptions can be found on pages 25-29 of *Amalgam Illness: Diagnosis and Treatment*[47] or in the medical literature[48].

Reviewing some of the key factors noted in standard medical texts is a good place to start. Medical textbooks are conservative in the material they present. Controversy regarding the nature of mercury poisoning or the lack of utility of certain tests to detect it may be due to the emotionally charged nature of this subject, not to medical or scientific questions.

Erethism is generally mentioned as the most common and important finding in mercury poisoning.

[47] see www.noamalgam.com for more information or to obtain the book.

[48] The review by Gerstner and Huff (Clinical Toxicology of Mercury, *Journal of Toxicology and Environmental Health* volume **2**, pages 491-526 (1997)) is particularly good. Also the descriptions in the 1948 and 1955 editions of *Textbook of Medicine* by Cecil are excellent though more modern editions have less useful descriptions. Any pre-1972 edition of *Diseases of Occupations* by Hunter also contains accurate and useful descriptions of what mercury poisoned people are like. For a firsthand description of mercury poisoning see http://www.stanford.edu/~bcalhoun/AStock.htm.

Clinical Toxicology, Ford, first edition, 2001: "Erethism is a constellation of signs and symptoms, including shyness, emotional lability, nervousness, insomnia, memory problems, and inability to concentrate. "

Merritt's Textbook of Neurology, ninth edition 1995: "the psychological changes which include timidity, seclusiveness and irritability have been referred to as 'erethism."

Textbook of Medicine, Cecil, seventh edition 1948: "Erethismus mercurialis, a peculiar psychic disturbance characterized by ready excitability and a strange shyness in the presence of strangers, a symptom of great interest and importance; insomnia; headache; vertigo; mental depression and dullness; and, rarely, hallucinations."

Erethism is the constellation of personality changes that appear early on in mercury poisoning, often before other symptoms are present.

Textbook of Clinical Neurology, Goetz, first edition, 1999: "Personality changes can develop before neurological signs appear. So-called "mercurial neurasthenia" may develop for weeks or months before the patient seeks treatment. This syndrome consists of extreme fatigue, hyperirritability, insomnia, pathological shyness, and depression."

I have included a description of mercury poisoning from an older textbook written in the days when occupational poisoning was common and doctors had to rely on clinical skills rather than lab tests to make good diagnoses.

Textbook of Medicine, Cecil, seventh edition 1948: "Tremors of the orbit, lips, tongue, fingers and limbs. These are usually moderately fine at first, but at intervals become a coarse and jerking. They may become very severe and in rare instances involve contractions of the limbs of such violence as to require restraint. The tremor is intentional and subsides during rest. Under observation it may increase and diminish, rhythmically recurring. When it is marked, the patient may require assistance in eating and in other activities. Weakness of both the flexor and extensor muscles of the hand and forearm has been reported, but marked paresis is rarely, if indeed ever, caused by mercury. Polyneuritis is said to result occasionally from exposure to mercury. No ataxia occurs and reflexes are not notably affected."

Modern medical textbooks also make it clear that mercury from "silver[49]" fillings is a potential hazard to both dental patients and the people who work in the dentist's office.

Harrison's textbook of internal medicine, 14th edition 1998: "Low-level exposure from dental amalgams may also be associated with adverse immunologic reactions."

Cecil Textbook of Medicine, 21st edition, 2000, "10% of dental offices have excessive mercury vapor levels; and accidental spillage can lead to mercury poisoning."

Textbooks and journal articles cover some of the basic biochemical phenomena underlying mercury's ability to affect the body in a wide variety of ways.

Textbook of Clinical Neurology, Goetz, first edition, 1999 lists the following in a table on the effects of different heavy metals under mercury: "Alters cell membranes; causes combination of: metabolic disturbance, disturbance of Ca^{+2} homeostasis, oxidative injury, aberrant protein phosphorylation."

In the journal *Endocrinology,* volume **89**, number 6, pages 1489-98 (1971) an article entitled Effect of Mercury on Response of Isolated Fat Cells to Insulin and Lipolytic Hormones by Jack M. George shows how mercury has physiological effects that can cause exactly the constellation of problems seen in type 2 non insulin dependent diabetes – elevated blood sugar, insulin resistance and weight gain.

One very important question is whether you have to continue to worry about mercury if a doctor has told you that you definitely don't have a problem with it, or that you have something else. Standard medical textbooks make it clear that mercury poisoning is hard to diagnose and easy to mistake for something else.

Cecil Textbook of Medicine, 21st edition, 2000: "With mild exposure, the manifestations are likely to be subtle and diagnosis is difficult. Insomnia, nervousness, mild tremor, impaired judgment and coordination, decreased mental efficiency, emotional lability, headache, fatigue, loss of sexual drive, and depression are early manifestations and are often mistakenly ascribed to psychogenic causes." In a separate section, *Cecil* also says "Chronic metal intoxication may also produce a dementing syndrome."

Clinical Management of Poisoning and Drug Overdose, third edition, 1998, Haddad, Shannon and Winchester, eds. "Chronic mercurialism may be misdiagnosed as Parkinsonism, depression or Alzheimer's disease."

Merritt's Textbook of Neurology, 9th edition, 1995: "Because of the prominent motor manifestations, confusion with ALS[50] can occur."

Because mercury poisoning is hard to diagnose, many doctors may order blood

[49] Silver amalgam fillings are 50% mercury, and approximately 30% silver plus 20% other materials. It would be more accurate to refer to them as mercury fillings.

[50] ALS is amylotrophic lateral sclerosis, also known as Lou Gherig's disease. In the United Kingdom it is referred to as motor neuron disease.

or urine tests for mercury. Textbooks point out that this is not helpful.

Cecil Textbook of Medicine, 21st edition, 2000, "Because of the body's metabolism of mercury, blood and urine levels may be unreliable."

Textbooks are also clear on the fact that some children are very sensitive to mercury and become poisoned at low levels that don't bother others.

Clinical Toxicology, Ford, first edition, 2001: "Acrodynia ... Although it is most often postulated that the etiology of this syndrome is an idiosyncratic hypersensitivity reaction to mercury because of the lack of correlation with mercury levels, many of the symptoms resemble recognized mercury poisoning."

With medical textbooks saying that mercury poisoning is difficult to diagnose and can cause a wide variety of problems, a thorough description of it is necessary.

Mercury's effects on thought and emotion cause the most troubling symptoms.

Mercury poisoning changes people's emotional responses to things. Irritability, argumentativeness, avoidant behavior, anxiety, shyness, being easily embarrassed, and a desire to avoid strangers lead to social withdrawal and makes it difficult for them to relate to others. The victim generally feel that their behavior patterns are appropriate even when this is clearly not so.

The argumentativeness may reach the point where it is impossible to hold a job or maintain most social relationships.

Fatigue, lassitude and depression further limit social interaction. The depression brings feelings of futility, hopelessness, discouragement, and impending doom. Minor problems become overwhelming and insurmountable, while these emotional changes themselves are frightening.

Feelings of enjoyment and happiness become less and less common. There is no motivation for important tasks, fun activities, or day to day routine.

Intelligence gradually deteriorates. Previously bright persons become dull and slow thinking. They suffer from a progressive decline specifically affecting short term memory as well as the faculties for logical reasoning. Thus their ability to do things like balance the checkbook, do math, or play chess suffers. Thoughts become heavy, repetitive and pedantic. Creative thinking becomes progressively more difficult. Verbal expression suffers from an inability to select the right words to convey their meaning.

"Brain fog" sets in with a poor ability to concentrate and pay attention.

Men frequently become quietly depressive and withdrawn while women more often become anxious, shy, fearful and high strung.

Symptoms come and go over a period of months or years. Life seems to progress in fits and starts, with periods of productivity and social enjoyment coming less and less often as poisoning progresses.

In severe cases, mercury poisoning can cause psychosis. Psychosis interferes with a person's thinking, emotions, memory, communication, behavior and interpretation of reality. When sufficiently impaired in all these areas that they can not meet the ordinary demands of life, a person is psychotic.

Psychotic people often have very disorganized behavior which may be childlike or infantile, an inappropriate mood for the situation they are in, poor ability to control their impulses, speak in an incoherent jumble of words, and may have delusions and hallucinations (usually without an understanding of the fact that these are not real). Thought processes are often quite disrupted, with the psychotic person's train of thought seeming perfectly logical to them but being at best very loosely connected as far as anyone else is concerned.

Mercury can impair the brain's electrical regulation and cause epilepsy (either grand mal or petit mal varieties). Poor electrical regulation that is not bad enough to cause epileptic seizures can cause impulsivity and sudden strong emotions not rationally related to what is going on. It can also cause sudden confusion or disorientation or loss of train of thought.

Mercury affects the senses. The most well known effect is numbness in the hands, feet, and around the mouth (the medical term for this is stocking and gloves parasthesia). There can be a generalized reduction in sensitivity to touch and pain which is often more pronounced in children.

The sense of smell becomes less acute. While the ability to perceive sound is not reduced the person loses the ability to pick meaningful sounds out of background. For example, they are not able to understand speech directed at them at a party, or if music is playing.

The ability to focus the eyes and to control the iris progressively deteriorates, as does convergence - the ability to bring both eyes to bear on nearby objects so as to see one object with depth perception rather than to have double vision. People with convergence problems have difficulty keeping their place while reading and find reading for long periods quite tiring.

Color vision may also become less acute and colors do not appear as bright.

Mercury poisoned children typically have learning or developmental disabilities and trouble relating to peers. The poor visual convergence and accommodation due to mercury poisoning is a common cause of dyslexia and reading difficulties.

Children who are poisoned in infancy or before birth may have hyperextensible joints, hypermobile hips, and may lay in crib with their feet up by their head.

Early physical symptoms include dizziness, tinnitus (ringing in the ears), insomnia and daytime drowsiness. The biological clock runs slow. The poisoned person's body wants to wake up late and stay up late. It takes them a long time to "get going" in the morning.

There may be a tendency towards diarrhea - often alternating with constipation, cold hands and feet, a tendency towards sweating (some people have the opposite symptom and do not sweat at all, which is more common in women), flushing or reddening of the skin - particularly on the face and neck. Some people blush frequently, but others do not blush at all.. Digestive disturbances are also common.

The skin becomes dry, athlete's foot and toenail fungus progress, and the insides of the ankles, particularly behind the ankle bone and a bit above it become dry, itchy, flaky and peel. This becomes annoying enough to keep the victim up at night. There may also be hair loss. Hair may thin out or it may fall out in patches leaving bald spots.

Tremors eventually appear. There may be twitching of the eyelids. Poor coordination of the lips and tongue may render speech more difficult to understand. Handwriting deteriorates and eventually becomes illegible.

Mercury affects the blood coagulation mechanism and leads to easy bruising and bleeding in some people.

Women with mercury problems often have menstrual problems, irregularities, and may suffer from infertility.

Mercury also interferes with the brain's regulation of water balance and causes excessive urination.

Mercury can reduce the function of the adrenal and thyroid glands. Typically it does this enough to make the person miserable, but not enough to make laboratory tests abnormal.

Mercury poisons the immune system, making fighting off minor illnesses more difficult. Immune dysregulation can also lead to increased allergy, asthma and other respiratory complaints. It is also a common trigger for autoimmune conditions such as lupus, multiple sclerosis or rheumatoid arthritis. In physiologic terms, mercury poisoning causes an imbalance between T helper type 1 and T helper type 2 cells along with reduced activity of natural killer (NK) cells. This is the same problem that zinc deficiency causes and is most likely due to an acquired zinc deficiency caused by mercury's derangement of mineral transport.

Heart racing (the medical term is tachycardia) is quite common in people with mercury poisoning. The heart rate may vary dramatically over a period of a few minutes for no apparent reason. Heart pain (angina) may occur. Doctors may hear intermittent heart murmurs and may find a flattened T wave or a prolonged QT interval on an EKG.

Mercury induced derangement of mineral transport almost invariably results in reduced body zinc levels (usually reflected by high hair zinc levels), poor zinc absorption, increased loss of zinc in the stool and urine, and difficulty concentrating zinc into cells where it is needed. People with low zinc due to mercury problems typically need levels of zinc supplementation that would be excessive in normal individuals. Mercury toxic people tolerate high level zinc supplementation for prolonged periods of many years. Deranged mineral transport also typically results in a need for large amounts of magnesium and molybdenum. Sometimes manganese or selenium is also needed, sometimes excessive selenium is retained. Copper levels are usually somewhat high and copper supplementation is usually harmful.

Blood and urine porphyrins are often elevated in mercury toxic people though this test is seldom performed

It is not unusual for MCV and MCH on a blood count (CBC) to be mildly elevated (suggesting B-12/folate deficiency anemia) in people with mercury poisoning, even when they have more than adequate levels of B-12.

The liver enzymes AST and ALT (also called SGOT and SGPT) may be mildly elevated when a "chemistry panel" test is done on someone with mercury poisoning, though this is not the rule.

Cholesterol may go up quite a bit in the earlier stages of mercury poisoning. In some people it later falls due to inhibited steroid synthesis, typically in conjuction with inability to make enough cortisol and DHEA for their body's needs.

Blood sugar may go up (and may be diagnosed as type 2 diabetes) due to mercury poisoning.

In chronic mercury poisoning there is seldom any evidence of the kidney problems which are present in acute poisoning.

The most common result from a person's doctor ordering standard tests like a CBC, chem panel, urinalysis and thyroid test on someone with a fairly serious mercury poisoning problem is that all the tests come back normal, or close enough that the doctor doesn't consider the results significant.

In certain circumstances mercury poisoning may be recognized in an 'acute' phase. For example, after placement of fillings which extend below the gumline, after placement or replacement of several fillings, after placement of a dissimilar metal crown or bridge over or against amalgam.. In this earlier phase there may be red blood cells in the urine, greatly elevated urinary porphyrins leading to pink - not red-tinged - urine, burning urination, as well as the above symptoms. Very high exposure levels cause nausea, loss of appetite and diarrhea. In general acute poisoning leads to much more pronounced physical symptoms and much less pronounced mental and emotional symptoms. As time progresses the mental and emotional symptoms slowly increase.

Other Toxic Elements: Signs and Symptoms

General Discussion

Mercury intoxication is sometimes complicated by lead, copper or arsenic. Nickel, aluminum, antimony, titanium or tin are often elevated in hair samples from mercury poisoning victims but do not appear to be a toxin in most of them. However, exposure is ubiquitous. Some symptoms more characteristic of these may help determine whether they are involved, or are the primary toxin.

Most of the heavy metals are capable of producing anorexia, fatigue and depression. Most of them also interfere with porphyrin synthesis (as do most liver diseases). Mercury and arsenic in particular cause a hypothalamic intoxication, affecting mood, emotions, and body functions such as sleeping, appetite, blood pressure, heart rate and thirst.

Most of the heavy metals and some of the essential nutritional elements affect energy production in the mitochondria. Since nerve cells, especially ones with long axons, require a lot of energy intoxication by these metals features substantial neurological and psychiatric symptoms. The exact neurological symptoms caused by the different metals overlap quite a bit. Because the overlap in symptoms is due to these metals causing the symptoms the same way, they are synergistic toxins. What may appear to be a mixed metal intoxication when first glancing at a hair test usually turns out to be mercury deranged mineral transport where some elements are high on the test but not inside the person. Occasionally it actually is a mixed metal problem.

The major issue when looking at a hair test is to decide whether the person has mercury, some other toxic element, or something other than one of the elements on the test. Since the test is an attempt to find out what is going on with a real human being, one way to approach this is to consider the kinds of problems one would expect high levels of the elements in the red to cause and compare that to what is going on with the person whose hair got tested. If the symptoms match the red zone results, then those results may be a good indicator of the problem. If the symptoms are unrelated to the red zone results, deranged mineral transport, and thus mercury, seem more likely to be the problem.

Retention of Other Toxics due to Mercury

Part of the derangement of mineral transport that mercury induces is to cause the body to retain higher levels of certain minerals than it might otherwise. Which elements this affects varies from individual to individual, but arsenic, antimony, nickel, tin and aluminum are elements many mercury toxic people retain too much of. These elements are often retained to the point where they too exert a toxic effect on the victim. In this case, while the underlying problem is mercury poisoning, reducing exposure to whatever other toxic elements that person is retaining will lead to their body levels of those toxins falling and cause significant improvements. However this is not a cure for their condition. Once the mercury is removed, these other toxins are excreted well and fall to low levels. It is common to see serial urine tests where mercury stops coming out in significant amounts for several months but antimony, tin, aluminum, arsenic, or other metals come pouring out at intermittent times. This occurs when enough mercury has been removed from the relevant body compartment to clear the mineral transport protein that handles that particular element. Which elements will be retained in a give individual, and the order in which they will come out is not predictable.

Titanium, zirconium and boron also follow this pattern of being retained by some people but whether they build up to significantly toxic levels is not clear. They do not usually appear to be relevant to the victim's health.

Aluminum and antimony are particular problems due to this mercury induced secondary toxin retention. Children absorb and retain antimony much more strongly than adults do. Antimony exposure is primarily through flame retardants on bedding and clothes. Most flame retardant bedding and bedclothes ARE NOT LABELED AS SUCH! By law they only need to be labeled if they are NOT flame retardant. Aluminum comes from processed cheese, most baking powders (and thus most bakery goods), cooking in uncoated aluminum pans (Teflon or any other coating prevents aluminum dissolution) and the majority of antiperspirants. Antiperspirants not containing aluminum often contain zirconium. The "natural antiperspirant crystals" are crystals of aluminum salts.

Aluminum (Al)

Aluminum elevation in hair is believed to reflect body burden. This is probably true under conditions of both normal and deranged mineral transport. Blood and urine aluminum levels are not believed to be representative of body burden.

Mercury induced derangement of mineral transport sometimes causes retention of aluminum and elevation of aluminum levels in hair. Eliminating dietary and other aluminum sources usually leads to reduced body levels and symptomatic improvement over a couple of months.

Aluminum may become elevated due to mercury induced derangement of mineral transport before the test meets a counting rule. Some suspicion regarding the presence of mercury is appropriate in a test with elevated aluminum, especially if accompanied by one or more of antimony, nickel tin or titanium.

Body aluminum stores are affected by intake, and may be reduced by restricting dietary intake.

Dietary aluminum interferes with the absorption of calcium, iron and phosphate.

Children absorb aluminum from food more readily than adults do. Children are also more sensitive to the toxic effects of aluminum than adults are, and become poisoned at lower levels.

People with kidney problems of the sort that elevate creatinine and BUN on blood tests will accumulate aluminum to higher levels than healthy people will since most aluminum is excreted by the kidneys.

Old people typically have poor kidney function and eat or are fed a diet that may be high in aluminum.

Aluminum can block the enzyme that makes alpha ketoglutarate in the Krebs cycle (this can also occur due to manganese deficiency). Since alpha ketoglutarate is used to get rid of ammonia formed from the metabolism of amino acids, this can contribute to elevations of blood ammonia which then cause impaired brain function, with poor memory, confusion and in extreme cases delirium.

Aluminum may be a factor in Alzheimer's disease and other dementias, causing confusion, disorientation, poor memory and poor coordination.

The most well known type of aluminum poisoning is dialysis encephalopathy, where people who receive dialysis treatments because of kidney failure get poisoned by aluminum in the water used, or in medicine given them to prevent excessive phosphorus absorption.

Dialysis encephalopathy, which is chronic severe aluminum poisoning, causes malaise, cognitive decline specifically including memory loss, halting speech, muscle jerks, clumsiness, reduced utilization of blood sugar and reduced levels of the neurotransmitter acetylcholine. When the poisoning becomes very severe, respiratory problems and degeneration of the bones (the medical term is osteomalacia) also occur.

Due to their greater sensitivity to aluminum, children receiving dialysis get this problem much more frequently than adult dialysis patients.

Aluminum toxicity may also cause problems with the heart muscle (cardiomyopathy) and an anemia where the red blood cells are too small (the medical term is microcytic anemia).

Aluminum is believed to be a possible trigger for sarcoidosis, an autoimmune condition.

Antimony (Sb)

Hair is a good measure for the body burden of antimony. Blood, urine and feces are not representative.

Mercury induced derangement of mineral transport causes retention of antimony in many people. This retention causes the antimony to build up to levels where it is also directly toxic, and people with this problem do improve significantly over a period of months if sources of antimony exposure are eliminated. This mercury induced retention does elevate hair antimony and often also causes the counting rules to indicate deranged mineral transport. Even if the counting rules are negative, some suspicion needs to remain that mercury may underlie the problem if there is no obvious source of past or present exposure to unusually large amounts of antimony or if there are other signs of mercury problems, e. g. elevated hair aluminum, nickel, tin or titanium.

Trivalent (+3) antimony are generally more toxic than pentavalent (+5) compounds. Trivalent compounds are also retained by the body for much longer than pentavalent compounds. Antimony trioxide and most industrially relevant compounds contain trivalent antimony. For example, the flame retardant in bedding and pajamas is trivalent antimony. The rarely used antimony medicinals contain pentavalent antimony.

Antimony in the trivalent +3 state concentrates in the liver, causing fatty degeneration which if severe may lead to jaundice. With high exposures liver damage and even failure may occur.

Antimony can suppress white blood cell levels and specifically affects neutrophils. Mercury and the use of DMSA for chelation can also suppress neutrophils and these suppressions can be synergistic. Immune suppression during administration of pentavalent antimony compounds has been studied[51].

With high antimony exposures there may be iron deficiency anemia as there is with arsenic.

Cardiotoxicity is the most significant long term risk associated with antimony exposure and in general is a problem everyone with toxic levels of antimony will experience. Antimony concentrates in the heart, and causes heart problems via damage to the heart itself. There are usually EKG changes, prolongation of the QT interval with flattening or inversion of T waves being common. Sometimes ST segment shifts and P wave changes are seen. The QRS complex is usually normal. Antimony predisposes people towards heart attacks and diseases of the heart muscle (cardiomyopathies). Antimonial cardiotoxicity may produce arrhythmias, myocardial depression and damage, fainting due to heart slowing (Stokes-Adams attacks), heart failure and cardiac arrest including sudden cardiac death. One recent research study found dramatic elevations of mercury and antimony in the heart tissue of people with idiopathic dilated cardiomyopathy but not other heart conditions[52]. Older literature describes heart and other problems observed in human poisoning by antimony in great detail[53]

Pentavalent antimony is believed to have a blood half life of about 3 weeks. Some is retained in the body indefinitely. Thus on removing someone from exposure improvements are to be expected over the first month or two, and by three months any residual problems will not resolve on their own, they will require chelation to correct.

Antimony may impair the ability of cells to concentrate magnesium and potassium at high levels of toxicity. It does not otherwise appear to cause deranged mineral transport. This may be part of how it affects heart performance.

When antimony impairs magnesium utilization, low body magnesium - which is usually represented by high hair magnesium - causes vertigo, disorientation, confusion, balance problems, nervousness, muscle tension, clumsiness, weakness, heart arrythmias, reduces sweating and makes people anger more easily.

Low body magnesium is often associated with hyperactivity, muscular tightness, cramps and twitching, menstrual cramps, depression, high blood pressure, arrythmias and cardiovascular problems.

Antimony may reduce men's testosterone levels with consequent lack of energy, motivation, well being and a general achiness and malaise. Sexual desire is seldom reduced.

Blood and urine porphyrins are often elevated though this test is seldom performed

Antimony problems cause digestive slowing and inefficiency with consequent gas, bloating, varying stool consistency, and colic.

Antimony can cause a rash, often a papular type that is similar to chicken pox, occasionally a more widespread and extensive rash.

---

[51] Wortmann, Aronson E, Byrd, Grever and Oster Herpes zoster and lymphopenia associated with sodium stibogluconate therapy for cutaneous leishmaniasis , *Clinics in Infectious Disease*, volume **27**, number 3 pages 509-12, 1998.

[52] Frustaci, Magnavita, Chimenti, Caldarulo, Sabbioni, Pietra, Cellini, Possati and Maseri, Marked elevation of myocardial trace elements in idiopathic dilated cardiomyopathy compared with secondary cardiac dysfunction. *Journal of the American College of Cardiology*, volume **33**, number 6, pages 1578-83, 1999.

[53] It is briefly reviewed and cited on page 28 of *Mineral Tolerance of Domestic Animals*, which may conveniently be viewed online at http://books.nap.edu/books/0309030226/html/28.html.

Children accumulate antimony much more readily than adults, attaining far higher levels at a fixed exposure.

People with elevated antimony are more prone to nosebleeds, especially spontaneous ones.

People with antimony problems are cross and peevish. Children don't like to be touched or looked at. Adults are sulky, sad and fretful.

In very poisoned people there may be weakening or loss of reflexes as well as reduced sensation (e.g. people with antimony problems are less sensitive to pain than would be expected) and muscle strength.

Antimony toxic people may do better on a high protein diet with control or restriction of carbohydrates.

As with most heavy metals, antimony poisoning causes fatigue, lethargy, malaise, headaches, aches and pains, and other nonspecific symptoms early on. These often are frustrating for the person's doctor to the point that the doctor wants to prescribe antidepressants to get rid of the troublesome patient they can't figure out.

Most of the relevant literature on antimony poisoning of human beings is old enough not to be accessible via PubMed on the internet[54]. There is much information on heart problems in humans exposed to antimony.

Chronic antimony and arsenic poisoning are similar, and these two toxins are highly synergistic.

[54] It is briefly reviewed on page 28 of *Mineral Tolerance of Domestic Animals*, which may be read online at http://books.nap.edu/books/0309030226/html/28.html#pagetop, or it found in many academic libraries as may the journal papers cited in it.

Arsenic (As)

Hair levels of arsenic are an accurate reflection of body burden in most cases. Exposures ended many years earlier may not be apparent, especially if treatments were used that are ineffective for detox[55] but which did mobilize the arsenic and allowed it to burrow deep into the tissues. Blood, urine and stool levels are notoriously inaccurate and do not reflect body burden at all except within a few days of exposure to large amounts or arsenic.

Arsenic poisoning causes the nonspecific symptoms characteristic of heavy metal problems: headaches, fatigue, restlessness, insomnia, drowsiness, vertigo and dizziness, listlessness, malaise, abdominal pain, aches and pains, and cognitive decline. These nonspecific complaints will often lead the frustrated health plan doctor to offer antidepressants and ignore physical causes for the problem – and physical symptoms of it.

Clinical Management of Poisoning and Drug Overdose (Haddad, Shannon and Winchester, editors, third edition 1998) states that "the exposure source may never be determined in up to 50% of arsenic poisoning cases." Yet, as with all heavy metals, the mainstream medicine approach to diagnosing arsenic is to ask about occupational exposures and do no testing if one is not obvious. Even when testing is done, it is typically blood or urine arsenic levels which most mainstream medical textbooks state are not diagnostically useful.

The most common symptom of arsenic poisoning is numbness, tingling, pins and needles or pain in the feet and sometimes hands (painful sensorimotor peripheral neuropathy in proper medical terminology). It often also involves the lower legs. This can involve weakness that may get to the point of sometimes

[55] For example, the use of chlorella, glutathione, N-acetylcysteine, whey protein or a "high sulfur" diet.

being unable to stand. There may also be numbness of the lips. Arsenic also makes people more susceptible to developing such neuropathy from alcohol consumption. Ford's *Clinical Toxicology*, page 786, says: "When exposure is not immediately apparent, patients with arsenical peripheral neuropathy have been misdiagnosed as having Guillain-Barré syndrome." There are many possible causes of peripheral neuropathy. Since it is a miserable condition, most of the causes are incurable, and hair arsenic levels are easy to check it makes sense to rule out arsenic and other heavy metals as a cause whenever someone has this unpleasant condition.

Many people with arsenic poisoning have diarrhea, headaches and skin problems. A fair number have white lines transversely across their fingernails (Mees' lines in proper medical terminology), hair loss, general weakness, and confusion. Some have nausea, vomiting, irritability and abnormal blood count results.

A common effect of arsenic problems that a doctor can easily check is that tendon reflexes are often diminished or lost.

Another symptom that arsenic can cause is heart racing and palpitations. When the heath care provider checks, the heart may be dilated, with accumulation of pericardial fluid. EKG abnormalities typically involve QT interval prolongation. QT prolongation can lead to potentially fatal complications called torsade de pointes-type ventricular arrythmia, and complete atrioventricular block. Since arsenic also increases hardening of the arteries (atherosclerosis) it can play a significant role in heart disease. In fact, arsenic induced atherosclerosis without other heart issues has been observed in laboratory animals at exposure levels some people do experience, and whether the animals developed atherosclerosis or not depended on their genetics, not on the amount of arsenic they consumed.

Like its chemical relative antimony, arsenic may impair the ability of cells to concentrate magnesium and potassium at high levels of toxicity. It does not otherwise appear to cause deranged mineral transport.

Low body magnesium - which is usually represented by high hair magnesium - causes vertigo, disorientation, confusion, depression, balance problems, nervousness or hyperactivity, muscle cramps twitching and tension (including menstrual cramps), clumsiness, weakness, heart arrythmias, high blood pressure, reduced sweating and makes people anger more easily. Red blood cell magnesium is a reasonable measure of magnesium adequacy since it does its job inside the cells. Serum, plasma and urine magnesium are not informative. Hair magnesium is sometimes high when body stores are low.

Arsenic interferes with the utilization of thiamin (vitamin B1), with the production of acetylcholine, and with the transfer of pyruvate into the mitochondria.

Reduced thiamin utilization impairs the conversion of food into energy and leads to fatigue, poor healing, and impaired thinking.

Reduced acetylcholine levels cause confusion, poor memory and muscular weakness.

Dietary carbohydrates are turned into pyruvate in order to be combined with oxygen and turned into energy in the Krebs citric acid cycle. With impaired transport of pyruvate into the mitochondria people feel horrid after eating too much carbohydrates, have blood sugar elevations, and have difficulty clearing lactic acid produced during intense exercise. People with impaired pyruvate utilization often feel much better on the Atkins diet. Poor pyruvate utilization can be misdiagnosed as type 2 diabetes. The peripheral neuropathy that arsenic causes may then be considered as being due to progression of the diabetes, giving both patient and health care provider false confidence they have arrived at the correct diagnosis and are doing all they can do to treat it.

Arsenic toxic people need a high protein diet, with limitation or restriction of carbohydrates.

In some people arsenic may affect adrenal and liver function. In these people there may be low blood sugar (hypoglycemia), often in combination with low blood pressure (below 100/67), anxiety, depression, nausea, vomiting and diarrhea.

Adrenally insufficient people need to eat some carbohydrate and may not tolerate Atkins or ketogenic style diets, nor do they tolerate fasting. Adrenally insufficient people who also do not do well with carbohydrates due to pyruvate elevation feel miserable all the time because their bodies can't use the food they eat.

People with too much arsenic typically have light complexions and may have a rash composed of small eruptions mostly on the areas where the clothes rub. These may appear similar to chicken pox or popped pimples – they do not have any kind of core or pus in them. The skin will flake and peel away from them. These respond well to a high vitamin content skin care product[56].

Arsenic can induce goiter in certain susceptible individuals. This is not common.

In some individuals, arsenic can cause seizures. In others, its effects on the brain do not bring on frank seizures, but rather impulsivity, emotional volatility and poor judgment.

The victim of arsenic poisoning experiences great anguish, and often has an agonized expression on their face. Restlessness leads the victim to change locations frequently. They like to travel. There is much generalized fear - most notably of death and of being alone. Victims are frightened, worried, and selfish. They may curse, rave, and act viciously. They may be self injurious and suicidal. Sometimes arsenic will have such a profound effect on the personality as to cause psychotic behavior patterns.

The victim has fearful dreams. The victim sleeps with their hands laid beyond the pillow, "above" the head, rather than on or under the pillow cradling their head, or by their body.

Arsenic Victims like to drink burning hot liquids. They crave sour or tart drinks and coffee, as well as milk.

Women may have a short menstrual cycle.

Arsenic affects vision. In more severe cases, there may be visual changes such as blurred vision, eye irritation, or optic neuropathy. Most people with arsenic poisoning are very sensitive to small changes in their glasses prescriptions, and may insist their glasses be remade time and again until the eye doctor gets frustrated and tells them the glasses are fine but they are crazy.

There are no distinctive laboratory test abnormalities for arsenic. It does not appear in urine or blood once exposure has ended. Iron deficiency anemia is sometimes caused by arsenic poisoning. Sometimes pyruvate is elevated in blood and urine, and occasionally blood sugar is also high. Blood and urine porphyrins are often elevated though this test is seldom ordered (even though most health plan labs perform it). Porphyrin elevations make people sensitive to the sun, and they burn easily. Platelets may also be low (reducing blood clotting ability) and neutrophils may be low reducing immune function.

Arsenic poisoning victims are more susceptible to outbreaks of warts, cold sores and shingles if they have been exposed to the relevant herpes virus.

Arsenic is a well known carcinogen. Having had cancer or being at risk for it is a good reason to check for it and other heavy metals.

---

[56] For example, Rejuvenex® skin cream available at www.lef.org, or any similar product.

Acute poisoning by arsine, the gaseous hydride of arsenic (AsH_3) is quite different than poisoning by arsenic salts. Arsine is highly toxic to the blood, destroying red blood cells and causing kidney damage by clogging the kidneys up with released hemoglobin.

Beryllium (Be)

It is relatively difficult to establish good reference ranges for every element under every circumstance. Despite the fact that the statistical definition of the *Doctor's Data* reference ranges (and the detection limit) would reasonably be expected to lead to a bar on the plot for beryllium half the time or so, it is quite unusual to see one.

It is not known whether hair beryllium is a good indicator of body burden.

Beryllium can cause cancer, especially if inhaled. Beryllium is also a suspected cause of sarcoidosis.

Beryllium can cause amplified immune responses and increase immune sensitization to other substances it is mixed with – or cause inappropriate immune sensitization to innocuous materials. In medical terms it is an adjuvant.

Individual sensitivity to beryllium varies tremendously. Most exposed workers never react to it. In towns where there were beryllium using factories there would be sporadic unexplained illnesses and deaths of family members or others who came into frequent contact with exposed workers. Eventually it became clear that the small amount of beryllium on the workers' clothing and person was enough to cause severe reactions in sensitive individuals even though all the workers in the factory were just fine. This problem has abated with the introduction of scrupulous industrial hygiene.

Almost everyone who reacts to beryllium has the HLA DP-β-1 gene. Beryllium disease is caused by immune sensitization to beryllium. The disease process does damage by having the body's own immune cells go out of control. In a very loose sense of the word, people become "allergic" to beryllium and even a very small exposure can wreak havoc in a sensitive person even though they used to tolerate large amounts just fine and everyone else around them continues to do so.

Beryllium disease is clinically indistinguishable from sarcoidosis. To differentiate the two, a test for beryllium sensitivity must be performed such as beryllium stimulated lymphocyte proliferation or a beryllium patch test.

Bismuth (Bi)

Hair levels of bismuth are a reasonably accurate guide to body burden when mineral transport is orderly. They may or may not be accurate when mineral transport is deranged, sometimes being falsely low, sometimes falsely high. In the face of deranged mineral transport, very elevated bismuth levels represent actual elevations of bismuth maybe 1/3 or 1/2 of the time, which is accurate often enough to merit concern for bismuth toxicity. Do note however that bismuth is a component of many cosmetic preparations and thus contamination is common.

Very elevated urinary bismuth after chelation in the presence of deranged mineral transport is strongly suggestive of bismuth toxicity even if hair levels are normal.

The kinetics of bismuth in the body are complex. It is known that some modest fraction, perhaps about 1%, of "inert" bismuth compounds such as bismuth subsalicylate (the active ingredient in remedies like Pepto-Bismol®) are absorbed into the bloodstream when it is taken by mouth. It is not clear how much more, if any, is absorbed by people who take it when they have ulcers. The half life of bismuth in the blood is about 3 weeks, yet the half life of bismuth in urine is significantly longer, about 6 and a half weeks. The slow onset and gradual build up of symptoms in people who are exposed, and their persistence after ceasing exposure, suggests the brain clears bismuth over a period of at least decades if it ever does, similarly to mercury.

The complex and poorly understood kinetics of bismuth in the human body make it clear that blood, urine and even hair levels of bismuth cannot be relied on to accurately reflect an elevated body burden in the affected tissues. Thus laboratory testing can not be relied on to rule out bismuth toxicity. In cases where bismuth exposure was significant, ended a long time ago (years), and the person has problems quite characteristic of bismuth poisoning it is irrational to assume that a low hair bismuth level means that they don't have a bismuth problem. In medical terms this is to say that hair testing has a significant false negative rate for bismuth toxicity.

Bismuth crosses the placenta during pregnancy and gets into the growing baby's body. Elevated hair bismuth is associated with the occasional case of childhood developmental disorders though this is of unclear significance presently.

Bismuth toxicity usually results from repeated low level exposures and occasionally from large acute ones. Mercury induced derangement of mineral transport may result in accumulation of bismuth from exposures that don't affect most people. Also, like mercury, the absorption and retention of bismuth appear to vary greatly from person to person so that occasional individuals will be found who were poisoned by using bismuth containing products[57] to the same extent that others do without developing problems.

Chronic bismuth toxicity will most often result in either neurological or personality issues, and in many women, polycystic ovaries. Very high levels of toxicity or large acute exposures may result in generalized inflammation of the mucous membranes inside the mouth (stomatitis is the medical term), a black line at the gums or black discoloration of the gums, excessive salivation and sometimes kidney damage or general irritation and inflammation of the skin. These problems do not occur during chronic intoxication especially of people whose bodies absorb and hold onto bismuth especially well.

In severe cases, neurological problems may include lack of energy; clumsiness; muscle tension, twitching or rigidity; generalized problems with brain function such as confusion, convulsions and even coma[58]; and tremor. These brain problems may also express themselves as psychotic behavior.

Bismuth toxicity may easily be confused with Parkinson's disease and similar neurodegenerative conditions.

In less severe cases the problems bismuth causes are vague and likely to be chalked up to "personality." People with bismuth problems complain about their condition a lot, are willing to discuss it extensively with relative strangers, express anguish and discontent, are somewhat "ornery" and "alienated" in a social sense, yet have a desire for company or companionship. They have restless sleep and are drowsy in the morning and a few hours after meals.

The neurological effects of bismuth toxicity can suddenly occur even years after exposure has ended. This sudden appearance often preceded by weeks or months of depression, anxiety, irritability and possibly trembling. Unless the toxicity is severe enough to cause

[57] Stomach soothing agents, or make-up, or diaper rash balm.

[58] Encephalopathy is the proper medical term for this symptom complex.

obvious tremors or convulsions this situation is quite likely to have a person's doctor decide they are a "nut case" and give them a psychiatric diagnosis out of frustration by the time the problems get bad enough that a fresh look at their condition should be taken and might arrive at an accurate diagnosis.

By impairing both hormones and metabolic energy production, bismuth may sometimes lead to weight gain though this is not the rule.

Bismuth toxicity may also interfere with hormone production at lower levels than are necessary to cause overt neurological problems.

Polycystic ovaries or similar bismuth induced hormone derangement's in women may result in obesity, excessive body hair, menstrual disturbances, and infertility. Sometimes there is high blood pressure or poor regulation of blood sugar. There is seldom masculinization or marked elevation of androgenic (male type) hormones.

Cadmium (Cd)

Hair cadmium levels accurately reflect body burden in the presence of orderly mineral transport but often do not when mineral transport is deranged. Blood and urine cadmium are also believed to accurately reflect body burden.

Note that artists who show vastly elevated cadmium on a hair test may have absorbed it or may simply have gotten a very small amount of pigment on their hair[59]. Many red and yellow pigments contain cadmium salts.

[59] Washing the hair at the laboratory doesn't help since oil based paints actually form an insoluble polymer on drying and there is no solvent that would remove it and leave the hair intact.

Low levels of iron and calcium in the diet increase cadmium absorption from the intestinal tract, as does a high dietary intake of thiols (sulfur compounds found in many foods). Supplementation with iron, calcium and also zinc reduces cadmium absorption, as does restriction of dietary thiol intake.

The half life of cadmium in people is estimated to be between 18 and 33 years. Because of this, cadmium can accumulate over the lifetime and even relatively low exposures can cause toxicity if continued for many years.

Cadmium interferes with the metabolism of iron, copper and zinc as well as vitamin D. Cadmium causes loss of calcium with the possibility of kidney stones and osteoporosis.

Cadmium can impair the body's ability to make blood sugar from protein or fats. People with this problem are lethargic and can't think clearly if they don't eat often enough and have adequate carbohydrate in their meals (e. g. they feel horrible on the Atkins diet).

Cadmium interferes with the function of the cell surface protein that exchanges sodium and potassium (the medical term is Na, K ATPase). People with cadmium problems usually feel better with a high potassium intake even if their serum potassium is mildly elevated.

Cadmium reduces children's intelligence.

Cadmium interferes with the function of the part of the adrenal gland that makes adrenaline (the medical term for this part of the gland is the adrenal medulla, and the medical term for adrenaline is epinephrine). Most of the other heavy metals affect the part of the adrenal gland that makes cortisol (the medical term is adrenal cortex) without affecting the adrenaline producing part.

Cadmium can cause enlargement of the heart and elevation of blood pressure. The blood pressure elevation is in part due to cadmium's interference with potassium handling and may improve with relatively high potassium intake from nutritional supplements or diet.

Cadmium interferes with the liver's production of heme groups and can cause toxic porphyria, an elevation of coproporphyrin and sometimes uroporphyrin in urine, blood and feces.

The effect of cadmium on the blood is seen only in more severe poisonings. It causes anemia due to reduced levels of red blood cells, as well as a reduction in the number of leukocytes (one kind of white blood cell) and an increase in the number of neutrophils (another kind of white blood cell).

Long term exposure to high levels of cadmium can result in such severe toxicity that kidney disease results and protein is found in the urine. Generally the first protein found is beta 2 microglobulin which is elevated in urine but not in the blood. Also, due to the effects on calcium metabolism, cadmium toxic people are at increased risk for kidney stones.

Vitamin C has been shown in animals to substantially decrease the toxic effects of dietary cadmium[60]. Zinc is also therapeutic.

Kidney stones, low red blood cell anemia, high blood pressure, heart problems, significant problems with osteoporosis, low sperm counts, lung and prostate cancer can be caused by cadmium and people with these conditions need to be screened for it.

In terms of symptoms cadmium causes, the rare occupational inhalation exposures cause a variety of respiratory symptoms, other exposures do not.

The subjective symptoms of cadmium problems include vertigo, headache, shivering, needing to eat carbohydrates, difficulty responding with appropriate excitement, energy and mood elevation to fun or stressful events, and pain in the lower back and legs.

[60] See *Mineral Tolerance of Domestic Animals* and references cited therein. This book can be read online at http://books.nap.edu/books/0309030226/html or found in many academic libraries.

Gadolinium (Gd)

It is unclear to what extent hair gadolinium levels accurately reflect body burden. However, the primary source of human exposure to gadolinium is through intravenous injection of contrast agent for MRI imaging and it is common for people who have had this to have vastly elevated hair gadolinium levels months or years later.

It is generally claimed that gadolinium from MRI contrast agent injections is rapidly cleared from the body and is completely safe. There is some evidence of safety for most people but no legitimate studies demonstrating that a small fraction of the population is not sensitive to gadolinium. There is no reasonable evidence regarding body clearance and specifically regarding interindividual variations in clearance or tolerance.

Thorium was used as an X-ray contrast agent for several decades before problems were recognized so there is no real reason to expect that if gadolinium causes problems that the medical profession or government regulators would have figured it out yet.

Chronic gadolinium toxicity mostly affects the liver. The liver makes many of the blood proteins necessary to form clots (scabs). In gadolinium poisoning, blood clotting is slowed, clotting and prothrombin time are elevated, and sometimes the liver enzymes ALT and AST (SGOT and SGPT) are elevated. In addition, gadolinium poisoning interferes with calcium metabolism in the heart and other smooth muscle.

Gadolinium toxicity may also contribute to anemia, though this is less than clear from studies to date.

This is not to say that one or two MRI contrast injections will necessarily cause toxicity – they most likely will not. Studies on people who received many contrast injections over months or years for research purposes have not identified obvious problems[61]. What these kinds of studies show is that most people are not affected in an obvious manner by a bunch of gadolinium contrast injections. What they fail to explore is whether there are a few people who are much more sensitive to gadolinium, or whether there are subtle or long term impairments from gadolinium administration.

Gold (Au)

It is not known whether hair gold levels accurately reflect body burden.

The most common source of gold toxicity is its use as an arthritis treatment.

Gold metal, gold jewelry and gold dental work do not lead to absorption of any significant amount of gold. Gold salts can be absorbed through the skin, by eating them, or by breathing their dust. When eaten, gold salts may produce abdominal cramps, sometimes with diarrhea.

Gold toxicity can cause certain blood disorders. One is too low a level of blood platelets (the medical term is thrombocytopenia), which are responsible for helping form blood clots. Another is too low a level of white blood cells (the medical term is leukopenia). Yet another is where new blood cells don't form fast enough and there just aren't enough blood cells to take care of things (the medical term is aplastic anemia).

61 This is discussed in *Neurology*, volume **48**, number 4, pages 832-5, 1997.

Gold toxicity affects the liver, reduces its function, and causes toxic porphyria. The porphyria results in a mild sun sensitivity.

The skin is also affected by gold toxicity, becoming itchy, inflamed, and flaking or peeling at various places, often but not always sites of gold exposure (the medical terminology for this is exfoliative dermatitis). This inflammation may extent to the mucous membranes inside the mouth (the medical term for this is stomatitis).

Gold toxicity may affect sight, making bright lights uncomfortable and causing double vision or other visual disturbances.

People with gold poisoning may suffer from depression, feel worthless, despair of ever improving, and think of suicide.

Lead (Pb)

Elevated hair lead levels correlate with body burden, but NOT with blood and urine levels. In the majority of cases, hair lead will be elevated if there is a problem. Unfortunately, hair lead is not always elevated in the presence of high body burdens when exposure ended long ago, so a low hair lead level does not rule out lead poisoning (nor does a low blood lead level – blood lead is much more likely to be falsely normal than is hair lead) if the suspected exposure is old and has ended. Thus if there are multiple hair tests I suggest considering the highest lead level observed to be the most accurate one. This also appears to be the most appropriate way to treat blood test results[62]. Also if an exposure was ended

62 See Recent Developments in Low-Level Lead Exposure and Intellectual Impairment in Children, Koller, Brown, Spurgeon and Levy, Environmental Health Perspectives, volume 112, number 9, pages 987-94, 2004. Downloadable on the web for free at http://ehp.niehs.nih.gov/members/2004/6941/6941.pdf.

months or a few years ago and a hair test is contemplated, the best sample to use is the ends of the hair that represent growth during the period of exposure rather than currently.

There are many laboratory test abnormalities that are distinctive for lead poisoning when someone is extremely toxic. Much misery and brain damage happens at lead levels below those needed to make the tests typically discussed in medical books be abnormal, for example, free erythrocyte protoporphyri, zinc protoporphyrin, basophilic stippling, aminolevulinic acid dehydratase, anemia (with low MCV and MCH[63]). Blood and urinary porphyrins will often elevate before other laboratory tests become abnormal. Hair lead levels are the first laboratory index measuring lead to become abnormal during chronic exposures, and are the last to normalize after exposure has ended.

Because on occasion no laboratory tests may be abnormal for a sick person with an old exposure, careful consideration of history, review of old medical records and lab tests, and possibly even a trial of chelation are necessary before deciding lead is not an issue if no other cause is apparent and the person's state is consistent with lead problems. If there was ever a lab test showing toxicity and the person has not chelated extensively, the lead is still there and they are still toxic, even many years later.

Blood lead levels are only meaningful during brief high exposures to lead. However, most medical research has used blood lead measurements without determining hair levels for historical reasons so blood lead levels will be discussed here despite the fact that hair lead levels are more reliable. Research using blood lead levels in large numbers of people has provided good overall information on what happens as lead levels rise even though the test is not very reliable on an individual basis.

The developing nervous system of the unborn baby, infant and young child is more susceptible to lead toxicity than older children or adults Early exposure to lead causes decline in cognitive function. There is no safe level of lead in this regard. Despite compelling research results, mainstream doctors consider children to be safe at a blood level up to 10 µg/dL even though studies show this corresponds to a loss of 7 IQ points[64]. Fifty years ago the supposedly safe level was 40. As the years passed and massive stacks of compelling research accumulated, the blood lead level considered safe was slowly reduced. Of course, the children from the 1950's – 1970's with a blood lead level of 39 who were considered just fine at the time still have the same cognitive impairment due to the lead which remains in them as they did back then when nobody knew any better.

Early lead exposure and increased lead body burden is a known contributor to later obesity and its attendant health problems. Increased lead exposure in childhood also results in reduced height after growth is complete.

Children develop brain inflammation and damage from lead much more easily than adults. Adults may develop nerve damage weakening the muscles which extend the fingers, hands and arms, which children do not. Generally the problems lead causes occur at the time of exposure and those problems remain constant from that point throughout life if proper detoxification is not performed. Thus an adult exposed in childhood may be of less than expected height, overweight, have a decrement in intellectual function or intelligence, and have a darker complexion than expected, but not have nerve problems, muscle wasting, etc.

[63] These abbreviations appear on laboratory test reports for a complete blood count (CBC). They stand for mean corpuscular volume and mean corpuscular hemoglobin.

[64] Due to improper use of statistics to exclude unfavorable data, the actual decline is more likely to be 12 points than 7. See footnote 62.

Adults are considered safe up to a blood lead level of 30 µg/dL. That is, children are considered three times more sensitive to the toxic effects of lead once it gets in them.

Unfortunately children also absorb lead much more readily than adults. Adults may absorb 5-10% of the amount of lead they consume, while very young children absorb 25-80% of it. Younger children absorb more, with those age 2 or under absorbing about half again as much as six year olds, who absorb 3 or 4 times more than adults (and 2 year olds absorb 5 or 6 times as much as adults). Dietary calcium levels strongly influence lead absorption. Those consuming less calcium absorb more lead. Thus children who do not consume dairy products for cultural or health reasons are at much greater risk than children living in the same location who do consume dairy products. Calcium taken as a nutritional supplement in forms other than carbonate or hydroxyapatite is effective at reducing lead absorption. Other things that affect absorption are magnesium, zinc and iron, which reduce it, and vitamin D and copper which increase it. Vitamins C and E decrease the toxic effects of lead, fluoride increases them.

All soluble lead salts taste sweet. Infants and small children will mouth whatever is available. They will then seek out and eat items that taste good, as lead containing items do.

People of all ages absorb inhaled lead quite well. Thus, sanding leaded paint or plaster in old construction for a do it yourself home painting or remodeling project is an easy way to get poisoned. Since children are much more sensitive to lead than adults and also absorb it better, it is not uncommon for children to have serious health problems following such projects while the adults are unaffected which may mislead the family and health care practitioners into not considering group exposure to lead as a potential problem that needs to be looked into.

Lead is stored in the bones, where it remains as a reservoir of toxin to keep the person poisoned for life. When lead is first absorbed blood levels are high, but they decline as the lead is taken up by the bones and internal organs. Since about 95% of the body's lead is stored in the bones and it only comes out slowly, it is not possible to treat a poisoned person effectively with only a few bouts of chelation.

Children are much more sensitive to the developmental and neurological effects of lead than adults are. Thus mildly toxic adults will have a variety of health complaints such as irritability, depression, fatigue, abdominal discomfort, poor sleep or reduced libido without much apparent brain involvement. Mildly toxic children will typically have relatively few health complaints with significant neurological issues such as reduced intelligence, learning disabilities, and developmental disorders such as ADHD, autism or pervasive developmental disorder. Children often have headaches, stomach aches and constipation early on in exposure. This very different presentation of lead induced problems between adults and children may again mislead family members and health care providers into not checking for common sources of lead exposure.

At higher levels of exposure children may suddenly have delirium, clumsiness (the medical term is ataxia), seizures, stupor, or coma. This can occur in a family setting of common exposure without anyone else having particularly notable problems at that time. The occurrence of unexplained instances of "heat stroke," "febrile seizures," or other surprising brain problems makes it wise to check for lead (and other heavy metals) using a hair analysis.

The classic picture of an adult with long standing lead problems is a somewhat overweight, tanned person with a dumb look on their face, modestly attention deficit, with a lot of abdominal cramping and digestive disturbances who likes to get up early in the morning, can't stay up late, and who may have lazy eye (the medical term is amblyopia) or a history of rough or violent behavior.

Lead is known to cause neurotransmitter imbalances, which is medical terminology for lead's ability to mess up the molecules the brain uses to handle information. Not surprisingly, lead can have profound effects on all aspects of brain function.

Lead toxicity may cause impairment of long term memory (versus short term memory for mercury) and of abstract and sophisticated thinking. Eye-hand (and foot) coordination are impaired, reaction time is slowed, and dizziness occurs. IQ falls as does the ability to pay attention. This becomes progressively worse as lead levels rise.

Lead can cause mood disorders. The behavioral effects of lead are apathy, depression, fatigue, confusion, irritability, sleep disturbances and difficulties controlling anger. Children especially are subject to impairments of intellectual function such as reduced intelligence, increased reaction time, increases in antisocial behavior and attention deficit hyperactivity disorder.

Studies by the US Centers for Disease Control over the last 50 years have shown that at present 2.2% of children have high enough blood lead levels to cause impairment – down from 4% in the 1970's.

Lead's emotional effects include a fear of being murdered, feeling worse at night, and better during exercise.

Lead causes subtle hearing loss, raising the auditory threshold. It also may also cause areas of decreased vision in the visual field, the most serious being amblyopia where central vision in one eye is seriously compromised. These problems typically originate in the brain rather than the eye and are characterized by a normal eye exam in the face of abnormal vision.

Lead may cause glaucoma, which is elevated pressure within the eye.

Lead may cause vertigo – dizziness and poor balance.

Lead causes high blood pressure. While uncommon, lead may cause EKG changes and other heart problems. These are reversible with chelation.

Lead can cause gout (and thus elevated uric acid on testing, which will show up before gout attacks do. Mercury poisoning often causes reduced uric acid levels). Dietary restriction of purine containing foods and avoidance of excessive protein intake improves gout.

Molybdenum is involved in the enzymes that make uric acid and sulfate. Lead does not elevate uric acid by interfering with molybdenum function and thus in a lead poisoned person, sulfate is often low while uric acid is normal or high and the person tolerates alcohol. With mercury (which does interfere with molybdenum), uric acid and sulfate are both reduced and there is relative alcohol intolerance.

High levels of lead cause anemia, and also basophilic stippling of blood cells which is checked for when a blood count lab test is done. The anemia due to lead involves reduced MCV and MCH[65].

Blood and urine porphyrins are often elevated due to lead's effects on the liver and on the blood. This test is seldom performed even though it is readily available from standard health plan labs.

High levels of lead cause abdominal pain. In severe cases there is sudden tensing and tightening of the abdomen with lots of pain which can be relieved by holding or pressing on it. Both acute and chronic lead poisoning are usually associated with constipation, but diarrhea will sometimes occur instead.

Lead impairs regulation of thyroid and adrenal function by the hypothalamus and pituitary (parts of the brain that control your hormones), leading to reduced levels of both thyroid hormone and cortisol. While these hormone levels may

[65] These abbreviations appear on laboratory test reports for a complete blood count (CBC). They stand for mean corpuscular volume and mean corpuscular hemoglobin.

be "within normal limits" they also may be inadequate to meet the person's requirements and leave them quite miserable. Since the problem is with central regulation, the most common thyroid test – TSH – is irrelevant and cannot be relied on. In addition, lead may reduce the conversion of the thyroid hormone T4 (thyroxine) to the more active form T3 (triiodothyronine) in the liver, leading to functionally inadequate thyroid hormone levels in the face of normal appearing test results. This can have a profound effect on neurological development in children.

Lead is also known to reduce testosterone levels in men though this appears not to be due to problems with central regulation. Thus in lead poisoning, LH and FSH go up as testosterone goes down, while in mercury poisoning LH, FSH and testosterone all go down. Again, many results that are within the laboratory's normal limits represent inadequate amounts for the specific individual in question and cause a lot of misery.

Lead decreases fertility in both men and women. Since lead is not cleared from the body without proper chelation, people exposed as children will have reduced fertility as adults. In women reduced fertility is expressed both as greater difficulty conceiving and also through increased miscarriages and stillbirths. Both sexes may have a decreased libido.

Hormone levels, hormone regulation and fertility do improve with chelation.

The normal ranges or reference ranges given for lead on hair tests, and on blood tests and urine tests done by "alternative" but not "mainstream" laboratories are derived from measuring lead levels in people who are not considered poisoned. For most toxic metals, the levels of exposure the general population experiences are relatively low and results where the bar is in the yellow range or barely peeking into the red range (on a *Doctor's Data* report, colors and ranges differ for other labs) are seldom significant. However, the amount of lead the average person is exposed to does cause significant problems. Lead is an element where yellow range results may well correspond to significant cognitive or health impairment and red range results are quite likely to.

Nickel (Ni)

It is not clear whether hair levels of nickel accurately reflect body burden or not when mineral transport is orderly, but nickel is one of the first elements that may become very elevated when mineral transport becomes deranged. This will sometimes occur before the tests meets one of the counting rules so a fair amount of suspicion regarding mercury is appropriate on a test showing very elevated nickel. Other elements that frequently become elevated by mineral transport derangement even before a counting rule is met are aluminum, antimony, tin and titanium.

Small amounts of nickel appear to be essential for the function of some liver enzymes, though a minimum daily requirement hasn't been established and there is no evidence anyone can get deficient on any known diet.

The half life of nickel in the bloodstream is two to three days. Nickel appears to be retained for somewhat longer in some organs, with the brain having a much longer but unknown retention time.

The anti-alcohol drug Antabuse (disulfiram) is an effective chelating agent for nickel and its routine use under normal circumstances is associated with a vast increase in brain nickel levels since it readily crosses the blood-brain barrier.

Much nickel is excreted the urine. Nickel is also excreted at high levels in sweat, and in hot climates toxicity may vary by season in the face of ongoing exposure.

Many people are allergic to nickel, and wearing nickel containing jewelry or other items can cause skin eruptions.

Nickel is a known carcinogen.

Nickel toxicity stunts growth. Nickel is poisonous to the blood and causes destruction of red blood cells.

The symptoms of nickel toxicity include weakness, fatigue, dizziness, headache, cough and shortness of breath, often worse in the morning.

Palladium (Pd)

It is not known whether hair palladium accurately reflects body burden.

It is common for people to be allergic to palladium. This allergy can cause skin inflammation, inflammation of the mouth, and other problems in the mouth when palladium containing alloys are used for dental work in allergic individuals.

Palladium poisoning can cause nervousness, fatigue, confusion, memory loss, dizziness, migraine headaches, allergies to other materials and impaired immune function.

Platinum (Pt)

It is relatively difficult to establish good reference ranges for every element under every circumstance. Despite the fact that the statistical definition of the *Doctor's Data* reference would reasonably be expected to lead to a bar on the plot for platinum fairly frequently, it is not common to see one.

It is not known whether hair platinum levels accurately reflect body burden.

People may become allergic to platinum, which is different from being poisoned by it. Platinum allergy can cause a runny nose, itchy eyes and skin, hives and asthma on exposure to platinum salts and platinum containing medications but not to the metal itself.

There can be nausea and vomiting with acute exposure but this is not common with chronic exposure.

Platinum toxicity often causes ringing in the ears (the medical term is tinnitus), trouble hearing, poor balance, weakness and fatigue. It will sometimes also cause numbness or tingling in the fingers and toes (the medical term is peripheral neuropathy), reduced reflexes, and a lessened sense of taste as well as colicky abdominal pain and constipation. It suppresses the immune system and reduces the number of platelets in the blood. Blood and urine porphyrins are often elevated in platinum toxicity. Sperm production may be depressed.

Weakness, tingling and painful sensations in the hands and feet, extending to the forearms and lower legs, can occur as with many other heavy metal toxicities.

Platinum toxicity may affect mood and emotions. Increased fear, anxiety, apprehension, an increased libido and an exaggerated sense of self importance are possible consequences of platinum poisoning.

Platinum may impair the ability of cells to concentrate magnesium and definitely impairs kidney retention and intestinal absorption of magnesium. Low body magnesium – which is usually represented by high hair magnesium – causes vertigo, disorientation, confusion, balance problems, nervousness, muscle tension, cramps and twitching, menstrual cramps, clumsiness, weakness, heart arrythmias, reduces sweating and makes people anger more easily. It is often associated with hyperactivity in children and depression in adults.

Platinum may cause leukemia or other cancers.

Silver (Ag)

Hair silver levels are believed to accurately reflect body burden in the presence of normal, orderly mineral transport. It is unknown whether blood, urine or stool silver levels reflect body burden. Since silver is a component of dental amalgam, and some people obtain sterling silver dental crowns and onlays (outside the United States) elevated stool silver in people with such dental work in place is not interpretable.

Silver poisoning is referred to as argyria when it is accompanied by a characteristic bluish discoloration of the skin. Skin pigmentation can occur but is not ubiquitous and is often absent in cases with substantial neuropsychiatric issues. The coloration is an unnatural bluish gray resulting from deposition of silver in the skin.

Elevated silver can cause substantial impairment of the intellect in some, often with apathy. The intellectual impairment can be to the extent of mental retardation. This is more often found in children exposed in infancy. Other people experience no intellectual decline, usually those exposed after infancy.

Adults with long-standing silver problems may have an intact intellect and may even be of superior intelligence if the exposure was after infancy. These people are dramatic, emotional, expressive and have rapid changes in how they feel, They are impulsive, their mind is bombarded with constant thoughts on things to do. They are very discontented with their life, in a hurry, and continuously in motion going places and doing things. They want everything to get done fast and desire immediate results from things, e. g. diets should cause rapid weight loss from the first day, medical treatments should lead to relief of symptoms within minutes. They are often fearful and tend to misunderstand people's intentions and to misperceive situations. They may be concerned that they are losing their mental faculties or that they are going crazy. Memory may be poor[66] to an extent it interferes with work and day to day life. They often have a serious interest in the performing arts and may participate in them professionally or as hobby.

Brain fog and headaches are common among people with silver poisoning.

People with silver poisoning will often have epilepsy of either the petit mal or grand mal variety[67].

Silver poisoned people often have binocular vision problems – that is, problems using their eyes together to perceive the world accurately. Most standard vision exams do not pick these up (behavioral optometrists such as might be found through www.oep.org, www.covd.org or www.nora.cc diagnose and treat these problems). Silver poisoned people have great sensitivity to small glasses prescription changes as is also seen with arsenic. They may need glasses remade several times and be told by the eye doctor that the glasses are fine but they are crazy.

With high levels of toxicity an anemia can occur where the red blood cells are small but have normal amounts of hemoglobin in them[68]. Enlargement of the heart may also occur, as well as generalized muscle weakness.

---

[66] All forms of memory are impaired unlike in mercury toxicity where short term memory is specifically affected, or aluminum, antimony and arsenic where the poor memory goes along with confusion, or lead where there is specific impairment of the long term memory.

[67] People have one or the other kind of seizure and rarely if ever will a given person end up having the other kind. Petit mal seizures don't involve any movement or dramatic action – the person simply freezes in place, such as pausing in the middle of a word and then continuing to speak as if nothing happened, or continues with some automatic activity like walking but has no conscious awareness for a few seconds. Grand mal seizures are the type where there is random muscle activity and the people can flop around, sometimes hurting themselves in the process.

[68] In proper medical terminology hematocrit is low and it is a hyperchromic, microcytic anemia.

With substantial exposures over a long period, but below the level at which obvious toxic effects appear, silver can cause liver and spleen tumors.

As with all heavy metals, with very high acute exposure kidney failure is possible.

Selenium and vitamin E improve the situation of silver toxic people. Silver is believed to antagonize copper and induces a copper deficiency. Copper supplementation may sometimes improve the toxic person's situation somewhat. Supplementation should be done carefully, with modest amounts (2-3 mg a day for an adult) and it must be kept in mind that copper supplementation makes people with most other heavy metal toxicities worse.

Thallium (Tl)

It is relatively difficult to establish good reference ranges for every element under every circumstance. Despite the fact that the statistical definition of the *Doctor's Data* reference ranges would reasonably be expected to lead to a bar on the plot for thallium most of the time, it is common not to see one.

Blood thallium levels are often falsely normal (low) in poisoned people. Urine levels in a 24 hour collection are somewhat more reliable but false normals are still common. Hair levels are often informative but no estimate is available at present regarding false normal thallium levels in hair from toxic people.

Most thallium salts are colorless, odorless, and tasteless.

Thallium is concentrated into certain cells because it binds to a specific site on a cell surface mineral transport protein that is normally used to take potassium up in exchange for sodium (the biochemical term is Na,K ATPase). It concentrates particularly well into the brain, kidney, and the part of the heart called the myocardium. In addition to binding to the sulfhydryl groups of various enzymes, thallium specifically attaches to vitamin B2 – riboflavin – and inactivates it. Without vitamin B2 fats can't be burned for energy and the main route that carbohydrates take into the energy producing Krebs cycle through pyruvate is also blocked. Citric acid, succinic acid, suberic acid and adipic acid go up on an organic acid test. Only protein can be easily used for energy.

Thallium is also reabsorbed in the kidney by the same mechanism that retains potassium, so people with low serum potassium or low potassium intake retain thallium and are more susceptible to its toxic effects.

The half life for thallium elimination from the body is variously given in the literature as 8 to 30 days. It apparently does not clear from the nervous system at a significant rate since neurological problems may persist for years.

The majority of excreted thallium leaves the body in the fluids the intestines secrete to help digest food. Lesser amounts are secreted in the bile and also are excreted in the urine. About twice as much thallium leaves the body in feces as in the urine.

The medical literature is full of descriptions of acute poisoning by large amounts of thallium. Typically the symptoms start 12-24 hours after the thallium is eaten. When smaller amounts are eaten the symptoms might not appear for 24-48 hours. The first symptoms are from the digestive tract – abdominal pain, nausea, diarrhea, perhaps vomiting, with constipation later. One to a few days later neurological symptoms appear. There is hypersensitivity to touch and pain, unusual phantom sensations, as well as muscle weakness that often begins in the soles of the feet and the big toe and slowly spreads up to the body. Movement may also become clumsy. The pain hypersensitivity can be so extreme that the brush of a feather can cause intense pain. The victim may salivate excessively. There may also be excessive thirst. At this point a variety of other problems may appear. Psychic disturbances such as delirium, restlessness, hallucinations and delusions are common. There may be eye turn or crossed eyes. Color perception may be altered or reduced. Blindness (due to optic neuritis) is frequent. Convulsions or seizures are occasionally seen. Liver and kidney damage often occur. About a week after the thallium is eaten the head hair falls out and the outer part of the eyebrows is lost. Children tend to develop the psychic disturbances more while adults are more prone to the weakness, hypersensitivity and heart regulation problems that come from impairment of the nerves that run the body.

Heart function is affected and victims are at risk of sudden cardiac death for at least several weeks. Heart racing is a common symptom as is elevated blood pressure.

Chronic toxicity does not have the digestive tract problems or dramatic onset of symptoms that acute toxicity does. The neurological, psychiatric and vision problems described above are prominent, hair loss does occur, and there can be heart problems and high blood pressure due to impairment of the nerves that run the heart and other internal organs (autonomic neuropathy in medical terminology). There are typically also sleep problems. Tremors sometimes occur as well. The neurological and psychiatric effects can be severe enough to lead to dementia or psychosis.

A slow blurring or dimness of vision is characteristic, with loss of both central vision and also mild constriction of the field of vision.

Thallium does cause toxic porphyria through its effect on the liver.

In chronic toxicity the symptoms may come on slowly and will usually be subtle rather than dramatic. Mild to moderate chronic toxicities are quite unlikely to be diagnosed (e.g. the combination of hair loss, heart racing and high blood pressure might be misdiagnosed as too much thyroid hormone).

Thorium (Th)

It is relatively difficult to establish good reference ranges for every element under every circumstance. Despite the fact that the statistical definition of the *Doctor's Data* reference would reasonably be expected to lead to a bar on the plot for thorium most of the time, it is unusual to see one.

It is not known how well hair thorium reflects body burden or exposure, however thorium is chemically similar to uranium which is known to be well represented by hair levels.

Thorium, a radioactive heavy element that was once considered as a fuel for nuclear breeder reactors to make atomic bomb materials is used in certain specific industrial applications. It was also injected into people in large amounts as a contrast agent – thorotrast – for medical x-rays for decades until it became clear that this caused liver cancer in many recipients.

Thorium which is absorbed into the circulation concentrates most strongly in the spleen. The liver holds the next highest concentrations of thorium, at about 14% that of the spleen, then comes bone marrow at about 4%. The most significant health risk from absorbed thorium is liver cancer.

Inhalation of thorium leads to increased risk of cancers of the lung, pancreas and blood.

Tin (Sn)

Hair tin levels are believed to accurately reflect body burden when mineral transport is normal. With deranged mineral transport hair tin levels often go up dramatically. There is some corresponding elevation of body burden in these cases due to its mercury induced retention. Trying to do something about the tin directly is useless, however when the mercury is cleared, the tin comes out on its own without any need to treat it directly.

Sometimes tin will go up due to deranged mineral transport in a test that doesn't meet a counting rule. In this case mercury should be considered as a possible problem, especially so if one or more of aluminum, antimony, nickel or titanium is also elevated for no apparent reason.

Interestingly, there is some evidence that ultratrace amounts of tin are essential for normal growth.

Tin has both inorganic and organic forms. The words "stannous" and "stannic" refer to tin salts in the +2 and +4 valence states, e. g. stannous fluoride as used in toothpaste. The inorganic forms are not particularly toxic and require high exposures to cause trouble. The organic forms of tin are quite toxic.

For inorganic tin, severe acute toxicity leads to skin, eye and digestive tract irritation, anemia, muscle weakness and in men testicular degeneration.

Chronic toxicity from inorganic tin results in weakness, easy fatigability, malaise, generalized pain, depression, profuse sweating, and possibly anemia.

High but subtoxic exposures cause reduced lifespan birth weight and increased miscarriages and infant mortality.

Aerosol exposure to tin occurs mostly in occupational settings with poor industrial hygiene. Such an exposure may result in a condition called stannosis, (a contraction of stannum + osis, Latin and Greek words which mean "tin disease"). This may cause coughing and shortness of breath. The cough may bring up mucus that is lemon yellow and sweet tasting. The person generally coughs more and feels worse in the evening. Typically a fair amount of tin containing dust can accumulate in the lungs without affecting their function much. For historical reasons regular doctors expect all tin intoxication to be due to industrial exposure and thus may miss the 99+% of current cases that are from another cause.

Inorganic tin can be released into beverages or foods from tin cans or from tin coated copper cooking utensils. Tin ingestion can produce nausea, vomiting, abdominal pain, and diarrhea as well as the typical symptoms described above.

Acute high level exposure to organic tin compounds causes brain swelling, increased intracranial pressure, and brain damage as well as inflammation of the biliary tract.

Both chronic and acute exposure to organic tin are able to cause brain damage resulting in clumsiness, disorientation, seizures, nystagmus, loss of vision and hearing, vertigo, loss of sensation or phantom pain in the hands and feet (parasthesias) and other nerve problems. There are also forgetfulness, lack of motivation, headaches, loss of libido and sleep disturbances.

Organic tin causes mood disorders which involve alternating bouts of rage and deep depression that lasts for hours to days. These mood disorders can last indefinitely.

Tin poisoning – from both organic and inorganic forms – can cause toxic porphyria as is seen with many heavy metals and other toxins which affect the liver.

Adequate copper and iron intake appear to be mildly protective against tin toxicity.

Titanium (Ti)

Hair titanium is believed to accurately reflect body burden when mineral transport is orderly.

It is not uncommon in people with mercury induced mineral transport derangement to have very elevated titanium levels. In this case, titanium's chemical cousin (congener - the element below it in the periodic table) zirconium is usually also quite elevated.

Elevations may be due to deranged mineral transport even if no counting rules are met. If there is no obvious source of the titanium, or if one or more of aluminum, antimony, nickel, tin or zirconium is also elevated then some suspicion regarding the possibility of mercury poisoning is appropriate.

People with surgically implanted pieces of titanium (plates, screws, staples, pins, clips, posts for dental implants) invariably have quite high hair titanium levels. There does not appear to be any pathology associated with this.

Titanium is generally believed to be nontoxic. There is however some possible association with lupus, eczema and Autoimmunity. Given the strong association of mercury with these conditions some caution should be used before ascribing them to elevated titanium on the basis of a high hair test level until careful thought has been given to ensuring that it is not an elevation induced by mercury's influence on mineral transport.

Tungsten (W)

Hair and urine tungsten levels are reliable measures of exposure as long as mineral transport is not deranged. Tungsten is not on the *Doctor's Data* Hair Element Profile, but it is on the Toxic Exposure Profile. It is also on the test by Trace Elements, Inc.

Tungsten interferes with molybdenum retention and utilization. This causes sensitivity to sulfites, aldehydes, alcohol, low production of uric acid, and sensitivity to foods high in purines (organ meats, sardines, anchovies, herring, mackerel, salmon, tuna and wild game). Tungsten may also promote seizures.

Molybdenum is protective against tungsten poisoning.

Tungsten will clear reasonably rapidly on its own if exposure is ended and adequate molybdenum is provided. Tungsten poisoning can be treated with standard chelating agents like DMPS, DMSA, EDTA and ALA.

When eaten, soluble tungsten salts are well absorbed. Tungsten concentrates in the bones and spleen. Most tungsten is excreted in the urine.

Uranium (U)

Hair uranium levels accurately reflect body stores in the metabolically active tissues (other than bone).

Uranium is stored in the bones for long periods.

Low dietary iron levels tend to cause elevated uranium levels. Increasing dietary iron lowers hair and body levels of uranium. Note that hair iron levels do NOT reflect body stores or dietary intake.

Uranium levels and exposure vary tremendously from place to place depending on local geology. Places where there is a lot of granite or certain sandstones have relatively high uranium, and people who drink water pumped up from underground in these locales, and eat locally grown produce tend to be high in uranium. Generally people in the southwestern United States are much higher than the rest of the country, with local exceptions, for example, New Hampshire, certain locations in Idaho, a few places in Oregon, many others where the geology consists mostly of specific types of metamorphic rocks.

People (this will primarily be health care providers) in a specific area can use hair uranium as one means of suggesting if people might benefit from supplementary iron by averaging the actual numerical measurements or uranium in hair from at least a dozen people who have adequate iron levels as determined by other means who have lived in the same place for at least 2 years. Anyone doing this would be wise to obtain a geological map from the United States Geological Survey (USGS) or online and verify that all the people in question live in the same kind of geological environment. Once this is done, anyone from that area (who has lived there 2 years or more) with over twice this level is a candidate for supplementary iron.

Hair uranium levels over 0.5 mcg/g (which is the same as parts per million) are typically associated with fatigue and other nonspecific symptoms. Hair levels below 0.25 mcg/g appear to not be significant for most people. The green/yellow/red color banding is not relevant to health or to whether someone is high or low for their geographic locale.

What's High and Low with Deranged Mineral Transport

There is no general rule about what essential and other elements go high or low when mineral transport becomes deranged due to the presence of too much mercury. However some things happen often enough to be worth mentioning – especially since they can be helpful in deciding whether a test that counts ambiguously looks "toxic" or not.

Often iodine, boron, calcium and cobalt go out of range on the tests with 4 or more red's. While lithium also often heads towards low red, this is less useful since in many localities it is low in many people due to the geology[69].

Uranium is often driven up by mercury induced derangement of mineral transport, but since the amount of uranium people are exposed to varies wildly with geographic locale much confusion can be caused by paying too close attention to an individual's uranium levels without knowing how the uranium came out for a lot of other people who live in the same general area. Also, the elevation of uranium is real, and typically corresponds to a need for more iron intake.

Often congeners that have adequately similar chemical properties track each other. Congeners are the elements that are in a column on the periodic table. The most obvious ones are the alkali metals sodium, potassium, and rubidium (Na, K and Rb), the alkaline earth metals magnesium, calcium, strontium and barium (Mg, Ca, Sr and Ba) the transition elements titanium and zirconium (Ti and Zr) and the group Va elements arsenic and antimony (As and Sb). While lithium is an alkali metal, beryllium an alkaline earth, and phosphorus and bismuth group Va metals their chemistry is different enough to not move with the other congeners during mineral transport derangement. None of the other congeners appear to move together.

In the toxic element section, deranged mineral transport often results in elevation of some of the following: aluminum, antimony, nickel, tin and titanium, with elevations of arsenic, bismuth and silver due to mineral transport derangement occurring but being less frequent. The elevations of aluminum, antimony and tin are known to be due to increased body retention due to the mineral transport problem and reducing intake often leads to modest symptomatic improvement.

69 For example, most of Washington state.

Meaning of Essential and Other Element Results

Calcium (Ca)

Calcium levels in hair often move in the opposite direction to body inventory. High hair calcium suggests calcium loss and wasting. High hair calcium will occasionally be due to very elevated serum calcium, parathyroid problems or excessive vitamin D levels but these are uncommon (serum calcium is part of most standard "chem panel" blood tests). Low hair calcium may suggest deficiency or may indicate low intake with adequate body levels. Diets which do not contain dairy products are typically much lower in calcium than the average American diet.

Thus the apparently paradoxical suggestion is made by people who know how to use hair tests that people with elevated hair calcium should consume more calcium.

Elevated sulfate is an uncommon cause for excessive excretion of calcium and hence low body stores which might lead to osteoporosis. Sulfate is not tested for in hair, it can be tested for in blood or in urine.

Low blood calcium leads to slowing of the heart and poor muscular contractions. High blood calcium leads to excessive muscular contraction (tetany). Calcium is a major component of bones and gives them their strength. Much calcium is found in the plaques in hardened arteries and is believed to end up there when the body isn't regulating calcium properly. Vitamin K is believed to be involved in regulating where the calcium goes. Supplementary vitamin K may help people with osteoporosis or with high hair calcium levels. Since vitamin K is most well known for assisting with blood clotting, people who bruise easily may also be having trouble with where their body is putting its calcium, especially if they also have elevated hair calcium.

Magnesium (Mg)

Magnesium in hair typically moves contrary to body burden in the presence orderly mineral transport. High hair magnesium suggests low body inventories and vice versa.

Body magnesium levels are usually low in the presence of deranged mineral transport regardless of hair magnesium levels.

High hair magnesium often indicates improper distribution of magnesium in the various parts of the body, often in addition to overall deficiency, sometimes improper distribution without an overall deficiency. Generally serum and urine magnesium will appear normal but red blood cell magnesium will be low (as will magnesium in other cells which are not used for lab tests).

Low magnesium in the presence of normal orderly mineral transport may sometimes suggest low body magnesium stores, but the correlation is not a strong one. Thus it suggests the possibility that there might be problems but can't be relied on without further investigation either through lab tests or through consideration of what is going on with the person the hair test is from.

Low body magnesium – which is usually represented by high hair magnesium – initially causes difficulty falling asleep, poor sleep, headaches, muscle spasm induced neck and back pain as well as restless legs, heart racxing, arrythmias and irritability. Increasingly worse deficiency causes disorientation, confusion, balance problems, nervousness, muscle tension, clumsiness, weakness, sometimes dizziness or vertigo, reduces sweating and makes people anger more easily.

Low body magnesium is often associated with hyperactivity, muscular tightness, cramps and twitching, menstrual cramps, depression, high blood pressure and cardiovascular problems.

People with low magnesium often find they need chiropractic adjustment and the adjustments don't hold well. They go back to the chiropractor monthly or more often. The dizziness which is sometimes associated with low magnesium usually is resolved by the chiropractic adjustment.

Low hair magnesium in combination with high hair zinc is a sign that more vitamin B6 is needed.

The kidney's conservation of magnesium (and hence the amount in the body) is reduced by medical issues including use of diuretic medications, amphotericin B, theophylline, the chemotherapy agent cisplatin and many toxic heavy metals with mercury being the worst offender, and by medical problems such as hyperthyroidism, hyperaldosteronism, diabetes, acidosis and low phosphate levels.

Lifestyle issues that reduce the kidney's conservation of magnesium and hence lower body inventory or require supplementation are the use of alcohol, fasting, being on a ketogenic diet such as the Atkins diet, and drinking tea (which contains theophylline).

Low plasma taurine leads to excessive loss of magnesium in urine with a reduction in body levels, as well as poor retention of magnesium in tissues such as heart muscle that need high levels of it.

Low body magnesium reduces the kidney's ability to conserve calcium and potassium.

Magnesium is also lost in sweat. People with low body magnesium levels do not sweat much, or at all if very low. Giving them supplementary magnesium often allows them to sweat.

Since low body magnesium usually shows up as high hair magnesium, the apparently paradoxical suggestion is made by people who know how to use hair tests that people with elevated hair magnesium should consume more magnesium.

Low magnesium leads to muscle spasms, and this includes heart muscle spasms. In addition, low magnesium can lead to arrythmias and EKG changes. Magnesium is a natural regulator of calcium channels and also reduces blood pressure. In pregnancy, low magnesium may lead to premature labor, pre-eclampsia or eclampsia. Unfortunately most prenatal vitamins contain grossly inadequate amounts of magnesium, almost invariably in the poorly absorbed magnesium oxide form.

The modern dietary intake of magnesium is much less than was prevalent throughout most of human history – almost nobody not taking supplementary magnesium gets too much of it. In addition, magnesium is not well absorbed from the intestines so excessive intake leads to diarrhea rather than high blood levels.

Sodium (Na)

Hair sodium does not reflect dietary intake or body burden even when mineral transport is normal and orderly.

In cystic fibrosis hair sodium is vastly elevated. Cystic fibrosis causes profound problems, and people who are more than a few years old and have not been diagnosed with it don't have it.

In celiac disease – an intestinal allergy to certain grains causing malabsorption – sodium is often much lower than potassium.

Low hair sodium (especially with potassium much higher than sodium) suggests inadequate production of aldosterone by the adrenal glands but this is a relatively weak correlation. Generally people with low aldosterone production will crave salt and eat a lot of it. Many heavy metals impair adrenal production of aldosterone and other hormones.

Potassium (K)

Hair potassium does not reflect dietary intake or body burden even when mineral transport is normal and orderly.

Low hair potassium suggests the possibility of low thyroid function (hypothyroidism is the proper medical term). Urine potassium generally correlates with dietary intake, and red blood cell potassium generally correlates with body inventory. Serum potassium is primarily a marker of adrenal/kidney interactions. High hair potassium may reflect excessive aldosterone activity.

Low body potassium can cause general weakness, fatigue, thirst, constipation, weak reflexes and heart beat irregularities, the most common of which is a racing heart (the medical term is tachycardia). Low potassium can be due to certain types of diuretic medications or to a diet low in vegetable matter. Mercury deranged mineral transport often causes functionally low body potassium levels by interfering with the concentration of potassium into cells by cell surface pumping proteins. However this is not reflected by the hair potassium level – red blood cell potassium is the relevant measure.

Relation of Calcium, Magnesium, Sodium, Potassium, Phosphorus and Other Elements

There is a varied literature on "metabolic typing" using hair analysis, and several labs that include typing information and supplement recommendations with their reports. The clearest and most comprehensive book I am aware of in this area is *Nutritional Balancing and Hair Mineral Analysis*[70].

Other books are *Trace Elements and Other Essential Nutrients*[71], and *The Strands of Health: A Guide to Understanding Hair Mineral Analysis*[72].

There is much discussion of "fast" and "slow" oxidizers or metabolizers in metabolic typing. However, definitions vary. Many of the definitions flow from misunderstandings of how the Krebs cycle works and result in there being 2 mutually exclusive sets of vitamins and foods that magically are perfect for the relevant type. These lists are not observed to either work well or to correlate with the supposed characteristics of fast and slow metabolism. In this section on essential element results interpretation I will stick to a few

---

[70] by Dr. Lawrence D. Wilson, ISBN 0962865745, available from www.drlwilson.com.

[71] by Dr. David L. Watts, ISBN 1-885676-15-8.

[72] by Rick Malter, Ph. D., ISBN 0-9660415-3-4.

straightforward concepts that are generally agreed to work by health care practitioners who have tried them but do not base their practice on hair analyses, and that are confirmed by the test results and experiences of people whose cases I am familiar with. Those who want more detail on this kind of approach will find it in the books cited above.

Elevations of both calcium and magnesium in hair is believed by some to correspond to parathyroid problems causing wasting of both elements. Many "alternative" health care practitioners believe this is due to excessive consumption of soda pop with its sugar and phosphate. While this is possible, "excessive' consumption is at least several cans or glasses per day, most every day on a long term basis. Even when this is true, careful investigation often does identify some other problem that is actually causing the calcium and magnesium to go up. These kinds of parathyorid problems do cause bone loss (the medical term is osteoporosis) and are worth taking seriously even though they are uncommon.

Sometimes excess parathyroid activity is accompanied by low hair iron as well.

Glucose intolerance – metabolic syndrome, or if severe, type 2 diabetes – can show up on a hair test in two ways. A low body inventory of chromium, zinc and manganese can lead to carbohydrate handling problems. When mineral transport is deranged, either high OR low values of these elements may indicate deficiency. Also, if the ratio of calcium to magnesium is not within reasonable limits it suggests glucose intolerance. Different labs cite different limits for this due to variation in hair test methodologies and standardization procedures. The normal range quoted by *Doctor's Data* is 4 to 30 and is not a function of age or sex. This ratio is given on the *Doctor's Data* Hair Element Profile test report as well as on reports by several other laboratories.

The ratio of calcium to phosphorus is believed to indicate whether someone is a "fast" or "slow" metabolizer. This ratio is listed on several labs' test reports, and for a *Doctor's Data* test has a reference range from 1 to 12. Fast metabolizers have a low ratio and slow metabolizers have a high ratio. People wishing to do "fine tuning" using this ratio may wish to consider 1 to 3 to be slightly fast and 6 to 12 to be slightly slow, with 3 to 6 being ideal. Fast metabolizers make lots of adrenaline, slow metabolizers do not.

Slow metabolizers often show lower hair toxic levels for a given body burden than fast metabolizers do. This is to say that someone with a calcium to phosphorus ratio below 1 may be quite toxic but have all the bars in the toxic element section of the hair test within the green zone.

Slow oxidizers are believed to benefit from B complex vitamins, protein, carbohydrates, table salt, and keeping fat intake low to moderate. Slow oxidizers need supplementary acid with meals to assist in the digestion of protein. Betaine HCl or acid form vitamin C (ascorbic acid) work well. Another indication for supplementary acid with meals is the CO_2 level on a blood chemistry panel – when CO_2 is below 22, additional acid is helpful.

Fast oxidizers are supposed to do well with a high fat diet and substantial amounts of vitamin D.

The ratios of zinc to copper and to cadmium are of no utility when mineral transport is deranged. When mineral transport is normal and orderly, maintaining the proper zinc to copper ratio balances metabolic activity, and maintaining an adequately high ratio of zinc to cadmium is protective against the toxic effects of cadmium.

Potassium and sodium up, with calcium and magnesium down means the person's adrenals are shot, making way too much adrenaline but not enough cortisol. This generally goes along with being thin, high strung, excitable but not very stress tolerant.

Calcium	245	300- 1200
Magnesium	23	35- 120
Sodium	60	12- 90
Potassium	21	8.0- 38

Calcium	177	125- 370
Magnesium	19	12- 30
Sodium	420	12- 90
Potassium	490	12- 40

Calcium	106	300- 1200
Magnesium	8	35- 120
Sodium	270	12- 90
Potassium	100	8- 38

This form is provided to Andrew Cutler to be used as an educational tool and does not constitute endorsement by *Doctor's Data*.

Potassium and sodium down, with calcium and magnesium up means the person's adrenals not making enough cortisol or adrenaline and their thyroid is somewhat low too. This generally goes along with sluggishness and weight gain.

The figure above shows examples of these two patterns – mild, moderate and severe. A mild pattern of this type indicates a level of stress that is no problem for a normal healthy person but may eventually wear down someone who has an underlying condition like toxicity.

These two situations are very common in people with heavy metal or other intoxications. Mercury is particularly insidious at inducing these kinds of problems.

Since the adrenal glands have a very wide dynamic range from person to person and from day to day, the usual blood tests typically won't identify problems until shortly before death. This has long been well known in the medical community, with *Williams' Textbook of Endocrinology*, 10th edition, having this to say:

"the patient may present with vague features of chronic adrenal insufficiency – weakness, tiredness, weight loss, nausea, intermittent vomiting, abdominal pain, ... Psychiatric symptoms may occur in long-standing cases and include memory impairment, depression, and psychosis. Patients may be inappropriately diagnosed as suffering from chronic fatigue syndrome or anorexia nervosa. "

Even very severe cases that actually do produce abnormal results on the standard tests are often misdiagnosed. When something is wrong your doctor can't figure out, a simple, inexpensive hair test makes a lot of sense Hair testing allows adrenal and thyroid problems to be picked up earlier than traditional tests, preventing people from undergoing much unnecessary suffering.

An elevation of hair potassium relative to sodium suggests excessive thyroid hormone levels or a deficiency of copper. The *Doctor's Data* report lists the ratio of sodium to potassium (Na/K) and gives a reference range of 0.5 to 10. Ratios below 0.5 suggest thyroid levels are too high, while ratios greater than 10 suggest thyroid levels are too low.

Low potassium compared to calcium suggests low thyroid or copper excess. The *Doctor's Data* report does not give this ratio. You can easily calculate it and the relevant reference range is K/Ca = 0.03 to 0.35. Again, the hair test will typically suggest the problem before the blood test results are abnormal. Much misery can be avoided by promoting normal healthy thyroid hormone levels for that particular individual rather than waiting until their levels are so out of whack that nobody could be healthy with that much or little thyroid hormone – which is what the blood tests tell you.

Also note that the blood test most commonly used depends on the body's normal feedback mechanism in the pituitary gland working properly, and many of the heavy metal toxins, especially mercury, are known to concentrate in the pituitary and poison it.

Exercise is believed to raise sodium and potassium while lowering calcium and magnesium. Exercise is helpful for people with elevation of calcium and magnesium relative to sodium and potassium (low thyroid and adrenal function, sluggishness, weight gain) and to not good idea except in small and carefully measured amounts for people with the opposite pattern of low calcium and magnesium compared versus high sodium and potassium (adrenals make tons of adrenaline but not enough cortisol, the person is thin, high strung, excitable and not stress tolerant).

An elevation of sodium relative to magnesium is believed to indicate excessive production of adrenaline. This leads to hyperactivity, anxiety, panic, and poor immune function. The reference range for this is 1 to 5.

When sodium is low compared to magnesium (Na/Mg less than 1) the body can't make enough adrenaline leading to poor mood and energy.

Having a high ratio of Ca/P along with a low ratio of Na/Mg suggests the need for supplementary vitamin B6.

When Ca/K and Na/Mg are elevated and Ca/P is low impaired immune function is likely.

Digestive enzymes are often helpful when either Ca/P is low, or when both Na/Mg and Ca/K are elevated.

One of the more popular laboratories suggests some form of adrenal cortex extract for everyone who gets a hair test without apparent regard for the results. Clinicians report that the adrenal cortex extract is usually quite helpful, too. Apparently the lab has learned by experience that most people who get a hair test do it late enough in their odyssey through the health care system that almost all of them are suffering from adrenal fatigue. Other labs have specific results that are indications for the use of adrenal cortex extract.

Vitamin E is also uniformly recommended by some laboratories and is usually helpful. Again, since hair tests are seldom ordered early in the investigation of a person's health condition it seems that the simple fact someone has been sick for a long time is an adequate indication for vitamin E without actually needing to check test results.

Labs that generate reports with their hair tests to suggest supplements often suggest copper, sometimes in frankly toxic amounts. Even in low amounts supplementary copper usually makes people worse, not better. Copper should uniformly be excluded from supplements in all cases of heavy metal problems other than zinc or molybdenum (as verified by blood or urine testing, not solely based on a hair test result). If you want to try copper supplementation, start low (500 mcg), increase every few days and be aware and alert for signs of copper problems such as diarrhea, malaise, fever, etc.

The more sophisticated books (and laboratories) discuss how toxic elements often do not show up in hair until after a reasonable program of supplementation is in place. The clearer way to express this is to say that when mineral transport is deranged, hair levels of toxic elements often do not reflect body burden. Once mineral transport has been normalized which can be done simply by supporting normal healthy metabolic activity through the use of appropriate nutritional supplements and dietary modification, the levels of the toxic elements in hair will then more accurately reflect how much of them is present in the body.

The major advantage of the relationships discussed here is that they appear to be valid indicators regardless of whether mineral transport is deranged or normal. Thus the hair test with deranged mineral transport may contain much valuable information even if the detailed levels of the various minerals are of little utility.

Copper (Cu)

Hair copper is believed to accurately reflect body inventory when mineral transport is normal. Very elevated levels of copper (high red) on a hair test are usually significant even in the presence of deranged mineral transport.

Women having regular menstrual cycles have slightly higher hair copper than men and non-cycling women. The use of birth control pills reduces hair copper slightly. The 18-50 year old women's reference range is most appropriate for women who are having regular menstrual cycles. Women who are not cycling (and are neither pregnant nor lactating – conditions that make hair test interpretation difficult) should be compared to the male range which is about 7 ppm lower.

There is much superstition in the hair testing community about elevated copper often being due to contamination. While occasionally elevated hair copper is due to contamination, often the supposed source of contamination (swimming pools treated with copper sulfate as an algaecide) also cause great systemic absorption of copper through the skin and the elevated hair levels actually DO reflect elevated body copper rather than being an artifact. Most laboratories have washing procedures which really do remove contaminants such as copper from swimming pools and only leave in the copper that got into the hair from the person's body when the hair was growing.

Very low hair copper will rarely indicate copper overload, so it may make sense to check urine or red blood cell copper for someone with copper symptoms and red-range low copper levels in hair.

The idea that occasionally low hair copper is paradoxical and reflects a high body burden is known to be relevant to rare genetic conditions. In Menkes disease hair copper is low. In Wilson's disease hair copper is variously reported as either low or normal. In both conditions body copper builds up to extremely high levels.

Since there are other defects of copper metabolism not yet described in the academic journal literature indexed in PubMed, it is possible that some of them share this aspect of Wilson's and Menke's diseases. Many defects of copper metabolism, some apparently inborn, do lead to elevated body AND HAIR copper. A genetically determined pattern of high hair copper among relatives in the absence of signs of any other mineral aberration, or the same mineral aberrations related to inadequate partition of certain minerals into bile is occasionally seen. It is appropriate to check the hair of relatives of anyone with very high or very low hair copper and copper overload symptoms who does not have another obvious explanation for why they accumulate copper.

Blood copper levels are actively controlled by the body for various reasons and do not increase due to copper toxicity until highly toxic levels are reached in other tissues and a life threatening toxic crisis is imminent.. Blood copper levels do however go up and down substantially in response to inflammatory processes. Hair copper is fairly reliable as an indicator of body copper levels but blood copper is irrelevant. Urine copper levels are also of little meaning and can only be considered significant if they are vastly elevated when someone collects a timed (e. g. 24 hour) urine specimen and is not taking any sort of chelating or copper mobilizing agents.

Blood copper levels (red blood cell, serum, plasma) are elevated by the mineral transport derangement characteristic of mercury poisoning, and also by anything such as infection or allergy that causes inflammation. Birth control pills and cancer also elevate blood copper. High blood copper levels do not necessarily mean toxicity, it is only reasonable to be concerned when blood or red blood cell copper levels are well outside the normal range (much greater than the 97.5th percentile on reports presented that way).

Most copper is excreted by the liver via the bile, some in intestinal fluids, and only 4% comes out in the urine. Thus chelating agents that greatly increase urinary copper levels have very little effect on the body inventory and are not useful for treating copper toxic people under most circumstances.

Any mild abnormality or impairment of liver function can lead to copper excretion being mildly reduced so that the copper building up to toxic levels since the physiologic and toxic ranges are so close together. Thus people with gall bladder problems or a history of hepatitis or other liver disease have a significant risk of developing copper problems. Many liver diseases and conditions cause reduced copper excretion. Since much mercury is also excreted in the bile, a finding of mixed copper + mercury intoxication is common among those with weak livers.

Medical conditions that cause copper to build up in the body, such as liver cirrhosis, do not affect hair copper in a straightforward manner. Despite this, most of the time elevated body stores of copper due to poor excretion by the liver does show up as elevated hair copper when the liver impairment is not severe by mainstream medical standards.

Copper absorption becomes more efficient in people who don't have enough of it. Zinc inhibits copper absorption as does vitamin C. Molybdenum modulates copper metabolism to reduce body copper levels. Iron has some modest effect of reducing copper uptake and toxicity. Supplements that increase bile flow will increase copper excretion. Alpha lipoic acid reduces excretion of copper in the bile.

The half life of copper in the body is about a month. As copper excretion in bile is reduced, the half life lengthens. Interventions to modify copper body burden take more or less a month (a half life) to show their beneficial effects.

The average daily intake of copper is 1-2 mg. Supplementation with 10 mg will lead to signs of toxicity within a week unless taken along with supplements such as zinc that are antagonistic to copper absorption.

Copper causes the growth of new blood vessels (neovascularization is the proper medical term). This is good for healing, getting over heart attacks and strokes, and growth. It is also the leading cause of blindness and kidney damage in diabetics and allows growth of cancer tumors. Supplementary copper should be avoided by diabetics, people with retinal or kidney disease and cancer patients unless testing clearly shows it is appropriate for their individual situation.

Of the essential elements, copper is the one closest to toxic levels when present in normal physiological amounts. Thus modest variations in intake and retention can cause it to build up to toxic levels in some sensitive individuals. A few people become toxic at an intake of 5 mg per day (for an adult).

The derangement of mineral transport that mercury causes usually favors the accumulation of copper. Mercury toxic people do very well on supplement regimens that most "natural health practitioners" believe will lead to symptomatic copper deficiency and do not develop it after months or years of therapy.

Most body copper is in the liver and the brain and these can become damaged and function improperly due to copper toxicity. These are also the targets of toxicity for mercury and many other heavy metals. Thus copper is a synergistic poison with other heavy metals and those who have a heavy metal problem usually do best to maintain low-normal copper levels.

The liver problems caused by copper do not normally cause elevation of liver enzymes or bilirubin on standard medical tests until the poisoning is quite severe.

Low adrenal function causes copper accumulation.

Body copper is lowered by progesterone, estrogens raise it. Elevated copper tends to go along with (and possibly contribute to) PMS and estrogen dominance.

The behavioral effects of copper poisoning are almost indistinguishable from those of mercury. The symptoms of copper poisoning are readily confused with those of mercury. The major difference between copper and mercury is that for menstruating women, copper leads to much more significant PMS problems than does mercury.

Elevated hair copper is believed by some practitioners who are experienced in the interpretation of hair tests to increase spaciness, detachment from reality, and goes along with intelligent, creative childlike and artistic personality traits. Copper elevation is also believed to encourage phobias.

Kayser-Fleischer rings, often mentioned in medical books, appear on the cornea only after someone is severely ill,– e. g. cirrhotic and psychotic because they are so poisoned – these rings are absent in most copper toxic people.

Low body copper can cause an anemia that may appear from the blood count to be an iron deficiency anemia (low hemoglobin, hematocrit, possibly reduced MCV and MCH[73]), but iron markers such as ferritin, transferrin, and total iron binding capacity will be normal. It also causes poor immune function (through reduction of the number of white blood cells), poor growth and healing of connective tissue, reduced skin and hair pigmentation, and poor formation of the catecholamine neurotransmitters which are needed to pay attention, get excited, enjoy yourself and be energetic.

In copper deficiency all measures of copper will be low, urine copper, serum and red blood cell copper, hair copper, and also the protein ceruloplasmin in serum will be towards the lower end of its range (since the body makes it as needed to transport copper).

Prenatal copper deficiency causes neurological and skeletal abnormalities due to impairment of phospholipid synthesis and collagen cross-linking.

Zinc (Zn)

Hair zinc does provide useful information about body zinc levels but interpretation can be complex since elevated hair zinc can indicate low body levels, and low zinc can too.

Low hair zinc does correlate well with low red blood cell and total body zinc. Substantially elevated hair zinc is usually a sign of zinc wasting and consequent LOW body zinc levels (and low blood zinc levels). That is, people with high hair zinc usually have low red blood cell zinc and low tissue levels of zinc. Rarely will high hair zinc be due to elevated body levels of zinc.

Plasma and serum zinc are reduced in acute illness and are not reliable indicators of body inventory.

When mineral transport is deranged, elevated hair zinc invariably indicates a low body burden and zinc supplementation is appropriate. In fact, body zinc is almost invariably low when mineral transport is deranged unless zinc is being supplemented at high levels.

[73] These abbreviations appear on laboratory test reports for a complete blood count (CBC). They stand for mean corpuscular volume and mean corpuscular hemoglobin.

When mineral transport is normal and zinc status is not clear from how the person is, determining zinc levels in red blood cells is the most accurate measure of the levels of zinc where it is needed – inside cells.

Vegetarian diets are low in zinc. High carbohydrate intake is believed to reduce zinc levels and high consumption of grains is known to hinder zinc absorption. Modern foods are relatively low in zinc compared to foods 50 years or more ago (due to changes in agricultural methods since that time), and sometimes are preserved by the use of EDTA or other agents to sequester whatever zinc is in them and render it unavailable. In addition, zinc absorption from the intestines is limited and saturable, which means a smaller fraction is absorbed as the amount taken increases. Thus as you increase the amount taken a lot the amount absorbed only increases a little. Because of this very few people will have high enough zinc levels to be a problem unless they take zinc supplements several times a day for months on end.

Given the meaning of the color bands on a *Doctor's Data* Hair Element Profile test, low or high red zone zinc levels – with either normal or deranged mineral transport – suggest that supplementing with zinc would be a very good idea while checking to make sure it is not one of the uncommon cases where body zinc is actually too high. If the results bar goes more than half way across the yellow zone to high or low, then it is worth considering zinc supplementation. If the results bar doesn't go very far into the yellow zone and you like things to be really very normal and ideal, then you might feel like using a bit of supplementary zinc.

High hair zinc (and also low hair magnesium) are signs that more vitamin B6 may be needed. Alkaline phosphatase (a blood test) below 50 also suggests this.

Significant amounts of zinc are excreted in the bile flow from the liver, and substantial amounts more are excreted in the intestinal secretions used for digestion of food. Zinc is absorbed better with protein, by pregnant women and by people with high cortisol levels (or who take prednisone like medicines). It is reduced by copper, iron, phosphate, calcium, fiber, a high intake of caffeine, and functionally inadequate adrenal hormone levels. Absorption is regulated by zinc intake, so that people who have little zinc in their diet absorb it well, while those who supplement absorb it poorly. Due tot his factor, great increases in supplementation lead to only small increases in absorption and dividing the dose of zinc up is far more effective at getting more absorbed than is increasing the amount given in a single dose. The liver's excretion of zinc (and copper) is reduced by alpha lipoic acid.

Much zinc is absorbed in the large intestine and people with chronic diarrhea can become zinc deficient.

The half life of zinc in the body is about 4 months. Interventions to affect zinc levels will thus typically have their effects gradually over a period of several months.

Zinc is found in muscles and internal organs, where it forms a part of many different enzymes.

At very high levels, zinc toxicity causes rapid heart beat, blurred vision, reduced body temperature, porphyria and anemia.

Zinc is one of the minerals that our bodies respond to the taste of. People without enough zinc in their bodies will find dilute zinc solutions neutral or even tasty, while those with adequate zinc levels will find them metallic tasting and unpleasant. People with excessive zinc levels will find even very dilute zinc solutions unpalatable.

Low zinc levels lead to rough dry flaky skin especially on the hands, feet and around body orifices. It also causes slow wound healing, poor immune function – specifically cellular immunity –

in the presence of normal levels of white blood cells and antibodies, poor appetite, reduced taste sensitivity, reduced sense of smell, fatigue and poor night vision. Low zinc also lead to impaired growth and development in children, and to birth defects and difficulties during labor and delivery. Severe forms of childhood zinc deficiency are described in standard medical textbooks. Current estimates are that about 10% of children have mild zinc deficiencies that they are suffering some negative effect from, e. g. that they have recurrent ear infections – a problem which can often be stopped with supplementary zinc.

Low zinc levels (and high hair zinc) are often associated with eating disorders like anorexia and bulimia. Zinc supplementation, e. g. 25 mg 3 times a day is often helpful in normalizing eating behavior in people with these conditions.

Zinc is essential for normal development before and immediately after birth. Just after birth, poor survival and growth and depressed function of the immune system are salient features of zinc deficiency.

Low zinc levels are known to cause impairment of neurological and psychiatric functioning which are corrected by zinc supplementation, and zinc supplementation is known to improve attention and behavior in ADHD.

The specific aberration of immune function that zinc deficiency causes is an imbalance between T helper type 1 and T helper type 2 cells along with reduced activity of natural killer (NK) cells. This is also commonly observed in mercury toxic people and is most likely a consequence of their poor zinc status due to mercury induced mineral transport derangement. This results in allergies, autoimmunity and a poor immune response to pathogens like viruses, bacteria yeast and fungus.

Mercury induced derangement of mineral transport almost invariably results in reduced body zinc levels (usually reflected by high hair zinc levels), poor zinc absorption, increased loss of zinc in the stool and urine, and difficulty concentrating zinc into cells where it is needed. People with low zinc due to mercury problems typically need levels of zinc supplementation that would be considered excessive in normal individuals and tolerate high level supplementation for prolonged periods of many years.

Other health conditions that increase the risk of zinc deficiency are any form of malabsorption (including chronic diarrhea and inflammatory bowel diseases), chronic liver disease, kidney disease, diabetes, sickle cell anemia, and cancer. Other deficiencies also usually occur along with zinc deficiency.

Manganese (Mn)

Hair manganese is believed to accurately reflect body stores when mineral transport is orderly, but not when it is deranged. During deranged mineral transport a moderately (but not vastly) elevated hair manganese can indicate LOW body stores of this element. Moderate elevation is up to four times the upper reference range level given on a *Doctor's Data* Hair Element Profile test report.

Manganese is needed for the conversion of ammonia to urea. If this process is impaired people develop high ammonia levels on eating protein which causes them to become delirious and confused in the same way that alcoholics with liver disease do.

Manganese is essential for energy production in the Krebs cycle and as a cofactor for many enzymes. It is necessary for the body to use carbohydrates properly as it is a cofactor of the pyruvate decarboxylase enzyme.

Manganese is essential for growth, bone formation, and especially for formation of the bones and structures of the inner ear that are responsible for proper hearing and sound perception. Manganese is also believed to be essential in the production of some mucopolysaccharides which are important in making the body's secretions (saliva, tears, intestinal fluids) function properly.

Dietary manganese is well absorbed. Manganese is concentrated in the mitochondria. Thus measurements of serum and plasma manganese are not a good reflection of body stores. Hair and red blood cells are reasonable measures when mineral transport is normal.

Dietary calcium, phosphorus and iron reduce uptake of manganese from food. Low levels of these nutrients increases manganese absorption.

Excretion of manganese is mainly in the stool via secretion in bile by the liver. Manganese levels go up about threefold in people with liver disease that interferes with this process.

Manganese can be a synergistic toxin with most heavy metals, e. g. Mercury, lead, arsenic and antimony. Because of this, someone suffering from toxic levels of one of these elements may be quite sensitive to manganese even if it is not very high and may not do well with normal levels of supplementation. Additionally, modestly elevated levels of two or more synergistic toxins can make someone very sick even if they would never notice the effect of one or the other.

Manganese is toxic due to its activity as an oxidation catalyst. Since it concentrates in the mitochondria a mild intoxication might be misdiagnosed as a mitochondrial problem.

Elevated manganese may be associated with alcoholism. In such cases the desire for alcohol can be lessened by reducing manganese levels.

There is a slow and insidious onset of symptoms in manganese toxicity. Too much manganese first causes headaches, insomnia, memory loss, emotional instability, depression and anxiety, and later progresses to apathy, fatigue, confusion, excessive sexual desire (and later impotence in males), then ultimately extends to tremor, slow and jerky movements, psychotic or bipolar like behaviors, abnormal gait, jerky eye movements (nystagmus is the proper medical term), lack of facial expression, generalized pain and especially joint pain, very small handwriting and whispering speech. Severe long term manganese poisoning can result in neurological signs like a peculiar staggering gait, clumsiness, tight muscles, slow movements, dementia and other issues. In this most severe form liver problems may be apparent on standard medical laboratory tests. This most severe form of manganese poisoning can be misdiagnosed as Parkinson's disease or Multiple Sclerosis or a variety of rare conditions.

Manganese toxicity can cause an elevation of serum calcium (which is routinely measured on a "chem panel," one of the blood tests doctors order early on when presented with a confusing case).

Elevated manganese may interfere with uptake of iodine by the thyroid. Free T3 and free T4 are useful tests to explore this issue. TSH is not useful for this purpose. The target range for best health is free T4 about midrange and free T3 in the upper third of its range.

Excess manganese reduces brain dopamine and serotonin levels, presumably by poisoning the neurons that make them.

Many people with manganese poisoning feel sore all over and touching is painful if not done gently.

In the earlier stages of manganese toxicity there is often poor appetite to the point of substantial weight loss. Later on weight gain is common though not uniform.

Manganese excess can resemble Parkinson's disease. In fact, the brain damage caused by manganese is quite similar to that caused by Parkinsonism.

Both cause damage to the globus pallidus. Manganese but not Parkinson's disease causes damage to the striatum, while Parkinson's disease causes damage to the substantia nigra and manganism does not.

Manganese toxicity's most prominent features are psychiatric and neurological. Liver problems are secondary and are rarely apparent on standard blood tests.

Manganese toxicity can masquerade as Multiple Sclerosis, Wilson's disease, Parkinson's disease, thyroid problems (hypothyroidism or myxedema are the medical terms), psychiatric problems, and other heavy metal exposures. It is important to rule out manganese (and other heavy metal) problems with a hair test since excess manganese has a good treatment outcome and many of the other conditions do not.

Doctors are taught in school that manganese toxicity only occurs due to inhalation exposure in the workplace. In this case there is difficulty breathing and eventually "manganese pneumonia." In the vast majority of actual cases exposure is not by inhalation and there are no respiratory symptoms.

People with frank manganese deficiency may have elevated ammonia levels – especially if they consume a high protein diet. Elevated ammonia leads to confusion, disorientation, malaise, and a variety of other neurological problems. If this is occurring, plasma ornithine and alpha ketoglutarate are low while arginine and glutamate are elevated. Glutamine may take any value, but manganese deficient people will tolerate supplementation with it poorly. While plasma ammonia is also elevated it is important to realize that false positives on this test are the rule rather than the exception unless done at a hospital with its own laboratory and by a phlebotomist who routinely performs the test and is aware of how crucial proper handling of the blood sample is (which starts with immediately putting it on ice). Manganese deficiency is also believed to be correlated with low cholesterol, weight loss, hair and nail changes, dermatitis and blood clotting irregularities due to low levels of vitamin K dependent plasma proteins.

When pyruvate decarboxylase activity is low due to inadequate manganese levels, a person has to burn a lot of amino acids (protein) along with their carbohydrate to make energy. This use of protein then produces ammonia which is handled poorly when inadequate manganese is present.

Manganese deficiency can lead to neurological issues, such as confusion, disorientation, easily aroused anger, nervousness, rapid pulse and possibly tremors (which can also be due to high manganese). The most easily identified problem is ringing in the ears – tinnitus – which improves on manganese supplementation.

Mn deficiency: reduced growth, skeletal abnormalities, reproductive function abnormalities, clumsiness (ataxia) in the newborn, abnormal inner ear bone formation.

Manganese deficiency can cause fatigue, lack of physical endurance, slow growth of fingernails and hair, poor bone and cartilage formation, skin irritation (dermatitis), weight loss, and reduced fertility. In children it can cause reduced growth and skeletal abnormalities. Manganese deficiency during pregnancy or in infancy can cause poor coordination due to impaired balance, hearing problems and abnormal brain function. Severe deficiency can cause birth defects.

Seizures may occur in severe manganese deficiency.

Low hair manganese is also associated with allergy and inflammation. However so many other things are also associated with this that allergy and inflammation are not reasonably suggestive of a manganese deficiency.

Chromium (Cr)

Hair chromium accurately reflects body inventories and functional status when mineral transport is orderly. During disordered mineral transport very elevated hair chromium is sometimes a sign of deficiency, or may have no significance.

Blood and urine chromium levels are not good indicators of body inventory since both are increased greatly by glucose and by insulin. Thus people who eat lots of carbohydrates, and type 2 (adult onset) diabetics will have relatively low body inventories at a given blood or urine level.

Chromium is essential for insulin function and sensitizes the body to insulin. Since elevated insulin is a logical result of reduced chromium levels, and causes increased chromium losses, a vicious cycle can occur whereby inadequate body chromium inventory causes further reductions in body inventory through increased excretion.

The form of chromium the body uses is a chelated +3 form (trivalent chromium, with the physiologically useful form being held on to by specific ligands). The form that is highly toxic is the +6 form (hexavalent chromium). Hair testing (and urine and routine blood testing) does not distinguish between these forms. Special blood tests to distinguish the two forms of chromium are available.

Most trivalent chromium is excreted in urine.

Most forms of trivalent chromium are poorly absorbed and retained, but natural Chromium from foodstuffs, and certain specific types available as supplements, are well absorbed and retained. Hexavalent chromium is far more toxic than trivalent chromium, and also it is more well absorbed than most forms of trivalent chromium.

Very high levels of trivalent chromium can cause impaired health which may be improved by supplementary vanadium, and vice versa.

Most laboratory tests, including a hair test, do not distinguish between the valence states of chromium. Indirect evidence of whether someone has the trivalent or hexavalent form in them can be had by comparing plasma and red blood cell chromium levels. Trivalent chromium mostly stays in the plasma. Hexavalent chromium mostly concentrates in the red blood cells. Both of these analyses are a reflection of recent exposure rather than body burden.

Low levels of the physiologically useful form of chromium lead to carbohydrate intolerance and eventually elevated blood sugar as is found in diabetes. This then causes elevated blood fatty acids, tingling or painful or numb hands and feet[74], poor brain function and poor utilization of protein.

Chromium toxicity causes nonspecific symptoms such as weight loss, fatigue and malaise. Chromium toxicity (both trivalent and hexavalent forms) may affect the liver and the blood. There is both anemia and a reduction in platelet cells which assist with clotting. If enough blood cell destruction occurs then the kidneys may also be harmed.

Exposure to the hexavalent form of chromium is known to increase the risk of cancer, but exposure to the trivalent form – the form in food and nutritional supplements – is known not to increase the risk of cancer, which is to say that hexavalent chromium is carcinogenic, but trivalent chromium is not.

---

[74] The medical term for this is peripheral neuropathy.

Vanadium (V)

Hair vanadium levels are believed to accurately reflect body burden in the presence of normal orderly mineral transport.

Vanadium has insulin like and muscle building actions. Toxic levels of vanadium can interfere with the kidney's ability to retain potassium, leading to low body levels of potassium which reduces heart, muscle and nerve performance.

The vanadyl form of vanadium used as a nutritional supplement (e.g. vanadyl sulfate) has less potential for toxicity than other forms.

Trace amounts of vanadium are essential for normal growth and metabolism.

High levels of trivalent chromium can cause impaired health which may be improved by supplementary vanadium, and chromium may help when vanadium is high.

Vanadium can inhibit the cell surface transport protein Na/K ATPase, and also stimulate cardiac adenylate cyclase. Since Thallium also inhibits Na/K ATPase vanadium and thallium are in theory synergistic toxins.

There is some possibility that selenium and vanadium are synergistic toxins when both are present in excessive amounts[75], but this is less than clear in the literature.

At high levels of toxicity, vanadium can cause anemia as well as toxic porphyria, similarly to most heavy metal toxins. Again, there is some potential for synergistic toxicity in these areas.

Despite the theoretical possibility of synergistic toxicity between vanadium and heavy metals, it is the uniform experience of people with heavy metal problems and of health care practitioners who treat such people that those with elevated blood sugar improve on the use of vanadium as a nutritional supplement without experiencing any negative effects.

The half life of vanadium from nutritional supplements in the blood and urine is approximately 16-24 hours. Retention in other body tissues not routinely sampled (aside from hair) may be much longer than this.

Molybdenum (Mo)

Hair molybdenum accurately reflects body stores when mineral transport is normal but not when it is deranged. However this does not necessarily mean that the amount of molybdenum present is adequate for what that specific person needs, and some people who are "within normal limits" but towards the lower end of them are still functionally deficient in molybdenum while others with the same amount of molybdenum in their systems are fine.

Tungsten is believed to be an antagonist of molybdenum retention or an antagonist of its proper utilization.

Elevated copper accelerates molybdenum clearance and reduces body inventory at a given level of dietary molybdenum intake. Molybdenum and copper reduce each other's body levels.

Molybdenum is used in several enzymes, sulfite oxidase, aldehyde oxidase and xanthine oxidase. Low molybdenum can cause low sulfate levels, low uric acid levels, and intolerance to alcohol. The low sulfate levels may cause reduced tolerance to salicylate and phenolic containing foods like berries, tomatoes, cucumbers and many fruits. Reduced clearance of sulfite can lead to sulfite sensitivity or simply to asthma from sulfite naturally produced in

[75] The amounts of selenium typically taken as supplements are not excessive and will not create synergistic toxicity with vanadium.

the body. Reduced aldehyde clearance leads to alcohol intolerance with red flushing and nausea.

Excess molybdenum can be toxic. In severe cases the symptoms may include weakness, diarrhea, and loss of hair pigmentation. It can cause gout, anemia (the hypochromic microcytic type, that is Hgb and MCV and MCH[76] are all low on blood count results), generalized pain, reduced appetite and gonadal atrophy which causes low sex hormone levels. Since molybdenum reduces copper absorption and is not itself particularly toxic, a high molybdenum intake may cause problems by inducing copper deficiency without directly producing molybdenum toxicity.

Molybdenum absorption is reduced by increasing dietary sulfate levels. High blood sulfate may impair molybdenum reabsorption in the kidneys, too, so those with elevated sulfate may need a lot of supplementary molybdenum.

Boron (B)

Hair levels of boron are believed to accurately reflect body burden when mineral transport is orderly, but contamination from "borax" type hand soaps, detergents, etc. can occur. Some bedding is made flame retardant with borax to avoid the use of the far more toxic antimony. Of course people have a pretty good likelihood of absorbing boron from these "contamination sources," so identifying one of them does not automatically mean that a high hair boron level does not signify high body levels of boron – there is a reasonable chance that hair boron is high because body levels are high due to the identified exposure source. The hair handling and washing procedures that most laboratories use really do prevent external contamination from being an issue in most cases.

When mercury poisoning has caused deranged mineral transport, boron is often elevated.

Boron's functions in the body are not well understood, but it is believed to be helpful in causing calcium to form bone.

Boron is primarily excreted through the urine.

Boron is not believed to be absorbed significantly through intact skin, but is absorbed surprisingly well through scabs, damaged skin, the eyes, and is very well absorbed if ingested.

High boron levels can increase the need for vitamin B2 (riboflavin) by increasing its excretion in the urine.

Boron toxicity causes poor appetite and weight loss, poor digestion, nausea, vomiting, diarrhea, clumsiness, tremors, convulsions, and an itchy red rash that is firm to the touch.

Iodine (I)

Hair iodine is believed to accurately reflect body burden when mineral transport is normal. Hair iodine often becomes falsely low or sometimes high when mineral transport is deranged. Urine levels of iodine are believed to reflect recent intake more than body inventory.

Iodine is well absorbed from the diet.

Iodine's major function is as a component of thyroid hormone, and iodine deficiency leads to goiter (swelling of the thyroid gland in the neck) along with low levels of thyroid hormone.

Extremely high levels of iodine may interfere with thyroid function, also leading to low levels of thyroid hormone[77]. Coughing, nasal discharge and increased tear formation may also occur. Iodide increases mucous secretion

[76] These abbreviations appear on laboratory test reports for a complete blood count (CBC). They stand for hemoglobin, mean corpuscular volume and mean corpuscular hemoglobin.

[77] The medical term is hypothyroidism.

and used to be used medicinally for this purpose. White blood cell levels and activity may also be mildly reduced, increasing susceptibility to infection. High iodine may also reduce iron and hemoglobin levels.

The signs of low thyroid hormone levels are a hairline moved a bit down on the forehead, sensitivity to cold, low energy and metabolism, constipation, dry skin, thin, dry hair that is slow growing and lightens in color moderately, loss of the outer third of the eyebrows, mental dullness, slowness and depression, as well as a particular kind of skin puffiness called myxedema.

Hair iodine will sometimes be elevated in the presence of hyperthyroidism, and occasionally in the presence of hypothyroidism.

Iodine deficiency during pregnancy or infancy can lead to impaired brain development with permanent cognitive deficits or mental retardation.

In adults, low iodine intake leads to goiter – swelling of the thyroid gland in the neck.

Lithium (Li)

Hair lithium is believed to accurately reflect body inventory when mineral transport is orderly.

Lithium varies greatly in people depending on where they live due to variations in the amount in their local soil and rocks and hence their food.

Lithium is calming and mood stabilizing. It is believed to have this effect by slowing the dephosphorylation of inositol and thus prolonging the action of the phosphoinositide second messenger pathway, which amplifies the effect of some brain chemicals like serotonin that are responsible for calmness and a sense of well being.

Very high levels of lithium, such as are common in therapy for bipolar disorder (manic depression) also inhibit thyroid function and lead to hypothyroidism. This is not a risk when using nutritional supplements or when the hair lithium level bar is not all the way to the edge of the chart.

Very high lithium levels may also adversely affect the heart.

While lithium is a congener of sodium, potassium, rubidium and cesium, it generally does not move along with them during mercury induced deranged mineral transport.

While there is no formally recognized need for dietary lithium, low levels are believed by some to correlate with poor mood and a lack of emotional control. Lithium is available as a nutritional supplement at levels far less than those used in psychiatry (thus lacking the risks and side effects) to ensure that people who are low in it receive adequate amounts.

Phosphorus (P)

Hair is believed to reflect phosphorus intake and body levels reasonably well if mineral transport is orderly.

Phosphorus levels are reduced in chronic liver diseases and elevated in dialysis or kidney disease. Thus low hair phosphorus suggests that liver health be evaluated and high phosphorus suggests checking the kidneys. Poorly characterized liver dysfunction is quite typical of heavy metal problems. Many liver problems do not cause any abnormalities on standard blood chemistry panels or liver function tests. Standard tests for kidney function are generally informative.

Low phosphorus can cause low energy, weakness, appetite loss, intolerance to fructose containing items such as table sugar or fruit or soda pop, and fatigue. Low phosphorus due to dietary deficiency is quite unusual since phosphorus is in most foodstuffs.

Some people believe that low serum phosphorus (as determined on the ubiquitous chem panel or metabolic panel blood test) suggests poor digestion of protein which is most often remedied by using additional acid, usually in the form of betaine hydrochloride or the acid form of vitamin C (ascorbic acid) with meals.

Exposure to elemental phosphorus in certain allotropic forms can cause severe toxicity (for example, 'fossy jaw,' which is extensive abscesses and dead areas in the lower jaw). This is a well understood risk in very limited and carefully controlled segments of the chemical and munitions industry.

Selenium (Se)

When mineral transport is orderly, hair selenium appears to accurately reflect at least intake and probably body burden as long as you don't use a selenium or sulfur containing anti-dandruff shampoo (for example, Selsun Blue® or the selenium sulfide containing version of Head and Shoulders®).

Contamination of hair samples by dandruff shampoos containing selenium or sulfur – which is sometimes contaminated with selenium – is quite common and accounts for most observed very high hair selenium levels.

Red blood cell levels of selenium are believed to represent body burden, while whole blood, plasma, serum and urine measurements are less accurate. Measuring glutathione peroxidase or comparing the levels of T3 and T4 are ways to determine whether the amount of selenium present is adequate to meet the body's needs.

About half of selenium excretion is in the urine. Somewhat less than half is excreted in the stool, and significant amounts are excreted in sweat. Some is excreted in breath and this becomes a larger fraction as body inventory increases to toxic levels. Little is excreted in bile except when other toxins such as arsenic or mercury are also present and it is excreted in association with them.

Copper increases selenium excretion in the bile and is useful for reducing high levels

The toxic level of selenium is only about 15 times higher than the level needed for adequate nutritional intake. Only copper has a narrower window between minimum required levels and toxicity. A few people start to experience mild symptoms of selenium toxicity at an intake as low as 750 mcg a day (on an adult basis).

Selenium is used in several enzymes. Three very important ones are the enzyme that converts the T4 form of thyroid hormone to the more active T3 form, glutathione peroxidase that uses glutathione to protect cell membranes from oxidative damage, and phospholipid hydroperoxide glutathione peroxidase which also protects cell membranes. The activity of all of these enzymes varies as body selenium inventory rises and falls.

Selenium deficiency causes aching muscles, heart muscle weakness and heart disease. It can also contribute to sensitivity to the toxic effects of arsenic and mercury and perhaps other heavy metals at lower than the usual toxic levels.

Increased vitamin E intake partly offsets the negative effects of mild selenium deficiency.

Breast milk has more selenium than standard FDA approved infant formula.

Mothers of infants with neural tube defects, and the infants, have low hair selenium. Women who miscarry repeatedly have low hair selenium.

At excessively high levels selenium itself is a toxic heavy metal.

The selenite form of selenium is quite toxic, strongly pro-oxidant and is used in laboratory experiments to induce cataracts in rats. Organic forms of selenium such as selenomethionine or selenium yeast are much less toxic.

Severe selenium toxicity causes liver damage and chemical hepatitis, hair loss, weight loss, anemia, distress of the stomach and intestinal tract, red pigmentation and structural changes in the hair and nails, alterations in mental status, numbness tingling or pain in the hands and feet[78], and a garlic smell to the breath.

Flax seed meal is helpful in selenium toxic animals but it is unclear which component is protective so use of, e. g. the more convenient flax seed oil may not be helpful.

All antioxidants offer some protective effects in selenium toxicity.

Strontium (Sr)

Hair strontium levels are an accurate reflection of body burden when mineral transport is normal. When mineral transport is deranged, strontium levels usually correlate with calcium.

Low strontium correlates with tooth decay, soft bones and heart disease. Supplementary strontium in appropriate modest quantities is known to improve bone density and strength.

Strontium is not very toxic. Toxicity usually requires intakes well above 1000 ppm in the diet. Toxicity causes muscle weakness, rickets and bone weakness. It may also weaken ligaments and cause muscle spasms. This seems to happen more in the legs and feet. Children are more susceptible to strontium toxicity than adults, and the lower the calcium content of the diet, the more toxic a given dietary level of strontium is.

The radioactive form of strontium, strontium 90, is an entirely different problem and the amounts necessary for serious negative effects or even death are far lower than will appear abnormal on a hair test.

Sulfur (S)

Sulfur takes many forms in the body and its chemistry is complex. Hair sulfur does not give any relevant information about any of the important forms of sulfur that relate to the need for dietary restrictions or the use of specific supplements.

Barium (Ba)

It is not known to what extent hair barium levels accurately reflect body burden when mineral transport is normal. When mineral transport is deranged it sometimes follows calcium, magnesium and strontium which are its chemical cousins (congeners).

Barium sulfate, as used for X-ray imaging studies, is quite insoluble and thus safe. Soluble barium salts are quite toxic.

Barium interferes with potassium metabolism. Blood potassium levels fall which can cause paralysis and heart failure.

When eaten, barium can cause stomach pain, nausea, vomiting and diarrhea.

On long term exposure barium can cause weakening of the heart muscle (the medical term is cardiomyopathy).

[78] The medical term is peripheral neuropathy.

Cobalt (Co)

It is not known whether cobalt levels correlate with body stores.

Hair cobalt is not generally believed to correlate particularly well with B-12 levels, low cobalt levels in hair do not suggest the possibility of functionally inadequate B-12 levels but most doctors will probably want to check anyway since this is easy to do and B-12 therapy is inexpensive. The correct laboratory markers for B-12 deficiency are elevations of MCV and MCH[79] on a blood count and elevation of methylmalonate in urine.

Elevated cobalt may interfere with uptake of iodine by the thyroid gland and reduce thyroid hormone levels.

Cobalt toxicity can lead to enlargement of the heart. It can also cause nausea, vomiting and ringing in the ears. At high levels it can also lead to impaired coordination, tremors and deafness.

Cobalt toxicity can also affect the blood. It can lead to polycythemia – an increase in the number of red blood cells, with increased RBC number, hemoglobin, hematocrit, and total blood volume.

The name cobalt comes from kobold, a German word for evil spirit, because of the health problems that would befall cobalt miners.

Iron (Fe)

Hair iron is not a good measure of body burden. It does not correlate well with other measures of iron status such as ferritin, transferrin, red blood cell levels, etc.. The blood test for ferritin is the most accurate and useful guide for body levels of iron, though it may be unreliable in the presence of inflammatory disease conditions. Measurements of serum iron or red blood cell iron are much less accurate reflections of body burden as they may be influenced by many factors.

Urine iron is also not well correlated to body burden, though occasionally very low urine iron will be seen in people who have high body iron levels to the point of iron overload toxicity. The most extreme form of this is a disease called hemochromatosis. Many people do not have this disease but do have enough iron to cause them problems, especially if they also have other toxic elements at moderate to high levels.

Iron is not well absorbed. Only a few percent of iron in vegetable matter is absorbed. 15-20% of iron bound to heme rings in animal derived food is absorbed. This is found in liver, blood, red meat from mammals or dark meat from birds, and in the darker colored or red juices from these.

Adequate stomach acid is essential for iron absorption. The amino acids histidine and valine increase the absorption of dietary iron, as does supplementary stomach acid (betaine hydrochloride), and as do ascorbic acid alone and in combination with vitamin E. Buffered forms of vitamin C may not increase iron absorption as effectively as acid form vitamin C does.

People who are deficient in iron do absorb it better than people who have enough.

Ferric (+3 valence state, rust red color) iron is very poorly absorbed. Ferrous iron (+2 valence state, greenish color) is the form of iron that is absorbed.

Children absorb iron much more readily than adults and can be poisoned by eating iron supplements or plant foods (garden fertilizer) containing iron.

[79] These abbreviations appear on laboratory test reports for a complete blood count (CBC). They stand for mean corpuscular volume and mean corpuscular hemoglobin.

Vitamin E, copper and phosphorus are somewhat protective against the physiological stress caused by excessive iron levels.

The body has a hard time excreting iron. Oral lactoferrin is quite effective at reducing body levels. Otherwise, donating blood often and avoiding red meat and dark green leafy vegetables as well as any iron containing supplement is important.

Up to 8 percent of men have a tendency to develop hemochromatosis, a genetically based iron overload disease. Women become susceptible to this after menopause (natural as well as medically or surgically induced).

Excess iron can cause all the typical nonspecific symptoms of any heavy metal problem, such as fatigue, malaise, liver problems, digestive disturbances, or psychiatric issues.

Iron deficiency occurs below a body inventory (adult) of 150-200 mg. Iron overload occurs above a body burden of 2.5 to 3 grams (2,500 to 3,000 mg). The ideal body inventory for optimal health is about 275-650 mg for an average sized adult, corresponding to a ferritin level of 30-70.

Iron deficiency causes fatigue, inflammation of the tongue and spooning of the fingernails.

Adequately severe iron deficiency causes anemia where hemoglobin, hematocrit, mean corpuscular volume and mean corpuscular hemoglobin (MCV an MCH) are all low. In medical terms this is referred to as a hypochromic microcytic anemia.

Iron deficient children sometimes have their hair become coarser and they also tend to mouth and chew everything available[80].

Menstruating women are more susceptible to iron deficiency than men due to blood loss.

Generally people with heavy metal toxicity do not tolerate iron well. Many people's bodies dump it – even to the point of anemia – to maintain health. People with heavy metal toxicity whose bodies do not dump iron tend to be more symptomatic and improve on anything that reduces body inventory.

Low hair iron is suggestive of lead exposure, but not a definitive sign of it. Generally any reduction in body iron stores caused by heavy metal toxicity is NOT reflected in hair iron levels.

The body's major use for iron is in the oxygen transport and storage proteins hemoglobin and myoglobin. It is also used in some enzymes. Low body iron typically causes the kind of weakness, spaciness, attention deficit and tiredness anyone experiences when at high altitudes.

Some hair testing experts believe that certain cancers sometimes cause elevation of hair iron, though very high iron is not often due to cancer. Also, hair iron is often normal in cancer patients. Due to this weak correlation and the frightening nature of cancer I would suggest that vastly elevated hair iron (red range) in the presence of normal mineral transport suggests the merits of reasonable routine cancer screening that might be relevant to the person's condition. Very high iron levels with deranged mineral transport should not be considered suggestive of cancer. People should not allow themselves to be frightened by the test result (or by an excited health care provider) but rather should consider this to be a factor such as smoking or occupational exposure to asbestos or certain chemicals that means such screening is a prudent part of routine medical care even though it is likely not to turn up anything.

---

[80] Pica is the medical term for this behavior.

Germanium (Ge)

It is not known whether hair levels of germanium accurately reflect body inventory.

It is not clear whether germanium has any metabolic role. Inorganic germanium is quite toxic, organic germanium is generally believed to be much less so and many believe it to be essentially safe. There is no particular reason to expect this widely held belief to turn out to be particularly accurate when it comes to long term toxic effects from chronic exposure.

Organic germanium is often used as a nutritional supplement as a therapy for fibromyalgia and is reported to work fairly well for some people. This therapeutic use is based in large part on the presumption of safety. Germanium is also sometimes used as an antitumor agent.

As with most heavy metals, germanium toxicity affects the mitochondria. It can cause muscle weakness, kidney problems, anemia and tingling, numbness and pain in the hands and feet[81].

Rubidium (Rb)

It is not known whether hair levels of rubidium accurately reflect body inventory, but they most likely do not do so as rubidium is a close chemical relative of sodium and potassium.

Rubidium levels typically correlate well with sodium and potassium during disordered mineral transport or in adrenal stress response patterns. Very high rubidium is not likely to be significant unless it is much higher than sodium and potassium.

Rubidium has a 30 to 60 day half life.

Rubidium is antidepressant, affecting the serotonin neurotransmitter system.

Rubidium lowers body potassium levels.

High levels of rubidium can induce goiter and reduce thyroid hormone levels by interfering with the uptake of iodine by the thyroid gland.

Rubidium is sometimes used as an alternative cancer treatment.

Zirconium (Zr)

Hair zirconium is believed to accurately reflect body burden when mineral transport is normal. During deranged mineral transport hair zirconium levels often rise dramatically and they usually do so in conjunction with the level of titanium, which is zirconium's chemical cousin (congener).

Most zirconium compounds are generally believed not to be toxic. However, zirconium is a suspected cause of sarcoidosis which is an autoimmune condition.

Silicon (Si)

Hair silicon is believed to correlate with body inventory when mineral transport is orderly.

Silicon improves skin, hair nails, joint and bone properties. Low silicon predisposes people to tendentious and bone weakness or decalcification. Low silicon may also be associated with cardiovascular disease.

[81] The medical term for this is peripheral neuropathy.

High silicon levels cause kidney stones in some animals. It is unclear if this occurs in humans.

Silicon is not determined on a *Doctor's Data* test, but it is determined on a *Trace Elements, Inc.* test.

A few comments on absorption

Many toxins are poorly absorbed if eaten, but well absorbed if inhaled. Lead is one of these so sanding old paint, or old wood previously painted with lead paint and later stripped is a potentially disastrous source of exposure to the sander and to any children curiously watching and trying to help.

This is also true of antimony, which has now replaced lead as the color fast white pigment in many paints. While poorly absorbed when eaten, it is much more readily absorbed when dust is inhaled. In addition, children absorb antimony much more readily than adults.

Tungsten is potentially absorbed by inhalation during certain types of electric arc welding as well as grinding of "hard metal" machine tool bits.

Arsenic is potentially absorbed from the dust generated in sawing preserved wood, or the smoke or ash from burning it.

Most of these metals are also absorbed somewhat through the skin. Thus playing on an unsealed wood deck or wood playset made of arsenic treated wood, especially without washing afterwards, will lead to significant arsenic absorption. In fact, it is an occupational safety requirement that workmen who install such decking wear gloves and wash afterwards.

Anyone who used Mercurochrome or Merthiolate as a topical antiseptic absorbed a few percent of it. Each 2 ounce bottle of these contained about 2 lethal doses of mercury if it were completely absorbed. Studies on absorption through intact skin in laboratory animals showed that about 2% made it through the skin and into the bloodstream. Mercurochrome and Merthiolate were seldom applied to intact skin so it is reasonable to assume that as a rule more than 2% was absorbed.

Thimerosal was used as a preservative in many eye, ear and nose drops as well as some skin creams and several percent of the amount applied was absorbed. For practical purposes everyone other than small children has had substantial amounts of mercury applied to them externally during their lifetimes simply because its use in health care products was so widespread.

Many hospital surface and hand cleaning products contained Thimerosal and routine use could lead to absorption of substantial amounts of mercury. Due to the nature of hospital work, nurses are much more at risk for this occupational mercury poisoning than doctors.

Antimony salts from flame retardant treated bedding and children's pajamas are absorbed through the skin and antimony routinely shows up at high levels in the hair of developmentally impaired children. These children usually show some improvement when bedding free of antimony is used.

Case Summaries and Example Hair Tests

You Need to be Sure to get the Right Test

Through a comedy of errors one particular family of a father, mother, and two sons got a wide variety of hair tests done with assorted errors regarding test ordered, age and sex that led to material which illustrates two points very well.

1. You need the right test – the Hair Element Profile. The Hair Toxic Element Exposure Profile is not a suitable substitute and will completely miss disordered mineral transport.

2. Errors in transmitting patient information to the lab (age, sex) can change the appearance of the results dramatically. In this case, the father was at first accidentally classified as a woman. Hair mineral reference ranges are age and sex specific, and in fact change quite a bit depending on classification. Simply knowing whether someone is an adult or child is not adequate - the actual age in years of the person must be what appears on the test report.

While not illustrated here, if a young child with a perfectly normal hair test result is given an age significantly older than he or she actually is, the test will appear to meet the "all low" counting rule indicating disordered mineral transport due to excessive mercury accumulation when nothing of the sort is going on.

We will start with the mother, a 50 year old woman has been sick for over 10 years. She did get the right test to start out with. She has been diagnosed with endometriosis, temporomandibular joint problems, vulvodynia, and interstitial cystitis. She has had a complete hysterectomy and three other pelvic surgeries.

The symptoms and problems she reported are multiple food allergies, chemical sensitivities, foot pain (burning and numbness), constipation, bowel pain, insomnia, chronic fatigue, heartburn, mouth sores, painful scalp, joint pain, pelvic pain and bloating, bladder pain with urgency and frequency, multiple pelvic hernias, adverse reaction to female hormone supplementation, tingling in the arms and legs, unexplained weight loss, back pain, memory problems, lack of focus and concentration, feeling of being overwhelmed, depression, lack of interest, feeling of being cold all the time (while living in a very hot climate), eye floaters, and development of astigmatism.

HAIR ELEMENTS

DOCTOR'S DATA INC

SEX: Female
AGE: 50

POTENTIALLY TOXIC ELEMENTS

TOXIC ELEMENTS	RESULT µg/g	REFERENCE RANGE	PERCENTILE 68th 95th
Aluminum	17	< 12	
Antimony	< 0.01	< 0.06	
Arsenic	0.01	< 0.09	
Beryllium	< 0.01	< 0.02	
Bismuth	0.077	< 0.13	
Cadmium	0.026	< 0.1	
Lead	0.27	< 1.0	
Mercury	0.61	< 1.1	
Platinum	< 0.003	< 0.005	
Thallium	< 0.001	< 0.01	
Thorium	< 0.001	< 0.005	
Uranium	0.045	< 0.06	
Nickel	0.06	< 0.4	
Silver	0.03	< 0.1	
Tin	0.05	< 0.3	
Titanium	0.62	< 1.0	
Total Toxic Representation			

ESSENTIAL AND OTHER ELEMENTS

ELEMENTS	RESULT µg/g	REFERENCE RANGE	PERCENTILE 2.5th 16th 50th 84th 97.5th
Calcium	303	475- 1500	
Magnesium	20	45- 180	
Sodium	34	24- 180	
Potassium	15	20- 80	
Copper	42	9.0- 24	
Zinc	230	130- 200	
Manganese	0.11	0.2- 0.65	
Chromium	0.21	0.2- 0.45	
Vanadium	0.016	0.018- 0.065	
Molybdenum	0.04	0.04- 0.1	
Boron	0.47	0.7- 4.0	
Iodine	0.49	0.25- 1.3	
Lithium	0.006	0.008- 0.04	
Phosphorus	232	350- 600	
Selenium	0.98	0.95- 1.7	
Strontium	0.31	1.0- 8.0	
Sulfur	49300	43000- 50000	
Barium	0.23	0.5- 5.0	
Cobalt	0.023	0.013- 0.05	
Iron	5.7	5.8- 14	
Germanium	0.047	0.045- 0.065	
Rubidium	0.013	0.03- 0.25	
Zirconium	0.73	0.04- 1.0	

SPECIMEN DATA

COMMENTS:

Date Collected: | Sample Size:
Date Received: | Sample Type:
Date Completed: | Hair Color:
Treatment:
Methodology: ICP-MS | Shampoo:

V06.99

RATIOS

ELEMENTS	RATIOS	EXPECTED RANGE
Ca/Mg	15.2	4- 30
Ca/P	1.31	1- 12
Na/K	2.27	0.5- 10
Zn/Cu	5.48	4- 20
Zn/Cd	> 999	> 800

This form is provided to Andrew Cutler to be used as an educational tool and does not constitute endorsement by *Doctor's Data.*

This woman's test is presented on the previous page and counts as follows:

Counting rule #1: 4 bars go right. This satisfies counting rule #1.

Counting rule #2: 19 bars go left. This does not meet counting rule #2.

Counting rule #3: 2 bars go into the red band. This does not meet counting rule #3.

Counting rule #4: 9 bars are contained entirely in the middle green + white stripe. This satisfies counting rule #4.

Counting rule #5 is irrelevant since another counting rule was met.

Note that the highest essential element is copper, and that while it doesn't touch the red zone, it is close and the average level of minerals on this test is very low. Thus there is some reasonable chance that copper is a problem as well as mercury. People who have problems with copper AND mercury often have them because their liver isn't putting these similar metals into bile properly, so liver health can be worth exploring as well as toxicity.

Since the essential elements are "all low," it is not surprising that the toxic elements are also all low. This does not mean that they are all present at low levels, only that the hair levels may be lower than body levels.

Note that pubic hair was used. All of the hair testing labs are quite particular about which hair to use and where to get it. This is because they were quite careful to control all possible variables in developing accurate reference ranges. However I have seen a number of hair tests using hair from various locations on the head, or pubic hair, and the test results do correlate well with what is happening with the people.

She chose to have her amalgam fillings replaced with materials not containing mercury and noticed the following improvements promptly: relief from burning feet, fatigue, heartburn, sensitive teeth and insomnia.

Now let us consider her oldest son. This 13 year old boy, son of the 50 year old woman just discussed and the 54 year old man on pages 142-145 and brother of the 11 year old boy on pages 138-142, has no diagnosis[82] despite the following problems: severe allergies since age 2 (treated with allergy shots, then drugs, then Chinese herbs which were the most effective), exceptional shyness, obsessiveness, and Mees lines on his nails.

For whatever reason, the parents and doctor thought a Hair Element Profile had been ordered, but a Hair Toxic Element Exposure Profile showed up. They were able to get *Doctor's Data* to report a Hair Element Profile based on the same sample[83]. As you can see if you take the time to pore over the two tests and compare the results numbers, they are exactly the same (the quoted reference ranges for the Hair Element Profile are where the green and yellow meet, those on the Hair Toxic Element Exposure Profile are where the yellow and red meet).

---

[82] Which is to say that he doesn't quite meet any specific checklist for a named condition, not that he is normal or healthy.

[83] If one test has been performed the laboratory can provide results for the other test without needing a new hair sample submitted if they are contacted by the ordering health care professional and asked to do this within thirty days of the original test. If more than thirty days elapse a new hair sample will be required.

You will see one other peculiarity – the total toxic exposure index on the toxic exposure profile is substantially higher than the Total Toxic Representation on the Hair Element Profile. I don't know why this is so. I haven't found the Total Toxic Representation to correlate at all with people's actual conditions so I ignore it. Whatever meaning it has is likely to be more relevant in cases where mineral transport is normal than in cases where it is deranged.

If you first look at the Hair Toxic Element Exposure Profile presented on the facing page, you say "Oh! This poor kid has way way way too much arsenic!" And if you check a few medical books or good websites you will say "Wow! He has the Mees' lines that the books say arsenic causes, too!" (Mees lines are transverse white lines across the fingernails – they are observed in people with various heavy metals as well as with zinc deficiency and widely reported in textbooks as a sign of arsenic intoxication).

Now if you go look at the Hair Element Profile on page 137 you will see that the Hair Toxic Element Exposure Profile gave the same information. The essential elements portion of the Hair Element Profile counts out as follows:

Counting rule #1: 11 bars go right. This does not meet rule #1.

Counting rule #2: 12 bars go left. This does not meet counting rule #2.

Counting rule #3: 1 bars go into the red bands. This does not meet counting rule #3.

Counting rule #4: 14 bars are contained entirely in the middle green + white stripe. This does not meet counting rule #4.

Counting rule #5 is not met.

This is not an uncommon test statistically. Mineral transport is ordely and normal. The test is nowhere near meeting the counting rules, so the levels given in the toxic element section can be considered representative of body levels. The arsenic is way high with nothing else in the toxic element section anywhere near it.

In this case, the Hair Toxic Element Exposure Profile and the Hair Element Profile tests both tell the same story – this boy has arsenic poisoning.

Based on these test results the parents chelated their son with alpha lipoic acid on a 3-4 hour protocol and found gradual improvement in all problem areas. For the first time since infancy he does not require any medications or herbs to control his allergies. He has also become more outgoing and his obsessiveness is much reduced.

The parents also gave their son a reasonable assortment of vitamins and minerals two or three times a day during this time period.

HAIR TOXIC ELEMENT EXPOSURE PROFILE

DOCTOR'S DATA INC

SEX: Male
AGE: 13

POTENTIALLY TOXIC ELEMENTS

TOXIC ELEMENTS	RESULT µg/g	REFERENCE RANGE	PERCENTILE 68th 95th
Arsenic	0.35	< 0.2	
Lead	0.96	< 6.0	
Mercury	0.16	< 2.0	
Cadmium	0.22	< 0.7	
Chromium	0.45	< 0.65	
Beryllium	< 0.01	< 0.05	
Cobalt	0.008	< 0.1	
Nickel	0.1	< 1.0	
Zinc	210	< 250	
Copper	30	< 50	
Thorium	< 0.001	< 0.01	
Thallium	< 0.001	< 0.02	
Barium	0.32	< 5.0	
Cesium	< 0.002	< 0.01	
Manganese	0.68	< 1.2	
Selenium	0.86	< 2.7	
Bismuth	0.043	< 0.18	
Vanadium	0.045	< 0.25	
Silver	0.1	< 0.5	
Antimony	0.037	< 0.17	
Palladium	< 0.004	< 0.01	
Aluminum	10	< 22	
Platinum	< 0.003	< 0.01	
Tungsten	< 0.01	< 0.02	
Tin	0.19	< 0.8	
Uranium	0.019	< 0.2	
Gold	0.01	< 0.4	
Tellurium	< 0.05	< 0.05	
Germanium	0.032	< 0.085	
Titanium	0.79	< 2.0	
Gadolinium	< 0.001	< 0.008	
Total Toxic Exposure Index			

SPECIMEN DATA

Comments:
Date Collected: 7/30/2003
Date Received:
Date Completed:

Methodology:
µg/g = ppm
Sample Size:

Sample Type:
Hair Color:
Treatment:
Shampoo:

Metals are listed in descending priority order based upon data from the Agency for Toxic Substances and Disease Registry, which considers not only the relative toxicity per gram metal, but also the frequency for occurrence of exposure.

This form is provided to Andrew Cutler to be used as an educational tool and does not constitute endorsement by *Doctor's Data.*

HAIR ELEMENTS

DOCTOR'S DATA INC

SEX: Male
AGE: 13

POTENTIALLY TOXIC ELEMENTS

TOXIC ELEMENTS	RESULT µg/g	REFERENCE RANGE	PERCENTILE 68th 95th
Aluminum	10	< 8.0	
Antimony	0.037	< 0.066	
Arsenic	0.35	< 0.08	
Beryllium	< 0.01	< 0.02	
Bismuth	0.043	< 0.06	
Cadmium	0.22	< 0.15	
Lead	0.96	< 1.5	
Mercury	0.16	< 0.4	
Platinum	< 0.003	< 0.005	
Thallium	< 0.001	< 0.01	
Thorium	< 0.001	< 0.005	
Uranium	0.019	< 0.06	
Nickel	0.1	< 0.4	
Silver	0.1	< 0.13	
Tin	0.19	< 0.3	
Titanium	0.79	< 1.0	
Total Toxic Representation			

ESSENTIAL AND OTHER ELEMENTS

ELEMENTS	RESULT µg/g	REFERENCE RANGE	PERCENTILE 2.5th 16th 50th 84th 97.5th
Calcium	360	200- 700	
Magnesium	13	18- 70	
Sodium	58	12- 90	
Potassium	20	10- 40	
Copper	30	10- 30	
Zinc	210	130- 200	
Manganese	0.68	0.15- 0.65	
Chromium	0.45	0.2- 0.4	
Vanadium	0.045	0.02- 0.075	
Molybdenum	0.034	0.031- 0.062	
Boron	2.8	0.34- 3.0	
Iodine	1.7	0.25- 1.3	
Lithium	< 0.004	0.007- 0.023	
Phosphorus	160	160- 250	
Selenium	0.86	0.95- 1.7	
Strontium	0.44	0.3- 3.2	
Sulfur	48200	44000- 52000	
Barium	0.32	0.17- 1.5	
Cobalt	0.008	0.013- 0.035	
Iron	7.0	5.8- 13	
Germanium	0.032	0.045- 0.065	
Rubidium	0.016	0.009- 0.09	
Zirconium	0.23	0.047- 0.7	

SPECIMEN DATA

COMMENTS:

Date Collected:	7/30/2003	Sample Size:	
Date Received:		Sample Type:	
Date Completed:		Hair Color:	
		Treatment:	
Methodology:	ICP-MS	Shampoo:	

V06.99

RATIOS

ELEMENTS	RATIOS	EXPECTED RANGE
Ca/Mg	27.7	4- 30
Ca/P	2.25	0.8- 8
Na/K	2.9	0.5- 10
Zn/Cu	7	4- 20
Zn/Cd	955	> 800

This form is provided to Andrew Cutler to be used as an educational tool and does not constitute endorsement by *Doctor's Data.*

This 11 year old boy is the younger son of the 50 year old woman whose situation is described on pages 132-134 and the 54 year old man on pages 142-145 and brother of the 13 year old boy whose case is discussed on pages 134-137. He has chronic constipation, vision problems involving focusing and seeing up close or in 3D, a nervous shudder that goes through his body when he gets excited, slurred speech, reads below grade level, has below average expressive and receptive language as well as fine and gross motor skills, is socially shunned by some at school, as ADD like symptoms, auditory sensitivity, physical sensitivity, fatigue, is thirsty a lot and grits his teeth all night long.

His parents began looking for answers when he failed his school vision test at age 9. The following year the boy did quite poorly at school and had a lot of social problems. The parents eventually decided to get a hair test for their son.

This child got the Hair Toxic Element Exposure Profile by accident at first. The test was re-plotted as a Hair Element Profile. Neither the Toxic Exposure Index nor the Hair Element Profile have ages given on them, and the reference ranges used are those for an adult. They did finally get a Hair Element Profile report plotted with proper reference ranges for an 11 year old boy.

The two Hair Element Profiles are shown on the two pages following the Hair Toxic Element Exposure Profile presented on the facing page, with the left hand page being the one using an adult reference ranges and the right hand page being the one using the correct reference ranges.

Looking at the Hair Toxic Element Exposure Profile on the facing page, this boy appears to have a mixed metal problem, with chromium, manganese, aluminum and maybe tin, arsenic and lead adding up to a very high Total Toxic Exposure Index.

Plotted on an adult reference range, the counting rules are not met in the Hair Element Profile shown on page 140 though it is suspicious. However, when replotted using age appropriate reference ranges as shown in the Hair Element Profile report on page 141 the counting rules ARE met, indicating a high probability of mercury deranged mineral transport. Note that the Total Toxic Representation bar changes size a lot depending on which test was ordered and which age was used for the reference ranges.

	Adult reference	Correct reference
Counting rule #1:	14	13
Counting rule #2:	9	10
Counting rule #3:	3	4
Counting rule #4:	13	14
Counting rule #5:	no	rule #3 met

For this boy the Hair Toxic Element Exposure Profile is misleading and of little utility. Replotted with the wrong reference ranges the test is still not very informative. Only when the data is plotted as a Hair Element Profile with age and sex appropriate reference ranges does it finally tell this boy's story. He is having trouble because he has mercury poisoning.

Prior to the hair test this boy had undergone vision therapy unsuccessfully. He also was not able to wear the glasses the doctor had prescribed to correct his convergence and accommodation

HAIR TOXIC ELEMENT EXPOSURE PROFILE

DOCTOR'S DATA INC.

SEX: Male
AGE:

POTENTIALLY TOXIC ELEMENTS

TOXIC ELEMENTS	RESULT µg/g	REFERENCE RANGE	PERCENTILE 68th 95th
Arsenic	0.14	< 0.2	
Lead	4.3	< 8.0	
Mercury	0.63	< 3.0	
Cadmium	0.25	< 0.7	
Chromium	0.83	< 0.75	
Beryllium	< 0.01	< 0.05	
Cobalt	0.017	< 0.1	
Nickel	0.2	< 1.0	
Zinc	150	< 270	
Copper	21	< 55	
Thorium	< 0.001	< 0.01	
Thallium	< 0.001	< 0.02	
Barium	0.84	< 5.0	
Cesium	< 0.002	< 0.01	
Manganese	1.4	< 1.5	
Selenium	0.79	< 2.7	
Bismuth	0.021	< 0.22	
Vanadium	0.072	< 0.2	
Silver	0.2	< 0.5	
Antimony	0.066	< 0.15	
Palladium	< 0.004	< 0.01	
Aluminum	17	< 19	
Platinum	< 0.003	< 0.01	
Tungsten	< 0.01	< 0.02	
Tin	0.62	< 0.8	
Uranium	0.003	< 0.2	
Gold	0.02	< 0.4	
Tellurium	< 0.05	< 0.05	
Germanium	0.032	< 0.085	
Titanium	1.0	< 2.0	
Gadolinium	0.002	< 0.008	
Total Toxic Exposure Index			

SPECIMEN DATA

Comments:
Date Collected: 7/14/2003
Date Received:
Date Completed:

Methodology:
µg/g = ppm
Sample Size:

Sample Type:
Hair Color:
Treatment:
Shampoo:

Metals are listed in descending priority order based upon data from the Agency for Toxic Substances and Disease Registry, which considers not only the relative toxicity per gram metal, but also the frequency for occurrence of exposure.

This form is provided to Andrew Cutler to be used as an educational tool and does not constitute endorsement by *Doctor's Data*.

HAIR ELEMENTS

DOCTOR'S DATA INC.

SEX: Male
AGE:

POTENTIALLY TOXIC ELEMENTS

TOXIC ELEMENTS	RESULT µg/g	REFERENCE RANGE	PERCENTILE 68th 95th
Aluminum	17	< 7.0	
Antimony	0.066	< 0.066	
Arsenic	0.14	< 0.08	
Beryllium	< 0.01	< 0.02	
Bismuth	0.021	< 0.06	
Cadmium	0.25	< 0.15	
Lead	4.3	< 2.0	
Mercury	0.63	< 1.1	
Platinum	< 0.003	< 0.005	
Thallium	< 0.001	< 0.01	
Thorium	< 0.001	< 0.005	
Uranium	0.003	< 0.06	
Nickel	0.2	< 0.4	
Silver	0.2	< 0.12	
Tin	0.62	< 0.3	
Titanium	1.0	< 1.0	
Total Toxic Representation			

ESSENTIAL AND OTHER ELEMENTS

ELEMENTS	RESULT µg/g	REFERENCE RANGE	PERCENTILE 2.5th 16th 50th 84th 97.5th
Calcium	230	200- 750	
Magnesium	6.0	25- 75	
Sodium	180	12- 90	
Potassium	160	9.0- 40	
Copper	21	10- 28	
Zinc	150	130- 200	
Manganese	1.4	0.15- 0.65	
Chromium	0.83	0.2- 0.4	
Vanadium	0.072	0.018- 0.065	
Molybdenum	0.054	0.025- 0.064	
Boron	0.78	0.4- 3.0	
Iodine	0.82	0.25- 1.3	
Lithium	0.012	0.007- 0.023	
Phosphorus	174	160- 250	
Selenium	0.79	0.95- 1.7	
Strontium	0.43	0.3- 3.5	
Sulfur	49100	44500- 52000	
Barium	0.84	0.16- 1.6	
Cobalt	0.017	0.013- 0.035	
Iron	15	5.4- 13	
Germanium	0.032	0.045- 0.065	
Rubidium	0.14	0.011- 0.12	
Zirconium	0.86	0.02- 0.44	

SPECIMEN DATA

COMMENTS:

Date Collected: 7/14/2003
Date Received:
Date Completed:
Methodology: ICP-MS

Sample Size:
Sample Type:
Hair Color:
Treatment:
Shampoo:

V06 99

RATIOS

ELEMENTS	RATIOS	EXPECTED RANGE
Ca/Mg	38.3	4- 30
Ca/P	1.32	0.8- 8
Na/K	1.13	0.5- 10
Zn/Cu	7.14	4- 20
Zn/Cd	600	> 800

This form is provided to Andrew Cutler to be used as an educational tool and does not constitute endorsement by *Doctor's Data.*

HAIR ELEMENTS

DOCTOR'S DATA INC

SEX: Male
AGE: 10

POTENTIALLY TOXIC ELEMENTS

TOXIC ELEMENTS	RESULT µg/g	REFERENCE RANGE
Aluminum	17	< 8.0
Antimony	0.066	< 0.066
Arsenic	0.14	< 0.080
Beryllium	< 0.01	< 0.020
Bismuth	0.021	< 0.12
Cadmium	0.25	< 0.15
Lead	4.3	< 1.0
Mercury	0.63	< 0.40
Platinum	< 0.003	< 0.005
Thallium	< 0.001	< 0.010
Thorium	< 0.001	< 0.005
Uranium	0.003	< 0.060
Nickel	0.20	< 0.40
Silver	0.20	< 0.13
Tin	0.62	< 0.30
Titanium	1.0	< 1.0
Total Toxic Representation		

PERCENTILE: 68th, 95th

ESSENTIAL AND OTHER ELEMENTS

ELEMENTS	RESULT µg/g	REFERENCE RANGE
Calcium	230	160- 500
Magnesium	6	12- 50
Sodium	180	12- 90
Potassium	160	10- 40
Copper	21	9.0- 30
Zinc	150	110- 190
Manganese	1.4	0.18- 0.60
Chromium	0.83	0.23- 0.50
Vanadium	0.072	0.025- 0.10
Molybdenum	0.054	0.040- 0.089
Boron	0.78	0.50- 3.5
Iodine	0.82	0.25- 1.3
Lithium	0.012	0.007- 0.023
Phosphorus	174	160- 250
Selenium	0.79	0.95- 1.7
Strontium	0.43	0.21- 2.1
Sulfur	49100	45500- 53000
Barium	0.84	0.19- 1.6
Cobalt	0.017	0.013- 0.035
Iron	15	6.0- 17
Germanium	0.032	0.045- 0.065
Rubidium	0.14	0.008- 0.080
Zirconium	0.86	0.060- 0.70

PERCENTILE: 2.5th, 16th, 50th, 84th, 97.5th

SPECIMEN DATA

COMMENTS:

Date Collected: 7/14/2003
Date Received:
Date Completed:
Methodology: ICP-MS

Sample Size:
Sample Type:
Hair Color:
Treatment:
Shampoo:

V06.99

RATIOS

ELEMENTS	RATIOS	EXPECTED RANGE
Ca/Mg	38.3	4- 30
Ca/P	1.32	0.8- 8
Na/K	1.13	0.5- 10
Zn/Cu	7.14	4- 20
Zn/Cd	600	> 800

This form is provided to Andrew Cutler to be used as an educational tool and does not constitute endorsement by *Doctor's Data.*

problems. After the hair test the parents chelated him some, his vision improved somewhat due to chelation alone, and he then tried vision therapy a second time with great success. The outcome of this was not only that he passed his school vision screening test, but also that he no longer misses rebounds or balls passed to him when playing basketball, and that his reading is greatly improved and his self confidence restored.

Chelation also improved other issues without the need for active therapy.

Chelation was done with alpha lipoic acid every 3-4 hours, and a reasonable assortment of vitamins and minerals were given 2 or 3 times a day. After many months of chelation the parents gave him a break from it because he was becoming impulsive, disinhibited and loud to the point they were receiving complaints from school. After a couple of months off chelation this settled down.

While there are residual issues that will require further chelation to resolve, much progress has been made and the parents feel like they have a different child.

Now let's discuss the man who is husband to the woman whose hair test is discussed on pages 129-131 and father to the two boys whose hair tests are discussed on pages 134-137 and from page 138 to this page. His problems express themselves as nasal congestion due to chronic allergies, stiffness and pain in the shoulder girdle and hip girdle on arising which diminishes as the day progresses, night sweats, frequent low grade fevers and tiredness. He was diagnosed with polymyalgia rheumatica and prescribed low doses of prednisone for it.

When his hair test was done, the forms were not filled out clearly and the laboratory initially issued the reports with female reference ranges. The report was re-issued with male reference ranges. In this particular case there is little effect on how it plots other than for the calcium, cadmium and lead.

This test was done using pubic hair. All of the hair testing labs are quite particular about which hair to use and where to get it. This is because they were quite careful to control all possible variables in developing accurate reference ranges. However I have seen a number of hair tests using hair from various locations on the head, or pubic hair, and the test results do seem to correlate well with what is happening with the people.

	Incorrect female reference test on page 143	Correct male reference test on page 144
Counting rule #1:	7	7
Counting rule #2:	16	16
Counting rule #3:	1	1
Counting rule #4:	11	13
Counting rule #5:	rule #4 met	not met

Note that this test when plotted using proper reference ranges is unusual but not suspicious or abnormal for mineral transport and thus mildly suggestive of mercury, in having seven bars to the right. When plotted using incorrect reference ranges it counts positive for deranged mineral transport.

This man does have very high selenium. At high levels selenium can be toxic. Since pubic hair is seldom washed with selenium containing dandruff shampoos, this does suggest that selenium problems be considered, possibly due to excessive intake or to unusually high retention. Pubic hair was used due to lack of adequate amounts of head hair. The man says he does not use dandruff shampoo and even if this is not an accurate recollection the amount of shampoo used is small.

HAIR ELEMENTS

DOCTOR'S DATA INC

SEX: Female
AGE: 54

POTENTIALLY TOXIC ELEMENTS

TOXIC ELEMENTS	RESULT µg/g	REFERENCE RANGE
Aluminum	21	< 12
Antimony	0.019	< 0.06
Arsenic	0.04	< 0.09
Beryllium	< 0.01	< 0.02
Bismuth	0.032	< 0.13
Cadmium	0.13	< 0.1
Lead	0.62	< 1.0
Mercury	0.24	< 1.1
Platinum	< 0.003	< 0.005
Thallium	< 0.001	< 0.01
Thorium	< 0.001	< 0.005
Uranium	0.007	< 0.06
Nickel	0.17	< 0.4
Silver	0.17	< 0.1
Tin	0.18	< 0.3
Titanium	1.1	< 1.0
Total Toxic Representation		

PERCENTILE: 68th, 95th

ESSENTIAL AND OTHER ELEMENTS

ELEMENTS	RESULT µg/g	REFERENCE RANGE
Calcium	457	475- 1500
Magnesium	29	45- 180
Sodium	39	24- 180
Potassium	16	20- 80
Copper	37	9.0- 24
Zinc	220	130- 200
Manganese	0.3	0.2- 0.65
Chromium	0.4	0.2- 0.45
Vanadium	0.027	0.018- 0.065
Molybdenum	0.057	0.04- 0.1
Boron	0.55	0.7- 4.0
Iodine	4.0	0.25- 1.3
Lithium	0.015	0.008- 0.04
Phosphorus	272	350- 600
Selenium	8.3	0.95- 1.7
Strontium	0.76	1.0- 8.0
Sulfur	43300	43000- 50000
Barium	0.59	0.5- 5.0
Cobalt	0.024	0.013- 0.05
Iron	8.7	5.8- 14
Germanium	0.063	0.045- 0.065
Rubidium	0.019	0.03- 0.25
Zirconium	2.4	0.04- 1.0

PERCENTILE: 2.5th, 16th, 50th, 84th, 97.5th

SPECIMEN DATA

COMMENTS:
Date Collected:
Date Received:
Date Completed
Methodology: ICP-MS

Sample Size:
Sample Type:
Hair Color:
Treatment:
Shampoo:

V06 99

RATIOS

ELEMENTS	RATIOS	EXPECTED RANGE
Ca/Mg	15.8	4- 30
Ca/P	1.68	1- 12
Na/K	2.44	0.5- 10
Zn/Cu	5.95	4- 20
Zn/Cd	> 999	> 800

This form is provided to Andrew Cutler to be used as an educational tool and does not constitute endorsement by *Doctor's Data.*

HAIR ELEMENTS

DOCTOR'S DATA INC

SEX: Male
AGE: 54

POTENTIALLY TOXIC ELEMENTS

TOXIC ELEMENTS	RESULT µg/g	REFERENCE RANGE	PERCENTILE 68th 95th
Aluminum	21	< 12	
Antimony	0.019	< 0.080	
Arsenic	0.040	< 0.12	
Beryllium	< 0.01	< 0.020	
Bismuth	0.032	< 0.080	
Cadmium	0.13	< 0.15	
Lead	0.62	< 2.0	
Mercury	0.24	< 1.1	
Platinum	< 0.003	< 0.005	
Thallium	< 0.001	< 0.010	
Thorium	< 0.001	< 0.005	
Uranium	0.007	< 0.060	
Nickel	0.17	< 0.40	
Silver	0.17	< 0.10	
Tin	0.18	< 0.30	
Titanium	1.1	< 1.0	
Total Toxic Representation			

ESSENTIAL AND OTHER ELEMENTS

ELEMENTS	RESULT µg/g	REFERENCE RANGE	PERCENTILE 2.5th 16th 50th 84th 97.5th
Calcium	457	375- 1100	
Magnesium	29	40- 140	
Sodium	39	24- 180	
Potassium	16	20- 80	
Copper	37	9.0- 26	
Zinc	220	120- 200	
Manganese	0.30	0.20- 0.65	
Chromium	0.40	0.20- 0.45	
Vanadium	0.027	0.018- 0.065	
Molybdenum	0.057	0.040- 0.10	
Boron	0.55	0.70- 4.0	
Iodine	4.0	0.25- 1.3	
Lithium	0.015	0.008- 0.040	
Phosphorus	272	250- 400	
Selenium	8.3	0.95- 1.7	
Strontium	0.76	1.0- 6.0	
Sulfur	43300	42000- 49000	
Barium	0.59	0.50- 3.0	
Cobalt	0.024	0.013- 0.050	
Iron	8.7	5.8- 14	
Germanium	0.063	0.045- 0.065	
Rubidium	0.019	0.030- 0.25	
Zirconium	2.4	0.040- 1.0	

SPECIMEN DATA

COMMENTS:

Date Collected:
Date Received:
Date Completed:

Methodology: ICP-MS

Sample Size:
Sample Type:
Hair Color:
Treatment:
Shampoo:

V06.99

RATIOS

ELEMENTS	RATIOS	EXPECTED RANGE
Ca/Mg	15.8	4- 30
Ca/P	1.68	0.8- 8
Na/K	2.44	0.5- 10
Zn/Cu	5.95	4- 20
Zn/Cd	> 999	> 800

This form is provided to Andrew Cutler to be used as an educational tool and does not constitute endorsement by *Doctor's Data.*

Metabolic Defects – Manganese

This 47 year old woman has a generally boisterous personality, the outgoing life of the party type of person. She is quite obese. She is chemically sensitive, allergic, has had adverse reactions to several drugs while hospitalized, and knows she was exposed to a wide variety of toxics while a young girl growing up on a farm. She has asthma, chronic bronchitis, fibromyalgia, chronic fatigue and hypothyroidism. She felt good in college on a low carbohydrate, high protein Atkins style diet. She had a two year period of unrelenting severe headaches. Her hair test shown on the facing page counts as follows:

Counting rule #1: 7 bars go right. This does not meet rule #1.

Counting rule #2: 16 bars go left. This does not meet counting rule #2.

Counting rule #3: 2 bars go into the red bands. This does not meet counting rule #3.

Counting rule #4: 18 bars are contained entirely in the middle green + white stripe. This does not meet counting rule #4.

Counting rule #5: is not met.

Her hair test exhibits orderly mineral transport thought it is unusual due to seven bars to the right. It shows vastly elevated arsenic (which often goes along with chemical sensitivity), and vastly elevated manganese. She and others agree that textbook and online descriptions of "manganese madness" are fairly good descriptions of her as long as one sticks to the milder symptoms. For example, her high school math teacher was constantly telling her to write bigger and spread her work out so it was possible to read – small handwriting being a characteristic sign of manganese toxicity.

Her Red Blood Cell Elements test, on page 147, shows very elevated manganese and very low molybdenum. On seeing this test she started supplementing molybdenum 1000 mcg a day and enjoyed dramatic relief of many problems.

At this point it is completely unclear whether there is any significance to the very elevated zirconium.

To date I have seen several of these high hair manganese tests for people with no apparent exposure, but who have manganese type problems. There is apparently some condition as yet unrecognized by modern medicine whereby a few people's bodies simply accumulate far too much manganese for their own good. I would estimate that the incidence of this problem among adults is between 0.5 and 2 percent of those people who have the kind of confusing medical histories where it becomes rational to order a Hair Element Profile to screen them for heavy metal problems.

When mineral transport is orderly, a red blood cell element test will typically mostly agree with what the hair shows. There may, however, be useful differences and it can be worth checking, which also allows the RBC test to be used to confirm the hair test and ensure there is no problem with external contamination. When mineral transport is unusual as in this case there may be differences, and in fact the low molybdenum picked up on the Red Blood Cell Elements test but not on the Hair Element Profile did lead to a very helpful intervention.

When mineral transport is disordered RBC tests most often come out showing low zinc, low-ish to low molybdenum and magnesium, and high-ish copper.

HAIR ELEMENTS

DOCTOR'S DATA INC.

SEX: Female
AGE: 47

POTENTIALLY TOXIC ELEMENTS

TOXIC ELEMENTS	RESULT µg/g	REFERENCE RANGE	PERCENTILE 68th 95th
Aluminum	8.4	< 7.0	
Antimony	0.048	< 0.05	
Arsenic	0.42	< 0.06	
Beryllium	< 0.01	< 0.02	
Bismuth	0.055	< 0.1	
Cadmium	0.098	< 0.1	
Lead	0.37	< 1.0	
Mercury	0.56	< 1.1	
Platinum	< 0.003	< 0.005	
Thallium	< 0.001	< 0.01	
Thorium	0.001	< 0.005	
Uranium	0.033	< 0.06	
Nickel	0.23	< 0.4	
Silver	0.24	< 0.15	
Tin	0.28	< 0.3	
Titanium	0.65	< 1.0	
Total Toxic Representation			

ESSENTIAL AND OTHER ELEMENTS

ELEMENTS	RESULT µg/g	REFERENCE RANGE	PERCENTILE 2.5th 16th 50th 84th 97.5th
Calcium	245	300- 1200	
Magnesium	23	35- 120	
Sodium	60	12- 90	
Potassium	21	8.0- 38	
Copper	13	12- 35	
Zinc	170	140- 220	
Manganese	2.5	0.15- 0.65	
Chromium	0.33	0.2- 0.4	
Vanadium	0.046	0.018- 0.065	
Molybdenum	0.025	0.028- 0.056	
Boron	0.51	0.3- 2.0	
Iodine	0.65	0.25- 1.3	
Lithium	0.01	0.007- 0.023	
Phosphorus	154	160- 250	
Selenium	0.95	0.95- 1.7	
Strontium	1.5	0.5- 7.6	
Sulfur	47300	44500- 52000	
Barium	0.28	0.26- 3.0	
Cobalt	0.021	0.013- 0.05	
Iron	7.7	5.4- 14	
Germanium	0.049	0.045- 0.065	
Rubidium	0.021	0.007- 0.096	
Zirconium	2.6	0.02- 0.42	

SPECIMEN DATA

COMMENTS:

Date Collected:
Date Received:
Date Completed:

Methodology: ICP-MS

Sample Size:
Sample Type:
Hair Color:
Treatment:
Shampoo:

V06.99

RATIOS

ELEMENTS	RATIOS	EXPECTED RANGE
Ca/Mg	10.7	4- 30
Ca/P	1.59	1- 12
Na/K	2.86	0.5- 10
Zn/Cu	13.1	4- 20
Zn/Cd	> 999	> 800

This form is provided to Andrew Cutler to be used as an educational tool and does not constitute endorsement by *Doctor's Data*.

RED BLOOD CELL ELEMENTS

DOCTOR'S DATA INC.

SEX: Female
AGE: 47

NUTRIENT ELEMENTS

ELEMENTS	RESULT µg/g	REFERENCE RANGE	PERCENTILE 2.5th 16th 50th 84th 97.5th
Calcium	15.0	8- 31.0	
Magnesium	49.0	36- 64.0	
Potassium	84.0	65- 95.0	
Phosphorus	605	480- 745	
Copper	0.71	0.52- 0.89	
Zinc	12.4	8- 14.5	
Iron	893	745- 1050	
Manganese	0.030	0.007- 0.03	
Chromium	0.019	0.012- 0.07	
Selenium	0.23	0.19- 0.38	
Boron	0.056	0.005- 0.11	
Vanadium	0.0011	.0001-0.002	
Molybdenum	< 0.0005	.0005-0.002	

POTENTIALLY TOXIC ELEMENTS

TOXIC ELEMENTS	RESULT µg/g	REFERENCE RANGE	PERCENTILE 95th 99th
Antimony	0.0010	< 0.005	
Arsenic	0.003	< 0.01	
Cadmium	< 0.0008	< 0.005	
Lead	0.036	< 0.09	
Mercury	0.003	< 0.01	

SPECIMEN DATA

Comments:
Date Collected: Methodology: ICP-MS
Date Received: µg/g = ppm
Date Completed:

V10.01

This form is provided to Andrew Cutler to be used as an educational tool and does not constitute endorsement by *Doctor's Data*.

Here is the case of a woman whose family history and hair tests of some family members suggests a genetically transmitted abberation of manganese metabolism. This 38 year old woman is tired, droopy and exhausted all the time. She has psoriasis which became much worse following her first pregnancy. She has insomnia which recently has gotten to the point where she falls asleep exhausted at 8 PM and awakens at 3 AM, unable to get back to sleep.

She has two children, a 10 year old son who is considered normal, and a 6 year old daughter who has epilepsy and pervasive developmental disorder though it is not severe. The daughter developed normally until regressing while two years old.

The mother's hair test is presented on the following page (page 146), and counts as follows:

Counting rule #1: 13 bars go right. This does not meet rule #1.

Counting rule #2: 10 bars go left. This does not meet counting rule #2.

Counting rule #3: 2 bars go into the red bands. This does not meet counting rule #3.

Counting rule #4: 17 bars are contained entirely in the middle green + white stripe. This does not meet counting rule #4.

Counting rule #5 is not met.

Mineral transport is orderly and normal. The toxic elements are generally nice and low. A uranium level of 0.2 is not at all significant and would not be expected to cause problems.

What would be expected to cause problems is the manganese level of 3.6, which is dramatically elevated. This woman's problems appear to be due to a mild case of manganese poisoning. No source of high level exposure to manganese has been identified.

Both of her children had hair tests. Both children's hair tests counted out normal, showing orderly mineral transport. The son's test however had dramatically elevated manganese like the mother's – 3.76 for the son versus 3.6 for the mother[84]. The daughter's Hair Element Profile showed normal manganese levels. In fact the hair test provides little clue as to what is going on with the daughter other than that the calcium – magnesium – sodium – potassium stress sign was present indicating that in fact she does have some sort of physiological problem underlying her condition. Other testing confirms this but does not clarify what the underlying problem is. The daughter's response to chelation is also ambiguous.

This overall situation suggests that there is some unknown genetic trait that, when inherited, causes people to accumulate excessive levels of manganese to the point of toxicity.

[84] The son's test is not presented because it was performed at *Great Smokies Diagnostic Laboratories* in 2001 and the elements that *Great Smokies* determines have changed since that time. I didn't think that having to explain how to read a different test format for a test that is no longer available would do anything other than cause confusion.

HAIR ELEMENTS

DOCTOR'S DATA INC.

SEX: Female
AGE: 38

POTENTIALLY TOXIC ELEMENTS

TOXIC ELEMENTS	RESULT µg/g	REFERENCE RANGE
Aluminum	3.3	< 7.0
Antimony	0.044	< 0.05
Arsenic	0.028	< 0.06
Beryllium	< 0.01	< 0.02
Bismuth	0.081	< 0.1
Cadmium	0.029	< 0.1
Lead	0.08	< 1.0
Mercury	0.23	< 1.1
Platinum	< 0.003	< 0.005
Thallium	< 0.001	< 0.01
Thorium	< 0.001	< 0.005
Uranium	0.2	< 0.06
Nickel	0.05	< 0.4
Silver	0.04	< 0.15
Tin	0.39	< 0.3
Titanium	0.93	< 1.0
Total Toxic Representation		

PERCENTILE: 68th, 95th

ESSENTIAL AND OTHER ELEMENTS

ELEMENTS	RESULT µg/g	REFERENCE RANGE
Calcium	892	300- 1200
Magnesium	140	35- 120
Sodium	180	12- 90
Potassium	15	8.0- 38
Copper	13	12- 35
Zinc	180	140- 220
Manganese	3.6	0.15- 0.65
Chromium	0.41	0.2- 0.4
Vanadium	0.018	0.018- 0.065
Molybdenum	0.034	0.028- 0.056
Boron	0.72	0.3- 2.0
Iodine	2.8	0.25- 1.3
Lithium	0.008	0.007- 0.023
Phosphorus	158	160- 250
Selenium	1.3	0.95- 1.7
Strontium	1.4	0.5- 7.6
Sulfur	50200	44500- 52000
Barium	0.74	0.26- 3.0
Cobalt	0.2	0.013- 0.05
Iron	11	5.4- 14
Germanium	0.051	0.045- 0.065
Rubidium	0.012	0.007- 0.096
Zirconium	0.12	0.02- 0.42

PERCENTILE: 2.5th, 16th, 50th, 84th, 97.5th

SPECIMEN DATA

COMMENTS:

Date Collected: | Sample Size:
Date Received: | Sample Type:
Date Completed: | Hair Color:
Treatment:
Methodology: ICP-MS | Shampoo:

V06.99

RATIOS

ELEMENTS	RATIOS	EXPECTED RANGE
Ca/Mg	6.37	4- 30
Ca/P	5.65	1- 12
Na/K	12	0.5- 10
Zn/Cu	13.8	4- 20
Zn/Cd	> 999	> 800

This form is provided to Andrew Cutler to be used as an educational tool and does not constitute endorsement by *Doctor's Data*.

Metabolic Defects – Copper

Hair tests for a family of five, a father, mother, daughter and identical twin autistic boys are presented here. With autistic twins the parents' lives and interests are consumed exclusively by their children. For practical reasons the entire family more or less stays on the same diet as the twins. On seeing the substantial improvement nutritional supplementation brought about in the twins' condition the rest of the family also tried some relevant supplements and found them helpful.

The mother has a history of constipation and frequent illness in childhood and craved sweets. She had many mercury (silver amalgam) fillings placed as a child. She majored in technical subjects at college. She was diagnosed with irritable bowel syndrome sometime after college. She is mildly chemically sensitive, feeling funny or anxious on exposure to fragrances, car exhaust or fresh paint. Caffeine and chocolate give her heart palpitations and make her irritable. She is on the specific carbohydrate diet[85] which has improved her sense of well being, and she takes nutritional supplements that she believes have helped keep her healthier. She characterizes herself as being a perfectionist, having a type A personality and having overly strong attention to detail. She is socially active and has many friends.

The mother's hair test is presented on page 152, facing her daughter's on page 153 for comparison. The mother's test counts as follows:

Counting rule #1: 17 bars go right. This does not meet rule #1.

Counting rule #2: 6 bars go left. This does not meet counting rule #2.

Counting rule #3: 2 bars go into the red bands. This does not meet counting rule #3.

Counting rule #4: 15 bars are contained entirely in the middle green + white stripe. This does not meet counting rule #4.

Counting rule #5 is not met.

Despite the fact that this test does not meet any counting rule, it is statistically suspicious due to only six bars going to the left.

The very elevated copper of 93 is significant and suggests some toxicity from this essential element. Copper elevations are generally significant even if mineral transport is deranged due to mercury problems.

The very elevated antimony and silver, and not quite very elevated tin are not easily interpretable since there may be derangement of mineral transport driving results to the high side (as is apparent when looking at the essential element section).

The unexpectedly low hair mercury level is consistent with a derangement of mineral transport that is not bad enough to set the counting rules off yet.

Both mercury and copper are primarily excreted in the stool via bile flow from the liver. With some impairment of this process, a mild mixed copper plus mercury toxicity is likely to result. Given the combination of suspicious mineral

---

[85] This diet is described in the book *Breaking the Vicious Cycle* by Elaine Gottschall. It is often used to control imbalances in the kinds of microorganisms in the intestine, and is very helpful for toxic people such as autistic children, chemically sensitive adults or people with celiac disease.

transport, very elevated hair copper, unexpectedly low hair mercury, and elevations of antimony, silver[86] and tin, it seems quite reasonable to suspect this woman is suffering from a mixed mercury-copper toxicity due to some abnormality in biliary metal excretion.

This possible interpretation of the mother's test as suggesting a mixed copper and mercury toxicity becomes even more reasonable when comparing her hair test to her daughter's hair test. The similarities of suspicious (mother) and clearly deranged (daughter) mineral transport, very elevated copper, very elevated antimony as often seen in deranged mineral transport, and unexpectedly low hair mercury all suggest the possibility of some genetic factor making transport of mercury and copper into bile somewhat different for this mother and daughter than for the general population.

This possible genetic issue suggests that all of the mother's blood relatives would be wise to have a hair test to ensure they do not also have this possibly genetic proclivity to accumulate too much copper and mercury.

Note the relation of calcium and magnesium to sodium and potassium. This suggests the possibility of adrenal problems relating to glucocorticoids and related hormones.

Note that this woman has significantly higher hair uranium than other family members and that she does not live in a geographical area where uranium typically is very elevated. This suggests the possibility that her body may need more iron than it has or that she may need a higher dietary intake of iron – a common situation for women of her age which can be easily explored through blood testing by determining ferritin and doing a blood count (CBC).

The daughter had a kidney defect called hydronephrosis that was identified before birth and treated with continuous low dose antibiotics until it was surgically corrected at 1 year old. She has no cavities and no mercury (silver amalgam) or other sort of dental fillings. She can tune her parents out and appear to be selectively attention deficit like any child, but her attention to detail with homework is quite good. She likes to do things right but is not a perfectionist. She is very social and makes new friends easily. At a younger age she was often shy and occasionally still is. The specific carbohydrate diet did improve her overall well being somewhat.

The daughter's hair test, shown page 153, counts as follows:

Counting rule #1: 15 bars go right. This does not meet rule #1.

Counting rule #2: 8 bars go left. This does not meet counting rule #2.

Counting rule #3: 6 bars go into the red bands. This meets counting rule #3.

Counting rule #4: 13 bars are contained entirely in the middle green + white stripe. This does not meet counting rule #4.

Counting rule #5 is irrelevant since another counting rule was met.

The very elevated copper of 57 is significant and suggests some toxicity from this essential element. In fact, the daughter's copper is more elevated compared to its normal range than the

[86] Sometimes elevated in people who are developing toxicity due to mercury exposure from mercury (silver amalgam) dental fillings. These fillings are typically 50% mercury, 30% silver and 20% other materials, primarily tin.

HAIR ELEMENTS

DOCTOR'S DATA INC

SEX: Female
AGE: 37

POTENTIALLY TOXIC ELEMENTS

TOXIC ELEMENTS	RESULT µg/g	REFERENCE RANGE	PERCENTILE 68th 95th
Aluminum	5.3	< 7.0	
Antimony	0.14	< 0.05	
Arsenic	0.034	< 0.06	
Beryllium	< 0.01	< 0.02	
Bismuth	0.2	< 0.1	
Cadmium	0.048	< 0.1	
Lead	0.36	< 1.0	
Mercury	0.08	< 1.1	
Platinum	< 0.003	< 0.005	
Thallium	< 0.001	< 0.01	
Thorium	< 0.001	< 0.005	
Uranium	0.23	< 0.06	
Nickel	0.41	< 0.4	
Silver	3.0	< 0.15	
Tin	0.83	< 0.3	
Titanium	0.81	< 1.0	
Total Toxic Representation			

ESSENTIAL AND OTHER ELEMENTS

ELEMENTS	RESULT µg/g	REFERENCE RANGE	PERCENTILE 2.5th 16th 50th 84th 97.5th
Calcium	2960	300- 1200	
Magnesium	220	35- 120	
Sodium	48	12- 90	
Potassium	15	8.0- 38	
Copper	93	12- 35	
Zinc	190	140- 220	
Manganese	0.49	0.15- 0.65	
Chromium	0.29	0.2- 0.4	
Vanadium	0.041	0.018- 0.065	
Molybdenum	0.084	0.028- 0.056	
Boron	0.34	0.3- 2.0	
Iodine	2.5	0.25- 1.3	
Lithium	0.015	0.007- 0.023	
Phosphorus	198	160- 250	
Selenium	0.97	0.95- 1.7	
Strontium	11	0.5- 7.6	
Sulfur	47400	44500- 52000	
Barium	5.7	0.26- 3.0	
Cobalt	0.051	0.013- 0.05	
Iron	14	5.4- 14	
Germanium	0.049	0.045- 0.065	
Rubidium	0.024	0.007- 0.096	
Zirconium	1.3	0.02- 0.42	

SPECIMEN DATA

COMMENTS:

Date Collected: Sample Size:
Date Received: Sample Type:
Date Completed: Hair Color:
Treatment:
Methodology: ICP-MS Shampoo:

V06.99

RATIOS

ELEMENTS	RATIOS	EXPECTED RANGE
Ca/Mg	13.5	4- 30
Ca/P	14.9	1- 12
Na/K	3.2	0.5- 10
Zn/Cu	2.04	4- 20
Zn/Cd	> 999	> 800

This form is provided to Andrew Cutler to be used as an educational tool and does not constitute endorsement by *Doctor's Data*.

HAIR ELEMENTS

DOCTOR'S DATA INC

SEX: Female
AGE: 5

POTENTIALLY TOXIC ELEMENTS

TOXIC ELEMENTS	RESULT µg/g	REFERENCE RANGE	PERCENTILE 68th 95th
Aluminum	9.1	< 8.0	
Antimony	1.2	< 0.066	
Arsenic	0.028	< 0.08	
Beryllium	< 0.01	< 0.02	
Bismuth	0.13	< 0.13	
Cadmium	0.08	< 0.1	
Lead	0.58	< 1.0	
Mercury	0.06	< 0.4	
Platinum	< 0.003	< 0.005	
Thallium	< 0.001	< 0.01	
Thorium	< 0.001	< 0.005	
Uranium	0.12	< 0.06	
Nickel	0.74	< 0.4	
Silver	0.79	< 0.2	
Tin	0.71	< 0.3	
Titanium	0.8	< 1.0	
Total Toxic Representation			

ESSENTIAL AND OTHER ELEMENTS

ELEMENTS	RESULT µg/g	REFERENCE RANGE	PERCENTILE 2.5th 16th 50th 84th 97.5th
Calcium	1760	140- 500	
Magnesium	990	15- 45	
Sodium	52	12- 90	
Potassium	21	9.0- 60	
Copper	57	10- 22	
Zinc	120	100- 190	
Manganese	0.39	0.2- 0.55	
Chromium	0.43	0.26- 0.5	
Vanadium	0.087	0.03- 0.1	
Molybdenum	0.091	0.048- 0.13	
Boron	0.68	0.35- 3.0	
Iodine	1.9	0.25- 1.3	
Lithium	0.011	0.007- 0.023	
Phosphorus	141	160- 250	
Selenium	0.87	0.95- 1.7	
Strontium	9.3	0.19- 2.0	
Sulfur	41500	45500- 53000	
Barium	5.3	0.16- 1.2	
Cobalt	0.069	0.013- 0.035	
Iron	14	8.0- 19	
Germanium	0.05	0.045- 0.065	
Rubidium	0.02	0.012- 0.16	
Zirconium	0.66	0.03- 1.0	

SPECIMEN DATA

COMMENTS:

Date Collected: | Sample Size:
Date Received: | Sample Type:
Date Completed: | Hair Color:
| Treatment:
Methodology: ICP-MS | Shampoo:

V06.99

RATIOS

ELEMENTS	RATIOS	EXPECTED RANGE
Ca/Mg	1.78	4- 30
Ca/P	12.5	1- 12
Na/K	2.48	0.5- 10
Zn/Cu	2.11	4- 20
Zn/Cd	> 999	> 800

This form is provided to Andrew Cutler to be used as an educational tool and does not constitute endorsement by *Doctor's Data.*

mother's is[87]. Even in the face of deranged mineral transport this usually does correspond to excessive body levels of copper and people with this become more healthy if they do something to reduce body copper stores.

The very elevated antimony is often due to deranged mineral transport causing the retention of antimony in the body.

The unexpectedly low hair mercury level is consistent with a derangement of mineral transport.

Both mercury and copper are primarily excreted in the stool via bile flow from the liver. With some impairment of this process, a mixed copper plus mercury toxicity is likely to result. Given the combination of deranged mineral transport and very elevated hair copper this girl is most likely suffering from such a problem.

Since the mother's test also suggests a mixed copper and mercury toxicity the possibility exists that some genetic factor is making transport of mercury and copper into bile somewhat different for this mother and daughter than for the general population.

This possible genetic issue suggests that all of the mother's blood relatives would be wise to have a hair test to ensure they do not also have this possibly genetic proclivity to accumulate too much copper and mercury.

Note the relation of calcium and magnesium to sodium and potassium. This suggests the possibility of adrenal problems relating to glucocorticoids and related hormones.

Note how the chemically similar elements magnesium, calcium, strontium and barium all move in the same direction by a similar amount. This is another sign suggestive of deranged mineral transport.

Note the out of range result for the calcium to magnesium ratio (Ca/Mg). This suggests the possibility of poor blood sugar regulation. This is common in the presence of adrenal problems.

Note that the ratio of calcium to phosphorus (Ca/P) is elevated and the relation of sodium to magnesium (Na/Mg) is quite low. This combination suggests a need for more vitamin B6.

The father is a successful salesman. He is overweight and has acid indigestion and irregularity at times but otherwise has no significant medical history or health complaints. He has no cavities and no mercury (silver amalgam) or other dental fillings. Caffeine has little effect on him. He drinks beer routinely to slow himself down and relax as his mind tends to race. He is somewhat attention deficit, his attention to detail can be sloppy and he also does not always track social conversation flow well. He is outgoing, will talk to anyone, makes acquaintances easily but he doesn't keep in touch with old friends or pursue seeing his friends outside of situations (such as work) that bring him into contact with them. He is a perfectionist which gets in the way of getting things done. He also had some

---

[87] For those who like technical discussions involving numbers and equations, the way to compare the z score or number of standard deviations the result is above or below the (geometric) mean for a log normal distribution such as all the essential elements show in hair is as follows:

Calculate the geometric mean as the square root of the product of the upper and lower reference range limits.

Calculate the sigma value ("standard deviation" as I am loosely but improperly using the term here) as the upper reference limit divided by the geometric mean just calculated.

The z score is the log of (actual result divided by geometric mean) divided by the log of the sigma value.

Compare the z scores to find out who is farthest out.

This z score is the same one you would look up in an error function table to determine percentile or fraction of results lying further out from the average than the given result.

trouble in college with not doing homework – in a manner his wife characterizes as "defiant." A simple program of nutritional supplementation has noticeably improved his attention, energy and well being.

The father's hair test, shown page 158, counts as follows:

Counting rule #1: 8 bars go right. This does not meet rule #1.

Counting rule #2: 15 bars go left. This does not meet counting rule #2.

Counting rule #3: 4 bars go into the red bands. This meets counting rule #3.

Counting rule #4: 13 bars are contained entirely in the middle green + white stripe. This does not meet counting rule #4.

Counting rule #5 is irrelevant since another counting rule was met.

The significant elevations of antimony and tin support the idea that this test shows deranged mineral transport.

The sodium to potassium (Na/K) ratio out of range suggests the possibility of thyroid problems which could be followed up on by appropriate blood tests[88].

This family also has identical twin autistic boys. The twin autistic boys' hair was tested after a large number of biochemical interventions were already in place. Their hair tests are shown face to face on pages 154 and 155. The twins' hair tests count as follows:

	First twin (pg 156)	Second twin (pg 157)
Counting rule #1: bars to right.	7	9
Counting rule #2: bars to left.	16	14
Counting rule #3: bars in red.	0	0
Counting rule #4: bars in middle	17	16

Both tests count as normal. One is quite ordinary and average, the other unusual but certainly far from meeting any counting rule. Other tests did however suggest mercury toxicity, and in addition hair testing of the rest of the family suggests a high likelihood that the twins do have a genetic susceptibility to retaining mercury until it accumulates to toxic levels. A positive response to chelation confirmed this. Further details of their hair test interpretation and other background can put this in context.

Tin is vastly elevated in the hair of both boys. While bismuth is very elevated (red band) for one, and only elevated (yellow band) for the other, the bars are almost the same length, they just end near the yellow/red boundary.

Antimony was vastly elevated in a stool test for both boys. Since mercury causes the excessive accumulation and retention of antimony this is somewhat suggestive of a long-standing mercury problem.

The hair tests are very similar, not unreasonable for identical twins raised in the same household.

One significant difference between the two hair tests is that the second twin has much higher uranium than the first. This suggests the possibility of iron deficiency and in fact the second twin does show substantially lower (but not abnormally low) red blood cell count, hemoglobin and hematocrit on a blood count.

Despite not meeting the counting rules both twins responded positively to chelation with great progress, suggesting that there is in fact some underlying

[88] Free T3, free T4 and TSH. TSH alone is not appropriate or useful.

HAIR ELEMENTS

06/99

DOCTOR'S DATA INC.

SEX: Male
AGE: 2

POTENTIALLY TOXIC ELEMENTS

TOXIC ELEMENTS	RESULT µg/g	REFERENCE RANGE	PERCENTILE (68th, 95th)
Aluminum	7.6	< 8	
Antimony	0.12	< 0.066	
Arsenic	0.1	< 0.08	
Beryllium	< 0.01	< 0.02	
Bismuth	0.57	< 0.13	
Cadmium	0.089	< 0.15	
Lead	0.72	< 1	
Mercury	0.28	< 0.4	
Platinum	< 0.003	< 0.005	
Thallium	< 0.001	< 0.01	
Thorium	< 0.001	< 0.005	
Uranium	0.048	< 0.06	
Nickel	0.12	< 0.4	
Silver	0.12	< 0.2	
Tin	2.9	< 0.3	
Titanium	1.1	< 1	
Total Toxic Representation			

ESSENTIAL AND OTHER ELEMENTS

ELEMENTS	RESULT µg/g	REFERENCE RANGE	PERCENTILE (2.5th, 16th, 50th, 84th, 97.5th)
Calcium	111	125- 370	
Magnesium	7	12- 30	
Sodium	22	12- 90	
Potassium	21	12- 40	
Copper	10	8- 16	
Zinc	130	100- 190	
Manganese	0.18	0.2- 0.55	
Chromium	0.34	0.26- 0.5	
Vanadium	0.038	0.03- 0.1	
Molybdenum	0.11	0.05- 0.13	
Boron	3.3	0.6- 4	
Iodine	0.95	0.25- 1.3	
Lithium	0.005	0.007- 0.023	
Phosphorus	162	160- 250	
Selenium	0.53	0.95- 1.7	
Strontium	0.18	0.16- 1	
Sulfur	50100	45500- 53000	
Barium	0.2	0.16- 0.8	
Cobalt	0.011	0.013- 0.035	
Iron	13	8- 19	
Germanium	0.065	0.045- 0.065	
Rubidium	0.023	0.016- 0.18	
Zirconium	0.51	0.04- 1	

SPECIMEN DATA

COMMENTS:
Date Collected:
Date Received:
Date Completed:

Methodology: ICP-MS

RATIOS

ELEMENTS	RATIOS	EXPECTED RANGE
Ca/Mg	15.1	4- 30
Ca/P	0.683	0.8- 8
Na/K	1.05	0.5- 10
Zn/Cu	12.2	4- 20
Zn/Cd	> 999	> 800

This form is provided to Andrew Cutler to be used as an educational tool and does not constitute endorsement by *Doctor's Data.*

HAIR ELEMENTS

06/99

DOCTOR'S DATA INC.

SEX: Male
AGE: 2

POTENTIALLY TOXIC ELEMENTS

TOXIC ELEMENTS	RESULT µg/g	REFERENCE RANGE	PERCENTILE (68th, 95th)
Aluminum	12	< 8	
Antimony	0.14	< 0.066	
Arsenic	0.14	< 0.08	
Beryllium	< 0.01	< 0.02	
Bismuth	0.67	< 0.13	
Cadmium	0.11	< 0.15	
Lead	1.1	< 1	
Mercury	0.15	< 0.4	
Platinum	< 0.003	< 0.005	
Thallium	< 0.001	< 0.01	
Thorium	< 0.001	< 0.005	
Uranium	0.22	< 0.06	
Nickel	0.25	< 0.4	
Silver	0.19	< 0.2	
Tin	4	< 0.3	
Titanium	0.99	< 1	
Total Toxic Representation			

ESSENTIAL AND OTHER ELEMENTS

ELEMENTS	RESULT µg/g	REFERENCE RANGE	PERCENTILE (2.5th, 16th, 50th, 84th, 97.5th)
Calcium	115	125- 370	
Magnesium	10	12- 30	
Sodium	20	12- 90	
Potassium	23	12- 40	
Copper	12	8- 16	
Zinc	100	100- 190	
Manganese	0.25	0.2- 0.55	
Chromium	0.56	0.26- 0.5	
Vanadium	0.058	0.03- 0.1	
Molybdenum	0.17	0.05- 0.13	
Boron	1.1	0.6- 4	
Iodine	0.92	0.25- 1.3	
Lithium	0.005	0.007- 0.023	
Phosphorus	142	160- 250	
Selenium	0.58	0.95- 1.7	
Strontium	0.24	0.16- 1	
Sulfur	51300	45500- 53000	
Barium	0.26	0.16- 0.8	
Cobalt	0.014	0.013- 0.035	
Iron	16	8- 19	
Germanium	0.063	0.045- 0.065	
Rubidium	0.023	0.016- 0.18	
Zirconium	0.55	0.04- 1	

SPECIMEN DATA

COMMENTS:
Date Collected:
Date Received:
Date Completed:

Methodology: ICP-MS

RATIOS

ELEMENTS	RATIOS	EXPECTED RANGE
Ca/Mg	12	4- 30
Ca/P	0.804	0.8- 8
Na/K	0.899	0.5- 10
Zn/Cu	8.74	4- 20
Zn/Cd	917	> 800

This form is provided to Andrew Cutler to be used as an educational tool and does not constitute endorsement by *Doctor's Data*.

HAIR ELEMENTS

DOCTOR'S DATA INC.

SEX: Male
AGE: 38

POTENTIALLY TOXIC ELEMENTS

TOXIC ELEMENTS	RESULT µg/g	REFERENCE RANGE	PERCENTILE 68th 95th
Aluminum	1.9	< 7.0	
Antimony	0.17	< 0.066	
Arsenic	0.025	< 0.08	
Beryllium	< 0.01	< 0.02	
Bismuth	0.021	< 0.06	
Cadmium	< 0.009	< 0.15	
Lead	0.08	< 2.0	
Mercury	1.6	< 1.1	
Platinum	< 0.003	< 0.005	
Thallium	< 0.001	< 0.01	
Thorium	< 0.001	< 0.005	
Uranium	0.085	< 0.06	
Nickel	0.07	< 0.4	
Silver	0.03	< 0.12	
Tin	0.82	< 0.3	
Titanium	0.41	< 1.0	
Total Toxic Representation			

ESSENTIAL AND OTHER ELEMENTS

ELEMENTS	RESULT µg/g	REFERENCE RANGE	PERCENTILE 2.5th 16th 50th 84th 97.5th
Calcium	260	200- 750	
Magnesium	26	25- 75	
Sodium	7.0	12- 90	
Potassium	23	9.0- 40	
Copper	22	10- 28	
Zinc	210	130- 200	
Manganese	0.07	0.15- 0.65	
Chromium	0.14	0.2- 0.4	
Vanadium	0.017	0.018- 0.065	
Molybdenum	0.068	0.025- 0.064	
Boron	0.78	0.4- 3.0	
Iodine	0.89	0.25- 1.3	
Lithium	< 0.004	0.007- 0.023	
Phosphorus	213	160- 250	
Selenium	1.1	0.95- 1.7	
Strontium	0.77	0.3- 3.5	
Sulfur	52000	44500- 52000	
Barium	0.46	0.16- 1.6	
Cobalt	0.004	0.013- 0.035	
Iron	6.6	5.4- 13	
Germanium	0.034	0.045- 0.065	
Rubidium	0.007	0.011- 0.12	
Zirconium	0.17	0.02- 0.44	

SPECIMEN DATA

COMMENTS:

Date Collected:
Date Received:
Date Completed:

Methodology: ICP-MS

Sample Size:
Sample Type:
Hair Color:
Treatment:
Shampoo:

V06.99

RATIOS

ELEMENTS	RATIOS	EXPECTED RANGE
Ca/Mg	10	4- 30
Ca/P	1.22	0.8- 8
Na/K	0.304	0.5- 10
Zn/Cu	9.55	4- 20
Zn/Cd	> 999	> 800

This form is provided to Andrew Cutler to be used as an educational tool and does not constitute endorsement by *Doctor's Data*.

heavy metal problem, with mercury by far the most likely candidate. They were receiving nutritional supplements and their diet was controlled to improve symptoms. Either of these factors could result in normalization of hair test results, or it is possible the hair tests would have been normal without any intervention – a false negative for heavy metal toxicity.

The ratio of calcium to phosphorus (Ca/P) is low for the first twin and very close to low for the second twin. This suggests both would do well taking digestive enzymes with meals.

The results of assorted other testing often abnormal in mercury poisoned people are of interest. Also of interest is the fact that when both twins are tested their results are almost always very close to each other – with identical twins living in the same environment and eating the same diet, there is a "proper" value for each biochemical variable that their body controls very very closely. Just being within normal limits on the laboratory test report is not good enough for being healthy. This same principle of course applies to all other human beings, but very few have an identical twin to compare test results with to figure out if their body has the right amount of each substance for good health.

Results both twins had which were very similar:

- Very low secretory IgA in stool (a special antibody that protects the intestine – it is often low due to functionally inadequate vitamin A levels)
- Low CO_2 on a blood chemistry panel.
- Very elevated hair tin and uranium.
- Very elevated fecal antimony.
- isoleucine, leucine and cysteine low with alanine and glutamine high on various plasma amino acid tests.
- Both twins had similar thyroid test results that suggest low thyroid function.
- Both boys are chemically sensitive like their mother.

There are enough unrelated abnormalities on laboratory testing to make it clear that the problem these twins have can not be genetic, it has to be due to some factor like heavy metal poisoning that affects a wide array of biochemical systems.

Some areas where the twins were different:

- One twin had low blood sugar on several tests during the first year along with obvious symptoms of low blood sugar (the medical term is hypoglycemia), the other twin did not have this.
- While both twins have some abnormalities on their EEG's (brain wave measurements), the twin with the low blood sugar problem also has a much more abnormal EEG and had seizures at night.
- The less hyperactive twin with the more abnormal EEG also had trouble feeding himself and enunciating words clearly.
- The twin with the more normal EEG is much more hyperactive than the other twin. The less hyperactive twin likes to mash his head into corners and drag it along the floor, the more hyperactive one does not. The more hyperactive twin is about 5 pounds heavier and an inch taller.

The twins are both sound sensitive and this was greatly improved first with auditory integration training using the Tomatis method and then by using *The Listening Program*[89].

The twins were delivered 4 and a half weeks early by cesarean section. They were formula fed.

The twins are currently fed a mostly organic and natural diet conforming to the requirements of the specific carbohydrate diet. The specific carbohydrate diet was tremendously helpful for improving their digestion and correcting their frequent bouts of diarrhea.

[89] Described at www.advancedbrain.com.

Sometimes Mercury is High and Mineral Transport is Normal

This 35 year old woman has a history of allergies, irritable bowel syndrome and colitis, weight loss and difficulty maintaining weight, chemical sensitivity, and continuing development of more and more food allergies and reactions as time goes on. She needs to eat a high protein diet to remain coherent and functional and can only tolerate organic foods. She found great relief from bowel problems on the Specific Carbohydrate Diet[90]. Her hair test, shown on the facing page, counts as follows:

Counting rule #1: 13 bars go right. This does not meet rule #1.

Counting rule #2: 10 bars go left. This does not meet counting rule #2.

Counting rule #3: 2 bars go into the red bands. This does not meet counting rule #3.

Counting rule #4: 18 bars are contained entirely in the middle green + white stripe. This does not meet counting rule #4.

Counting rule #5: is not met.

In fact, the essential elements part of this test is quite ordinary and average. It indicates orderly and normal mineral transport.

It is unlikely most people count the essential element section before noticing that the mercury results bar goes all the way across the page, and the result itself is 14. Some hair tests are easy to interpret because the person's mineral transport proteins are tolerant of mercury, their mineral transport is not deranged, but they get exposed to a really large amount of mercury and it shows up nicely in their hair.

A few percent of the people with mercury problems nowadays have hair tests that look like this. Most of the people with mercury problems have hair tests showing disordered mineral transport and low hair mercury.

This woman had her amalgam fillings removed (though they are unlikely to have been the problem, but if left in during chelation could have become one) and chelated with oral DMPS every 8 hours on alternate weeks for 3 months and then switched to DMPS + ALA every 3 hours 3 days per week thereafter.

She also obtained substantial symptomatic relief taking hyrocortisone 10-20 mg every day[91].

---

[90] This diet is described in the book *Breaking the Vicious Cycle* by Elaine Gottschall.

[91] For a discussion of this use for hydrocortisone, see Jeffries, *Safe Uses of Cortisone*, Charles C Thomas Publisher, 1981.

HAIR ELEMENTS

DOCTOR'S DATA INC

SEX: Female
AGE: 35

POTENTIALLY TOXIC ELEMENTS

TOXIC ELEMENTS	RESULT µg/g	REFERENCE RANGE	PERCENTILE 68th 95th
Aluminum	3.2	< 7.0	
Antimony	0.019	< 0.05	
Arsenic	0.1	< 0.06	
Beryllium	< 0.01	< 0.02	
Bismuth	0.023	< 0.1	
Cadmium	0.057	< 0.1	
Lead	0.31	< 1.0	
Mercury	14	< 1.1	
Platinum	< 0.003	< 0.005	
Thallium	< 0.001	< 0.01	
Thorium	< 0.001	< 0.005	
Uranium	0.03	< 0.06	
Nickel	0.07	< 0.4	
Silver	0.11	< 0.15	
Tin	0.26	< 0.3	
Titanium	0.9	< 1.0	
Total Toxic Representation			

ESSENTIAL AND OTHER ELEMENTS

ELEMENTS	RESULT µg/g	REFERENCE RANGE	PERCENTILE 2.5th 16th 50th 84th 97.5th
Calcium	448	300- 1200	
Magnesium	37	35- 120	
Sodium	24	12- 90	
Potassium	21	8.0- 38	
Copper	21	12- 35	
Zinc	270	140- 220	
Manganese	0.65	0.15- 0.65	
Chromium	0.33	0.2- 0.4	
Vanadium	0.031	0.018- 0.065	
Molybdenum	< 0.01	0.028- 0.056	
Boron	0.81	0.3- 2.0	
Iodine	1.3	0.25- 1.3	
Lithium	0.005	0.007- 0.023	
Phosphorus	243	160- 250	
Selenium	1.4	0.95- 1.7	
Strontium	0.83	0.5- 7.6	
Sulfur	49900	44500- 52000	
Barium	0.47	0.26- 3.0	
Cobalt	0.007	0.013- 0.05	
Iron	12	5.4- 14	
Germanium	0.037	0.045- 0.065	
Rubidium	0.024	0.007- 0.096	
Zirconium	0.22	0.02- 0.42	

SPECIMEN DATA

COMMENTS:
Date Collected:
Date Received:
Date Completed:
Methodology: ICP-MS

Sample Size:
Sample Type:
Hair Color:
Treatment:
Shampoo:

V06.99

RATIOS

ELEMENTS	RATIOS	EXPECTED RANGE
Ca/Mg	12.1	4- 30
Ca/P	1.84	1- 12
Na/K	1.14	0.5- 10
Zn/Cu	12.9	4- 20
Zn/Cd	> 999	> 800

This form is provided to Andrew Cutler to be used as an educational tool and does not constitute endorsement by *Doctor's Data.*

Mercury can run in the Family

Here is a set of hair tests from most of the members of a three generation family that is experiencing some problems due to mercury. They all had a high fish consumption, averaging five days a week and eating lots of swordfish and tuna along with other fish.

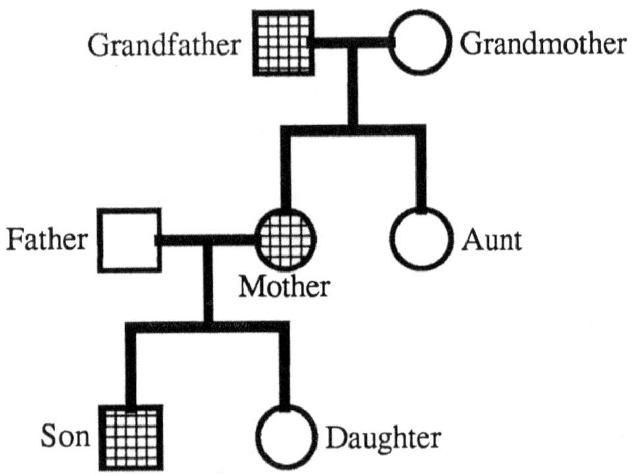

The family tree is shown above. There are hair tests for the grandfather and aunt, while the father, mother, son and daughter all have two hair tests a year apart. The gridwork shading shows the trait of very high hair mercury, presumably due to unusual retention of mercury to high levels.

The grandfather is retired. He has always eaten fish at least five times a week. He has high cholesterol, high trigylcerides (fat in the blood), Meuniere's disease (whistling in one ear), and some prostate issues. He is considered in good health for his age. He is quite rigid and stubborn and this is increasing with age. He does not have any mercury (silver amalgam) dental fillings.

The son was diagnosed with severe autism at 20 months due to being nonverbal, having no social skills, very limited receptive language, sensory issues, engaging in self stimulatory behaviors and in highly repetitive play. He also had growth failure, failure to thrive, anxiety, lost his physical ability to do things he used to be able to do[92], had poor mouth coordination and difficulty eating, had a low temperature and dry hair as well as allergies and food intolerances. He has about 10 spider nevi[93] on his arms and hands. His mother ate seafood at least 5 times a week during pregnancy and he ate it about twice a week from 6 to 20 months. He received the full recommended complement of childhood vaccinations through age 20 months.

The mother instituted treatment for her son by removing seafood from his diet, then later removing eggs, dairy, soy and gluten as well as the materials excluded on the Feingold diet. She tired the specific carbohydrate diet for about 3 months as well. She introduced yogurt as the only exception to the dairy exclusion. She eventually settled on a diet including all the above exclusions aside from the SCD, and in addition followed a four day rotation diet.

The mother is relatively small, 4'11" and 90#. She reports good health, does not have dental mercury (silver amalgam) fillings, and ate seafood on average 5 times a week until she became pregnant. During pregnancy and breast feeding she continued to eat seafood twice a week. She stopped eating seafood completely when her son was about one and a half. Despite her reporting good health she does complain of fatigue, lack of energy and lack of motivation. She is not very social and does have some mood swings. She had a mild depression after the birth of her son.

[92] The proper medical term is apraxia.

[93] The medical term for this is telangiectasia.

Note that the mother was lactating at the time of both hair tests. Thus the results must be interpreted with caution.

The father ate the same diet as the mother. He also reports good health and has no dental mercury (silver amalgam) fillings. He does have some back and neck pain due to an automobile accident about 12 years ago, and sometimes joint pain in his hands or mild carpal tunnel syndrome – he is a software manager and this is reasonably common in that line of work. Despite reporting good health, he had repetitive nosebleeds four years ago and would feel dizzy on standing up from lying down, and he presently has fatigue, reduced focus and concentration and a poor memory.

The daughter is still breast feeding and also eats solid foods, which were started at 6 months. She had no direct exposure to mercury from seafood but was exposed in the womb and during the first seven months of life through the mother's breast milk. She received standard vaccinations up to 5 months of age and none since then. She was put on the GFCF diet at age 7 months. Periodic reintroduction of dairy shows her to be intolerant of it with a distended belly, crying and tantrums.

The aunt was diagnosed with diabetes insipidus[94] at 7 years old, then was found to have complete deficiency of all hormones due to a nonfunctioning pituitary gland[95]. A brain tumor was discovered at age 10 and treated with radiation. She then received daily injections of growth hormone until age 14. She continues to take a variety of hormones: thyroid, adrenal, female hormones, and ADH for the diabetes insipidus. Growth hormone injections were re-started at age 27 due to excessive weight gain despite strict dieting. She continues to eat a very carefully regulated and very low calorie diet due to constant weight gain. She only had one mercury (silver amalgam) dental filling and that was placed at age 12. The symptoms she currently complains of are being overweight, depression, anxiety, sleep problems, poor memory, poor ability to focus or pay attention, low body temperature and always being cold. She is able to work full time.

The grandmother reports no significant health problems and is doing very well for someone her age. She has many mercury (silver amalgam) dental fillings, ate lots of fish lifelong until recently when she reduced her intake along with her husband. She is very social, full of energy, has no depression, no anxiety, perhaps a slight age related decrease in memory, and very seldom catches infectious diseases. A hair test was not performed on the grandmother because she perms and dyes her hair and there is no reason to believe she has any health problems it would be relevant to.

The serial hair tests were performed almost exactly a year apart.

During this year the father and the son chelated using alpha lipoic acid on a 3-4 hour schedule. The mother and daughter did not chelate. The exclusion of seafood from the diet for the father, mother, son, daughter and grandfather began shortly after the first hair test.

The hair tests are presented on pages 165-174, and the way they count is presented in a table on the next page.

[94] This is the medical term for a condition where extremely large amounts of urine are produced and the victim can easily die of dehydration. It is often due to lack of a specific hormone called ADH (antidiuretic hormone) that is released by the pituitary gland.

[95] Mercury is known from autopsy studies to concentrate in the pituitary gland and nearby brain structures.

	Grandfather	Father		Mother		Daughter		Son		Aunt
Test at age	62	32	33	32	33	0	1	1	2	27
Counting rule #1	8	7	11	9	11	11	13	13	10	6
Counting rule #2	15	16	12	14	12	12	10	12	13	17
Counting rule #3	1	3	1	1	3	3	1	4	4	4
Counting rule #4	18	15	13	18	18	12	13	14	6	13
Counting rule met	none	none	none	none	none	#5	none	#3	#3,#4	#3
Was mercury very elevated?	yes	no	no	yes	no	no	no	yes	no	no
Test on page	165	166	167	168	169	170	171	172	173	174

Note the very elevated aluminum on both children's tests both times.

The father's first hair test would be considered unusual on two counts (3 reds and 7 to the right) while the second test is unusual but on only one count (13 in the middle band).

The son's hair tests met the counting rules for deranged mineral transport both times, but the second test shows reasonably normal amounts of mercury while the first test shows very elevated mercury.

The daughter's first test met the counting rules while the second did not, though it remains unusual.

While the mother's first test counts out with quite ordinary and average mineral transport but the second one counts as unusual, it is not reasonable to make much interpretation of this since she was lactating during both tests and lactation has a significant but not well characterized effect on mineral transport. Given that neither test indicates abnormal mineral transport by meeting a counting rule it is most reasonable to interpret these tests as suggesting normal orderly mineral transport and thus that the toxic element levels in the hair (including the mercury) are an accurate reflection of body burden.

The first test for the father, mother, son and daughter, and the tests for the aunt and grandfather were done at a time when they were all equally exposed to methylmercury from seafood. Yet the grandfather, mother and son's tests all have very elevated hair mercury while the aunt, father and daughter do not. This suggests the possibility of some genetically transmitted trait that causes them to accumulate high levels of fish (methyl) mercury. With the son, daughter and aunt having deranged mineral transport and the father's hair test showing unusual mineral transport it is clear that they were all being exposed to more mercury than was good for them biochemically by their diet.

Also note the stress pattern shown by the calcium, magnesium, sodium and potassium in the son's age 1 hair test, and suggested in the daughter's age 0 test.

HAIR ELEMENTS

DOCTOR'S DATA INC.

SEX: Male
AGE: 62

POTENTIALLY TOXIC ELEMENTS

TOXIC ELEMENTS	RESULT µg/g	REFERENCE RANGE
Aluminum	2.8	< 7.0
Antimony	0.029	< 0.066
Arsenic	0.06	< 0.08
Beryllium	< 0.01	< 0.02
Bismuth	0.025	< 0.06
Cadmium	0.029	< 0.15
Lead	0.42	< 2.0
Mercury	8.9	< 1.1
Platinum	< 0.003	< 0.005
Thallium	< 0.001	< 0.01
Thorium	< 0.001	< 0.005
Uranium	0.011	< 0.06
Nickel	0.21	< 0.4
Silver	0.01	< 0.12
Tin	0.13	< 0.3
Titanium	0.46	< 1.0
Total Toxic Representation		

ESSENTIAL AND OTHER ELEMENTS

ELEMENTS	RESULT µg/g	REFERENCE RANGE
Calcium	322	200- 750
Magnesium	56	25- 75
Sodium	13	12- 90
Potassium	< 3	9.0- 40
Copper	14	10- 28
Zinc	200	130- 200
Manganese	0.08	0.15- 0.65
Chromium	0.29	0.2- 0.4
Vanadium	0.023	0.018- 0.065
Molybdenum	0.029	0.025- 0.064
Boron	0.9	0.4- 3.0
Iodine	0.54	0.25- 1.3
Lithium	0.005	0.007- 0.023
Phosphorus	189	160- 250
Selenium	0.7	0.95- 1.7
Strontium	1.2	0.3- 3.5
Sulfur	48700	44500- 52000
Barium	0.43	0.16- 1.6
Cobalt	0.02	0.013- 0.035
Iron	8.0	5.4- 13
Germanium	0.058	0.045- 0.065
Rubidium	0.007	0.011- 0.12
Zirconium	0.15	0.02- 0.44

SPECIMEN DATA

COMMENTS:
Date Collected:
Date Received:
Date Completed:
Methodology: ICP-MS

Sample Size:
Sample Type:
Hair Color:
Treatment:
Shampoo:

V06.99

RATIOS

ELEMENTS	RATIOS	EXPECTED RANGE
Ca/Mg	5.75	4- 30
Ca/P	1.7	0.8- 8
Na/K	4.33	0.5- 10
Zn/Cu	14.3	4- 20
Zn/Cd	> 999	> 800

This form is provided to Andrew Cutler to be used as an educational tool and does not constitute endorsement by *Doctor's Data*.

HAIR ELEMENTS

DOCTOR'S DATA INC.

SEX: Male
AGE: 32

POTENTIALLY TOXIC ELEMENTS

TOXIC ELEMENTS	RESULT µg/g	REFERENCE RANGE
Aluminum	1.4	< 7.0
Antimony	< 0.01	< 0.066
Arsenic	0.052	< 0.08
Beryllium	< 0.01	< 0.02
Bismuth	< 0.002	< 0.06
Cadmium	0.013	< 0.15
Lead	0.15	< 2.0
Mercury	2.2	< 1.1
Platinum	< 0.003	< 0.005
Thallium	< 0.001	< 0.01
Thorium	< 0.001	< 0.005
Uranium	0.013	< 0.06
Nickel	0.07	< 0.4
Silver	< 0.006	< 0.12
Tin	0.05	< 0.3
Titanium	0.51	< 1.0
Total Toxic Representation		

PERCENTILE: 68th, 95th

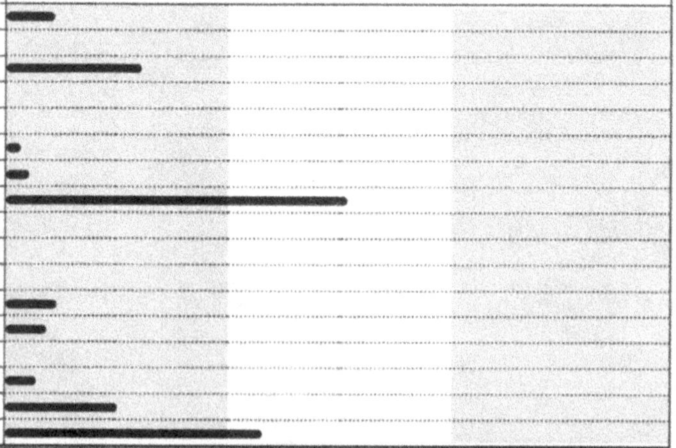

ESSENTIAL AND OTHER ELEMENTS

ELEMENTS	RESULT µg/g	REFERENCE RANGE
Calcium	522	200- 750
Magnesium	62	25- 75
Sodium	37	12- 90
Potassium	4.0	9.0- 40
Copper	12	10- 28
Zinc	190	130- 200
Manganese	0.05	0.15- 0.65
Chromium	0.22	0.2- 0.4
Vanadium	0.013	0.018- 0.065
Molybdenum	0.037	0.025- 0.064
Boron	1.2	0.4- 3.0
Iodine	0.26	0.25- 1.3
Lithium	0.004	0.007- 0.023
Phosphorus	307	160- 250
Selenium	1.0	0.95- 1.7
Strontium	0.95	0.3- 3.5
Sulfur	50900	44500- 52000
Barium	0.3	0.16- 1.6
Cobalt	0.007	0.013- 0.035
Iron	6.6	5.4- 13
Germanium	0.066	0.045- 0.065
Rubidium	0.005	0.011- 0.12
Zirconium	0.02	0.02- 0.44

PERCENTILE: 2.5th, 16th, 50th, 84th, 97.5th

SPECIMEN DATA

COMMENTS:
Date Collected:
Date Received:
Date Completed:
Methodology: ICP-MS

Sample Size:
Sample Type:
Hair Color:
Treatment:
Shampoo:

V06.99

RATIOS

ELEMENTS	RATIOS	EXPECTED RANGE
Ca/Mg	8.42	4- 30
Ca/P	1.7	0.8- 8
Na/K	9.25	0.5- 10
Zn/Cu	15.8	4- 20
Zn/Cd	> 999	> 800

This form is provided to Andrew Cutler to be used as an educational tool and does not constitute endorsement by *Doctor's Data.*

HAIR ELEMENTS

DOCTOR'S DATA INC.

SEX: Male
AGE: 33

POTENTIALLY TOXIC ELEMENTS

TOXIC ELEMENTS	RESULT µg/g	REFERENCE RANGE	PERCENTILE 68th 95th
Aluminum	2.5	< 7.0	
Antimony	< 0.01	< 0.066	
Arsenic	< 0.01	< 0.080	
Beryllium	< 0.01	< 0.020	
Bismuth	0.008	< 0.060	
Cadmium	0.030	< 0.15	
Lead	0.22	< 2.0	
Mercury	0.06	< 1.1	
Platinum	< 0.003	< 0.005	
Thallium	< 0.001	< 0.010	
Thorium	< 0.001	< 0.005	
Uranium	0.037	< 0.060	
Nickel	0.11	< 0.40	
Silver	< 0.006	< 0.12	
Tin	0.35	< 0.30	
Titanium	0.84	< 1.0	
Total Toxic Representation			

ESSENTIAL AND OTHER ELEMENTS

ELEMENTS	RESULT µg/g	REFERENCE RANGE	PERCENTILE 2.5th 16th 50th 84th 97.5th
Calcium	800	200- 750	
Magnesium	110	25- 75	
Sodium	44	12- 90	
Potassium	8	9- 40	
Copper	15	10- 28	
Zinc	230	130- 200	
Manganese	0.09	0.15- 0.65	
Chromium	0.25	0.20- 0.40	
Vanadium	0.030	0.018- 0.065	
Molybdenum	0.076	0.025- 0.064	
Boron	1.8	0.40- 3.0	
Iodine	0.53	0.25- 1.3	
Lithium	0.005	0.007- 0.023	
Phosphorus	222	160- 250	
Selenium	1.5	0.95- 1.7	
Strontium	2.9	0.30- 3.5	
Sulfur	47800	44500- 52000	
Barium	1.2	0.16- 1.6	
Cobalt	0.004	0.013- 0.035	
Iron	6.5	5.4- 13	
Germanium	0.036	0.045- 0.065	
Rubidium	0.008	0.011- 0.12	
Zirconium	0.041	0.020- 0.44	

SPECIMEN DATA

COMMENTS:

Date Collected:
Date Received:
Date Completed:

Methodology: ICP-MS

Sample Size:
Sample Type:
Hair Color:
Treatment:
Shampoo:

V06 99

RATIOS

ELEMENTS	RATIOS	EXPECTED RANGE
Ca/Mg	7.27	4- 30
Ca/P	3.6	0.8- 8
Na/K	5.5	0.5- 10
Zn/Cu	15.3	4- 20
Zn/Cd	> 999	> 800

This form is provided to Andrew Cutler to be used as an educational tool and does not constitute endorsement by *Doctor's Data.*

HAIR ELEMENTS

DOCTOR'S DATA INC.

SEX: Female
AGE: 32

POTENTIALLY TOXIC ELEMENTS

TOXIC ELEMENTS	RESULT µg/g	REFERENCE RANGE	PERCENTILE 68th 95th
Aluminum	2.1	< 7.0	
Antimony	0.015	< 0.05	
Arsenic	0.036	< 0.06	
Beryllium	< 0.01	< 0.02	
Bismuth	0.012	< 0.1	
Cadmium	0.049	< 0.1	
Lead	0.13	< 1.0	
Mercury	5.7	< 1.1	
Platinum	< 0.003	< 0.005	
Thallium	< 0.001	< 0.01	
Thorium	< 0.001	< 0.005	
Uranium	0.027	< 0.06	
Nickel	0.15	< 0.4	
Silver	< 0.006	< 0.15	
Tin	0.22	< 0.3	
Titanium	0.4	< 1.0	
Total Toxic Representation			

ESSENTIAL AND OTHER ELEMENTS

ELEMENTS	RESULT µg/g	REFERENCE RANGE	PERCENTILE 2.5th 16th 50th 84th 97.5th
Calcium	452	300- 1200	
Magnesium	77	35- 120	
Sodium	27	12- 90	
Potassium	13	8.0- 38	
Copper	14	12- 35	
Zinc	190	140- 220	
Manganese	0.05	0.15- 0.65	
Chromium	0.28	0.2- 0.4	
Vanadium	0.017	0.018- 0.065	
Molybdenum	0.032	0.028- 0.056	
Boron	0.78	0.3- 2.0	
Iodine	0.92	0.25- 1.3	
Lithium	0.006	0.007- 0.023	
Phosphorus	221	160- 250	
Selenium	1.1	0.95- 1.7	
Strontium	2.6	0.5- 7.6	
Sulfur	49500	44500- 52000	
Barium	1.0	0.26- 3.0	
Cobalt	0.012	0.013- 0.05	
Iron	6.6	5.4- 14	
Germanium	0.084	0.045- 0.065	
Rubidium	0.02	0.007- 0.096	
Zirconium	0.036	0.02- 0.42	

SPECIMEN DATA

COMMENTS:

Date Collected:
Date Received:
Date Completed:

Methodology: ICP-MS

Sample Size:
Sample Type:
Hair Color:
Treatment:
Shampoo:

V06.99

RATIOS

ELEMENTS	RATIOS	EXPECTED RANGE
Ca/Mg	5.87	4- 30
Ca/P	2.05	1- 12
Na/K	2.08	0.5- 10
Zn/Cu	13.6	4- 20
Zn/Cd	> 999	> 800

This form is provided to Andrew Cutler to be used as an educational tool and does not constitute endorsement by *Doctor's Data*.

HAIR ELEMENTS

DOCTOR'S DATA INC.

SEX: Female
AGE: 33

POTENTIALLY TOXIC ELEMENTS

TOXIC ELEMENTS	RESULT µg/g	REFERENCE RANGE	PERCENTILE 68th 95th
Aluminum	3.1	< 7.0	
Antimony	< 0.01	< 0.050	
Arsenic	0.010	< 0.060	
Beryllium	< 0.01	< 0.020	
Bismuth	0.012	< 0.10	
Cadmium	0.026	< 0.10	
Lead	0.11	< 1.0	
Mercury	0.05	< 1.1	
Platinum	< 0.003	< 0.005	
Thallium	< 0.001	< 0.010	
Thorium	< 0.001	< 0.005	
Uranium	0.023	< 0.060	
Nickel	0.10	< 0.40	
Silver	0.02	< 0.15	
Tin	0.12	< 0.30	
Titanium	0.56	< 1.0	
Total Toxic Representation			

ESSENTIAL AND OTHER ELEMENTS

ELEMENTS	RESULT µg/g	REFERENCE RANGE	PERCENTILE 2.5th 16th 50th 84th 97.5th
Calcium	793	300- 1200	
Magnesium	90	35- 120	
Sodium	14	12- 90	
Potassium	5	8- 38	
Copper	15	12- 35	
Zinc	220	140- 220	
Manganese	0.09	0.15- 0.65	
Chromium	0.35	0.20- 0.40	
Vanadium	0.053	0.018- 0.065	
Molybdenum	0.016	0.028- 0.056	
Boron	1.0	0.30- 2.0	
Iodine	0.39	0.25- 1.3	
Lithium	0.008	0.007- 0.023	
Phosphorus	171	160- 250	
Selenium	1.0	0.95- 1.7	
Strontium	4.1	0.50- 7.6	
Sulfur	46600	44500- 52000	
Barium	1.5	0.26- 3.0	
Cobalt	0.010	0.013- 0.050	
Iron	12	5.4- 14	
Germanium	0.057	0.045- 0.065	
Rubidium	< 0.003	0.007- 0.096	
Zirconium	0.098	0.020- 0.42	

SPECIMEN DATA

COMMENTS:

Date Collected: Sample Size:
Date Received: Sample Type:
Date Completed: Hair Color:
Treatment:
Methodology: ICP-MS Shampoo:

V06.99

RATIOS

ELEMENTS	RATIOS	EXPECTED RANGE
Ca/Mg	8.81	4- 30
Ca/P	4.64	1- 12
Na/K	2.8	0.5- 10
Zn/Cu	14.7	4- 20
Zn/Cd	> 999	> 800

This form is provided to Andrew Cutler to be used as an educational tool and does not constitute endorsement by *Doctor's Data*.

HAIR ELEMENTS

DOCTOR'S DATA INC.

SEX: Female
AGE: 0

POTENTIALLY TOXIC ELEMENTS

TOXIC ELEMENTS	RESULT µg/g	REFERENCE RANGE	PERCENTILE (68th, 95th)
Aluminum	33	< 8.0	
Antimony	0.08	< 0.066	
Arsenic	0.042	< 0.08	
Beryllium	< 0.01	< 0.02	
Bismuth	0.014	< 0.13	
Cadmium	0.043	< 0.1	
Lead	0.54	< 1.0	
Mercury	1.2	< 0.4	
Platinum	< 0.003	< 0.005	
Thallium	< 0.001	< 0.01	
Thorium	< 0.001	< 0.005	
Uranium	0.039	< 0.06	
Nickel	0.16	< 0.4	
Silver	0.13	< 0.2	
Tin	0.41	< 0.3	
Titanium	0.61	< 1.0	
Total Toxic Representation			

ESSENTIAL AND OTHER ELEMENTS

ELEMENTS	RESULT µg/g	REFERENCE RANGE	PERCENTILE (2.5th, 16th, 50th, 84th, 97.5th)
Calcium	178	140- 500	
Magnesium	10	15- 45	
Sodium	78	12- 90	
Potassium	190	9.0- 60	
Copper	11	10- 22	
Zinc	32	100- 190	
Manganese	0.06	0.2- 0.55	
Chromium	0.55	0.26- 0.5	
Vanadium	0.055	0.03- 0.1	
Molybdenum	0.2	0.048- 0.13	
Boron	4.3	0.35- 3.0	
Iodine	6.4	0.25- 1.3	
Lithium	0.012	0.007- 0.023	
Phosphorus	199	160- 250	
Selenium	0.84	0.95- 1.7	
Strontium	0.31	0.19- 2.0	
Sulfur	46400	45500- 53000	
Barium	0.32	0.16- 1.2	
Cobalt	0.008	0.013- 0.035	
Iron	12	8.0- 19	
Germanium	0.057	0.045- 0.065	
Rubidium	0.2	0.012- 0.16	
Zirconium	0.079	0.03- 1.0	

SPECIMEN DATA

COMMENTS:

Date Collected:

Date Received:

Date Completed:

Methodology: ICP-MS

Sample Size:

Sample Type:

Hair Color:

Treatment:

Shampoo:

V06.99

RATIOS

ELEMENTS	RATIOS	EXPECTED RANGE
Ca/Mg	17.8	4- 30
Ca/P	0.894	1- 12
Na/K	0.411	0.5- 10
Zn/Cu	2.91	4- 20
Zn/Cd	744	> 800

This form is provided to Andrew Cutler to be used as an educational tool and does not constitute endorsement by *Doctor's Data.*

HAIR ELEMENTS

DOCTOR'S DATA INC.

SEX: Female
AGE: 1

POTENTIALLY TOXIC ELEMENTS

TOXIC ELEMENTS	RESULT µg/g	REFERENCE RANGE
Aluminum	28	< 8.0
Antimony	0.024	< 0.066
Arsenic	0.043	< 0.080
Beryllium	0.010	< 0.020
Bismuth	0.010	< 0.13
Cadmium	0.028	< 0.10
Lead	0.28	< 1.0
Mercury	0.15	< 0.40
Platinum	< 0.003	< 0.005
Thallium	0.002	< 0.010
Thorium	0.002	< 0.005
Uranium	0.077	< 0.060
Nickel	0.08	< 0.40
Silver	0.02	< 0.20
Tin	0.54	< 0.30
Titanium	1.2	< 1.0
Total Toxic Representation		

PERCENTILE: 68th, 95th

ESSENTIAL AND OTHER ELEMENTS

ELEMENTS	RESULT µg/g	REFERENCE RANGE
Calcium	346	140- 500
Magnesium	36	15- 45
Sodium	36	12- 90
Potassium	74	9- 60
Copper	14	10- 22
Zinc	70	100- 190
Manganese	0.08	0.20- 0.55
Chromium	0.49	0.26- 0.50
Vanadium	0.092	0.030- 0.10
Molybdenum	0.11	0.048- 0.13
Boron	6.6	0.35- 3.0
Iodine	1.7	0.25- 1.3
Lithium	0.014	0.007- 0.023
Phosphorus	147	160- 250
Selenium	1.1	0.95- 1.7
Strontium	1.3	0.19- 2.0
Sulfur	44500	45500- 53000
Barium	0.97	0.16- 1.2
Cobalt	0.009	0.013- 0.035
Iron	6.2	8.0- 19
Germanium	0.043	0.045- 0.065
Rubidium	0.12	0.012- 0.16
Zirconium	0.072	0.030- 1.0

PERCENTILE: 2.5th, 16th, 50th, 84th, 97.5th

SPECIMEN DATA

COMMENTS:

Date Collected:
Date Received:
Date Completed:

Methodology: ICP-MS

Sample Size:
Sample Type:
Hair Color:
Treatment:
Shampoo:

V06.99

RATIOS

ELEMENTS	RATIOS	EXPECTED RANGE
Ca/Mg	9.61	4- 30
Ca/P	2.35	1- 12
Na/K	0.486	0.5- 10
Zn/Cu	5	4- 20
Zn/Cd	> 999	> 800

This form is provided to Andrew Cutler to be used as an educational tool and does not constitute endorsement by *Doctor's Data*.

HAIR ELEMENTS

DOCTOR'S DATA INC

SEX: Male
AGE: 1

POTENTIALLY TOXIC ELEMENTS

TOXIC ELEMENTS	RESULT µg/g	REFERENCE RANGE	PERCENTILE (68th, 95th)
Aluminum	36	< 8.0	
Antimony	0.091	< 0.066	
Arsenic	0.063	< 0.08	
Beryllium	< 0.01	< 0.02	
Bismuth	0.029	< 0.13	
Cadmium	0.022	< 0.15	
Lead	0.46	< 1.0	
Mercury	5.9	< 0.4	
Platinum	< 0.003	< 0.005	
Thallium	< 0.001	< 0.01	
Thorium	< 0.001	< 0.005	
Uranium	0.077	< 0.06	
Nickel	0.14	< 0.4	
Silver	0.12	< 0.2	
Tin	0.38	< 0.3	
Titanium	0.73	< 1.0	
Total Toxic Representation			

ESSENTIAL AND OTHER ELEMENTS

ELEMENTS	RESULT µg/g	REFERENCE RANGE	PERCENTILE (2.5th, 16th, 50th, 84th, 97.5th)
Calcium	140	125- 370	
Magnesium	13	12- 30	
Sodium	94	12- 90	
Potassium	190	12- 40	
Copper	21	8.0- 16	
Zinc	27	100- 190	
Manganese	0.05	0.2- 0.55	
Chromium	0.52	0.26- 0.5	
Vanadium	0.065	0.03- 0.1	
Molybdenum	0.12	0.05- 0.13	
Boron	3.0	0.6- 4.0	
Iodine	5.5	0.25- 1.3	
Lithium	0.021	0.007- 0.023	
Phosphorus	250	160- 250	
Selenium	1.1	0.95- 1.7	
Strontium	0.3	0.16- 1.0	
Sulfur	52400	45500- 53000	
Barium	0.19	0.16- 0.8	
Cobalt	0.011	0.013- 0.035	
Iron	8.9	8.0- 19	
Germanium	0.068	0.045- 0.065	
Rubidium	0.23	0.016- 0.18	
Zirconium	0.18	0.04- 1.0	

SPECIMEN DATA

COMMENTS:

Date Collected:
Date Received:
Date Completed:

Methodology: ICP-MS

Sample Size:
Sample Type:
Hair Color:
Treatment:
Shampoo:

V06.99

RATIOS

ELEMENTS	RATIOS	EXPECTED RANGE
Ca/Mg	10.8	4- 30
Ca/P	0.56	0.8- 8
Na/K	0.495	0.5- 10
Zn/Cu	1.29	4- 20
Zn/Cd	> 999	> 800

This form is provided to Andrew Cutler to be used as an educational tool and does not constitute endorsement by *Doctor's Data.*

HAIR ELEMENTS

DOCTOR'S DATA INC.

SEX: Male
AGE: 2

POTENTIALLY TOXIC ELEMENTS

TOXIC ELEMENTS	RESULT µg/g	REFERENCE RANGE	PERCENTILE 68th 95th
Aluminum	34	< 8.0	
Antimony	0.041	< 0.066	
Arsenic	0.055	< 0.080	
Beryllium	< 0.01	< 0.020	
Bismuth	0.013	< 0.13	
Cadmium	0.072	< 0.15	
Lead	0.45	< 1.0	
Mercury	0.26	< 0.40	
Platinum	< 0.003	< 0.005	
Thallium	< 0.001	< 0.010	
Thorium	< 0.001	< 0.005	
Uranium	0.056	< 0.060	
Nickel	0.06	< 0.40	
Silver	0.05	< 0.20	
Tin	0.42	< 0.30	
Titanium	2.2	< 1.0	
Total Toxic Representation			

ESSENTIAL AND OTHER ELEMENTS

ELEMENTS	RESULT µg/g	REFERENCE RANGE	PERCENTILE 2.5th 16th 50th 84th 97.5th
Calcium	120	125- 370	
Magnesium	18	12- 30	
Sodium	9	12- 90	
Potassium	< 3	12- 40	
Copper	17	8.0- 16	
Zinc	77	100- 190	
Manganese	0.05	0.20- 0.55	
Chromium	0.55	0.26- 0.50	
Vanadium	0.11	0.030- 0.10	
Molybdenum	0.34	0.050- 0.13	
Boron	5.6	0.60- 4.0	
Iodine	1.4	0.25- 1.3	
Lithium	< 0.004	0.007- 0.023	
Phosphorus	210	160- 250	
Selenium	0.79	0.95- 1.7	
Strontium	0.43	0.16- 1.0	
Sulfur	56100	45500- 53000	
Barium	0.27	0.16- 0.80	
Cobalt	0.006	0.013- 0.035	
Iron	4.8	8.0- 19	
Germanium	0.028	0.045- 0.065	
Rubidium	0.009	0.016- 0.18	
Zirconium	0.094	0.040- 1.0	

SPECIMEN DATA

COMMENTS:

Date Collected:
Date Received:
Date Completed:

Methodology: ICP-MS

Sample Size:
Sample Type:
Hair Color:
Treatment:
Shampoo:

V06.99

RATIOS

ELEMENTS	RATIOS	EXPECTED RANGE
Ca/Mg	6.67	4- 30
Ca/P	0.571	0.8- 8
Na/K	3	0.5- 10
Zn/Cu	4.53	4- 20
Zn/Cd	> 999	> 800

This form is provided to Andrew Cutler to be used as an educational tool and does not constitute endorsement by *Doctor's Data*.

HAIR ELEMENTS

DOCTOR'S DATA INC

SEX: Female
AGE: 27

POTENTIALLY TOXIC ELEMENTS

TOXIC ELEMENTS	RESULT µg/g	REFERENCE RANGE	PERCENTILE (68th, 95th)
Aluminum	3.5	< 7.0	
Antimony	0.013	< 0.050	
Arsenic	0.032	< 0.060	
Beryllium	0.010	< 0.020	
Bismuth	0.020	< 0.10	
Cadmium	< 0.009	< 0.10	
Lead	0.65	< 1.0	
Mercury	1.4	< 1.1	
Platinum	< 0.003	< 0.005	
Thallium	< 0.001	< 0.010	
Thorium	< 0.001	< 0.005	
Uranium	0.046	< 0.060	
Nickel	0.05	< 0.40	
Silver	0.04	< 0.15	
Tin	0.57	< 0.30	
Titanium	1.1	< 1.0	
Total Toxic Representation			

ESSENTIAL AND OTHER ELEMENTS

ELEMENTS	RESULT µg/g	REFERENCE RANGE	PERCENTILE (2.5th, 16th, 50th, 84th, 97.5th)
Calcium	296	300- 1200	
Magnesium	24	35- 120	
Sodium	32	12- 90	
Potassium	15	8- 38	
Copper	14	12- 35	
Zinc	210	140- 220	
Manganese	0.08	0.15- 0.65	
Chromium	0.27	0.20- 0.40	
Vanadium	0.058	0.018- 0.065	
Molybdenum	0.020	0.028- 0.056	
Boron	1.1	0.30- 2.0	
Iodine	0.12	0.25- 1.3	
Lithium	< 0.004	0.007- 0.023	
Phosphorus	214	160- 250	
Selenium	0.84	0.95- 1.7	
Strontium	0.85	0.50- 7.6	
Sulfur	52600	44500- 52000	
Barium	0.25	0.26- 3.0	
Cobalt	0.002	0.013- 0.050	
Iron	6.1	5.4- 14	
Germanium	0.037	0.045- 0.065	
Rubidium	0.022	0.007- 0.096	
Zirconium	0.12	0.020- 0.42	

SPECIMEN DATA

COMMENTS:

Date Collected:
Date Received:
Date Completed:
Methodology: ICP-MS

Sample Size:
Sample Type:
Hair Color:
Treatment:
Shampoo:

V06.99

RATIOS

ELEMENTS	RATIOS	EXPECTED RANGE
Ca/Mg	12.3	4- 30
Ca/P	1.38	1- 12
Na/K	2.13	0.5- 10
Zn/Cu	15	4- 20
Zn/Cd	> 999	> 800

This form is provided to Andrew Cutler to be used as an educational tool and does not constitute endorsement by *Doctor's Data*.

Hair Tests do not improve without Chelation

This boy has not been diagnosed with any developmental disorder but clearly has problems. From age 2 to 4 he would refuse to play with or be in groups of children. He was unable to sleep for more than two or three hours straight from four months to four and a half years of age. He is extremely sensitive to light, heat and sound, and has episodes of dilated pupils. He is sensitive to certain smells, such as rubbing alcohol and scented candles[96] He wets the bed and is often constipated. He becomes hyperactive, spacey and irritable when eating foods containing casein or gluten, and also becomes hyperactive from foods containing salicylates and phenols as well as from B vitamins, transdermal glutathione and supplementary chromium.

He has low blood sugar (the medical term is hypoglycemia), which was a problem until age 4, went away and returned at age 5. The hypoglycemia causes continuous hunger, sugar craving, night waking and walking, and being very emotional. This improved when vanadium was given as a nutritional supplement.

Despite this he is gifted intellectually and his motor skills are age appropriate.

He had low plasma cysteine and sulfate, an accelerated rate of glycination, sulfation and slow glucuronidation, all markers of impaired liver health. He also had low zinc on blood testing. In addition he has a high rate of metabolizing caffeine (as determined by laboratory testing) which is consistent with "chemical sensitivity" due to excessive activity of the liver's phase 1 cytochrome P450 mixed function oxidase system.

The mother thinks the following description of the effects of low (but not zero) adrenal function fit her son to a T: "thin, hyperactive after excitement or stress, not very tolerant of stress, generally obsessive and compulsive and somewhat emotionally disturbed but without any clearly relevant psychiatric diagnosis."

Some of his symptoms became much worse on a trial of seven cycles of chelation and also much worse on a trial of increasing dietary precursors of cysteine. Other symptoms did improve greatly on chelation. However, new symptoms that appeared were being afraid to walk up stairs, hiding under tables, and becoming obsessive and compulsive. Due to this the family feeds him tolerated nutritional supplements and excludes things he has problems with from his diet but has not attempted any kind of heavy metal detoxification therapy beyond the chelation trial.

	First test (page 176)	Second test (page 177)
Counting rule #1: bars to right.	11	10
Counting rule #2: bars to left.	12	13
Counting rule #3: bars in red.	2	2
Counting rule #4: bars in middle	16	14

Counting rule #5 is not met by either test. Both tests count normal and in fact fall in the not uncommon category. Whether the arsenic and aluminum are elevated enough to be significant is unclear.

Both the first and the second test have a low Ca/P, suggesting a need for digestive enzymes. He is in fact given these and they are found to be helpful.

The sodium to potassium ratio suggests it would be worthwhile to check his thyroid, blood tests for this are also ambiguous but suggest that he may not have as much thyroid hormone as his body needs.

[96] This phenomenon is often referred to as "chemical sensitivity" in alternative medical practice.

HAIR ELEMENTS

06/99

DOCTOR'S DATA INC

SEX: Male
AGE: 4

POTENTIALLY TOXIC ELEMENTS

TOXIC ELEMENTS	RESULT µg/g	REFERENCE RANGE
Aluminum	20	< 8.0
Antimony	0.078	< 0.066
Arsenic	0.2	< 0.08
Beryllium	< 0.01	< 0.02
Bismuth	0.025	< 0.13
Cadmium	0.043	< 0.15
Lead	1.2	< 1.0
Mercury	0.21	< 0.4
Platinum	< 0.003	< 0.005
Thallium	< 0.001	< 0.01
Thorium	< 0.001	< 0.005
Uranium	0.083	< 0.06
Nickel	0.16	< 0.4
Silver	0.1	< 0.2
Tin	0.76	< 0.3
Titanium	0.57	< 1.0
Total Toxic Representation		

PERCENTILE: 68th, 95th

ESSENTIAL AND OTHER ELEMENTS

ELEMENTS	RESULT µg/g	REFERENCE RANGE
Calcium	121	125- 370
Magnesium	19	12- 30
Sodium	64	12- 90
Potassium	76	12- 40
Copper	13	8.0- 16
Zinc	36	100- 190
Manganese	0.11	0.2- 0.55
Chromium	0.41	0.26- 0.5
Vanadium	0.063	0.03- 0.1
Molybdenum	0.1	0.05- 0.13
Boron	4.7	0.6- 4.0
Iodine	0.33	0.25- 1.3
Lithium	0.007	0.007- 0.023
Phosphorus	220	160- 250
Selenium	1.2	0.95- 1.7
Strontium	0.06	0.16- 1.0
Sulfur	49400	45500- 53000
Barium	0.13	0.16- 0.8
Cobalt	0.014	0.013- 0.035
Iron	11	8.0- 19
Germanium	0.045	0.045- 0.065
Rubidium	0.072	0.016- 0.18
Zirconium	0.075	0.04- 1.0

PERCENTILE: 2.5th, 16th, 50th, 84th, 97.5th

SPECIMEN DATA

COMMENTS:

Date Collected: 10/21/2001
Date Received:
Date Completed:
Methodology: ICP-MS

Sample Size:
Sample Type:
Hair Color:
Treatment:
Shampoo:

RATIOS

ELEMENTS	RATIOS	EXPECTED RANGE
Ca/Mg	6.46	4- 30
Ca/P	0.549	0.8- 8
Na/K	0.845	0.5- 10
Zn/Cu	2.75	4- 20
Zn/Cd	834	> 800

This form is provided to Andrew Cutler to be used as an educational tool and does not constitute endorsement by *Doctor's Data.*

HAIR ELEMENTS

DOCTOR'S DATA INC.

SEX: Male
AGE: 6

POTENTIALLY TOXIC ELEMENTS

TOXIC ELEMENTS	RESULT µg/g	REFERENCE RANGE
Aluminum	24	< 8.0
Antimony	0.033	< 0.066
Arsenic	0.23	< 0.08
Beryllium	< 0.01	< 0.02
Bismuth	< 0.002	< 0.12
Cadmium	0.086	< 0.15
Lead	1.1	< 1.0
Mercury	0.29	< 0.4
Platinum	< 0.003	< 0.005
Thallium	< 0.001	< 0.01
Thorium	< 0.001	< 0.005
Uranium	0.015	< 0.06
Nickel	0.12	< 0.4
Silver	0.11	< 0.13
Tin	0.69	< 0.3
Titanium	1.6	< 1.0
Total Toxic Representation		

PERCENTILE: 68th, 95th

ESSENTIAL AND OTHER ELEMENTS

ELEMENTS	RESULT µg/g	REFERENCE RANGE
Calcium	85.0	160- 500
Magnesium	42	12- 50
Sodium	36	12- 90
Potassium	60	10- 40
Copper	11	9.0- 30
Zinc	63	110- 190
Manganese	0.11	0.18- 0.6
Chromium	0.39	0.23- 0.5
Vanadium	0.073	0.025- 0.1
Molybdenum	0.087	0.04- 0.089
Boron	3.7	0.5- 3.5
Iodine	0.54	0.25- 1.3
Lithium	0.013	0.007- 0.023
Phosphorus	203	160- 250
Selenium	0.73	0.95- 1.7
Strontium	0.27	0.21- 2.1
Sulfur	48700	45500- 53000
Barium	0.1	0.19- 1.6
Cobalt	0.009	0.013- 0.035
Iron	8.9	6.0- 17
Germanium	0.037	0.045- 0.065
Rubidium	0.094	0.008- 0.08
Zirconium	0.16	0.06- 0.7

PERCENTILE: 2.5th, 16th, 50th, 84th, 97.5th

SPECIMEN DATA

COMMENTS:

Date Collected: 6/4/2003
Date Received:
Date Completed:
Methodology: ICP-MS

Sample Size:
Sample Type:
Hair Color:
Treatment:
Shampoo:

V06.99

RATIOS

ELEMENTS	RATIOS	EXPECTED RANGE
Ca/Mg	2.02	4- 30
Ca/P	0.419	0.8- 8
Na/K	0.6	0.5- 10
Zn/Cu	5.73	4- 20
Zn/Cd	733	> 800

This form is provided to Andrew Cutler to be used as an educational tool and does not constitute endorsement by *Doctor' s Data.*

The next case is a young woman who had three hair tests, about a year apart. She did not chelate between the first two tests, but did chelate between the second and third tests. This young woman had a history of vaccine reactions as an infant, strange skin eruptions at age two tentatively diagnosed as Moluscum Contagiosum, poor muscle tone, generalized lethargy, autism, developmental delay, moderately low intellectual functioning, and epileptic seizures. She has a deficiency of the immune system protein IgA which has a population incidence of about 1 in 600, is believed to often be due to viral problems in infancy and correlates with developing lots of food allergies and reactivities as this woman has. Her pupils are always extremely dilated[97] regardless of light level and she does not sweat. She was a very light sleeper until melatonin was introduced. She hyperventilates for brief periods at least once a day. She has a distended abdomen and passes unusually large stools, with frequent constipation. She has numbness and lack of pain sensation in her hands and feet[98]. She is quite fearful of new situations. She has trouble finding words to express herself, and will often answer questions about things she has read by going to the page and pointing to a specific sentence. She has little grasp of subtlety and inference. She is unable to form social bonds.

She takes carbamazepine for seizures and Armor thyroid for hypothyroidism. She is also on a very restricted diet that is free of gluten, casein (GFCF diet, excluding dairy and most grains) and all known allergens as well as foods high in copper or phenolic compounds.

She also takes a very long list of nutritional supplements which improve her overall condition greatly.

Diet and supplements were much more effective at reducing the incidence of seizures than medications were.

She has had three hair tests done approximately one year apart. The tests are shown on the following pages and count as shown in the table below:

	Test 1 (pg 179)	Test 2 (pg 180)	Test 3 (pg 181)
Counting rule #1: bars to right.	13	13	17
Counting rule #2: bars to left.	10	10	6
Counting rule #3: bars in red.	4	4	1
Counting rule #4: bars in middle	13	14	15

The first two tests satisfy counting rule #3, the third test does not meet any counting rule but does look suspicious due to only 6 bars going left. No test meets counting rule #5.

Many interventions, but no chelation was done between the first and second tests. As you can see, the different essential elements changed around but the test continued to meet a counting rule and also continued to look "messed up." Proper chelation was conducted between the second and third test, resulting in great symptomatic improvement – reduction in seizures, ability to tolerate supplements she had reacted to in the past, better choice of words, more energy, disappearance of hand and foot numbness – plus the test looking a lot more normal, and no longer meeting a counting rule though it does count as a "suspicious" test.

The very elevated nickel on the first test is most likely an artifact of deranged mineral transport, not an accurate reflection of how much nickel was in her body at that time.

What looks something like the calcium magnesium sodium potassium "stress pattern" on the first test changes to reflect more of a sodium/potassium imbalance suggesting thyroid problems, which was consistent with both her situation and with blood tests showing her thyroid function to be low.

The calcium to phosphorus (Ca/P) ratio on the first test indicates she is a "slow metabolizer" and would benefit from taking digestive enzymes with meals, but the Ca/P ratio is normal on the second and third tests.

[97] The medical term for this is mydriasis.

[98] The medical term is stocking and glove parasthesias.

HAIR ELEMENTS

THE GREAT PLAINS LABORATORY, INC

SEX: Female
AGE: 16

POTENTIALLY TOXIC ELEMENTS

TOXIC ELEMENTS	RESULT µg/g	REFERENCE RANGE
Aluminum	9.3	< 8.0
Antimony	0.012	< .066
Arsenic	0.024	< 0.08
Beryllium	< 0.01	< 0.02
Bismuth	0.008	< 0.1
Cadmium	0.12	< 0.1
Lead	0.3	< 1.0
Mercury	0.44	< 0.4
Platinum	< 0.003	< 0.005
Thallium	< 0.001	< 0.01
Thorium	< 0.001	< 0.005
Uranium	0.069	< 0.06
Nickel	1.5	< 0.4
Silver	0.56	< 0.15
Tin	0.28	< 0.3
Titanium	0.65	< 1.0
Total Toxic Representation		

Percentile: 68th, 95th

ESSENTIAL AND OTHER ELEMENTS

ELEMENTS	RESULT µg/g	REFERENCE RANGE
Calcium	2380	350- 1000
Magnesium	240	35- 120
Sodium	12	12- 90
Potassium	11	7.0- 36
Copper	49	12- 35
Zinc	170	130- 220
Manganese	0.85	0.15- 0.65
Chromium	0.76	0.2- 0.4
Vanadium	0.32	0.02- 0.075
Molybdenum	0.081	0.029- 0.058
Boron	1.1	0.3- 1.5
Iodine	0.93	0.25- 1.3
Lithium	0.009	0.007- 0.023
Phosphorus	172	160- 250
Selenium	0.75	0.95- 1.7
Strontium	3.7	0.86- 6.2
Sulfur	44400	44000- 52000
Barium	2.5	0.34- 2.5
Cobalt	0.04	0.013- 0.05
Iron	14	5.8- 15
Germanium	0.042	0.045- 0.065
Rubidium	< 0.003	0.006- 0.06
Zirconium	0.12	0.025- 0.5

Percentile: 2.5th, 16th, 50th, 84th, 97.5th

SPECIMEN DATA

COMMENTS:

Date Collected:
Date Received:
Date Completed: 10/16/2001
Methodology: ICP-MS

Sample Size:
Sample Type:
Hair Color:
Treatment:
Shampoo:

RATIOS

ELEMENTS	RATIOS	EXPECTED RANGE
Ca/Mg	10	4- 30
Ca/P	13.9	1- 12
Na/K	1.06	0.5- 10
Zn/Cu	3.5	4- 20
Zn/Cd	> 999	> 800

ANALYZED BY DOCTOR'S DATA, INC. CLIA ID NO: 14D0646470 LABORATORY DIRECTOR: James T. Hicks, MD, Ph.D., FCAP

This form is provided to Andrew Cutler to be used as an educational tool and does not constitute endorsement by *Doctor's Data*.

HAIR ELEMENTS

DOCTOR'S DATA INC.

SEX: Female
AGE: 17

POTENTIALLY TOXIC ELEMENTS

TOXIC ELEMENTS	RESULT µg/g	REFERENCE RANGE	PERCENTILE (68th, 95th)
Aluminum	2.4	[illegible]	
Antimony	0.026	[illegible]	
Arsenic	0.042	[illegible]	
Beryllium	< 0.01	< 0.02	
Bismuth	0.013	[illegible]	
Cadmium	0.069	[illegible]	
Lead	0.24	[illegible]	
Mercury	0.53	[illegible]	
Platinum	< 0.003	< 0.005	
Thallium	0.001	< 0.02	
Thorium	< 0.001	< 0.003	
Uranium	0.058	< 0.08	
Nickel	0.49	[illegible]	
Silver	0.12	< 0.16	
Tin	0.21	[illegible]	
Titanium	0.5	[illegible]	
Total Toxic Representation			

ESSENTIAL AND OTHER ELEMENTS

ELEMENTS	RESULT µg/g	REFERENCE RANGE	PERCENTILE (2.5th, 16th, 50th, 84th, 97.5th)
Calcium	1620	[illegible]	
Magnesium	97	[illegible]	
Sodium	11	[illegible]	
Potassium	< 3	[illegible]	
Copper	33	[illegible]	
Zinc	220	130- 220	
Manganese	0.02	[illegible]	
Chromium	0.39	0.2- 0.4	
Vanadium	0.077	[illegible]	
Molybdenum	0.071	[illegible]	
Boron	0.41	[illegible]	
Iodine	0.04	[illegible]	
Lithium	0.011	[illegible]	
Phosphorus	201	[illegible]	
Selenium	1.2	0.95- 1.7	
Strontium	4.8	[illegible]	
Sulfur	47600	44000- 52000	
Barium	1.6	[illegible]	
Cobalt	0.01	[illegible]	
Iron	16	[illegible]	
Germanium	0.05	0.045- 0.065	
Rubidium	< 0.003	[illegible]	
Zirconium	0.2	[illegible]	

SPECIMEN DATA

COMMENTS:

Date Collected
Date Received
Date Completed 6/13/2002
Methodology ICP-MS

Sample Size
Sample Type
Hair Color
Treatment
Shampoo

RATIOS

ELEMENTS	RATIOS	EXPECTED RANGE
Ca/Mg	16.7	[illegible]
Ca/P	8.06	[illegible]
Na/K	3.67	[illegible]
Zn/Cu	6.67	[illegible]
Zn/Cd	> 999	[illegible]

This form is provided to Andrew Cutler to be used as an educational tool and does not constitute endorsement by *Doctor's Data.*

HAIR ELEMENTS

DOCTOR'S DATA INC

SEX: Female
AGE: 18

POTENTIALLY TOXIC ELEMENTS

TOXIC ELEMENTS	RESULT µg/g	REFERENCE RANGE	PERCENTILE 68th 95th
Aluminum	2.8	< 7.0	
Antimony	0.016	< 0.05	
Arsenic	0.038	< 0.06	
Beryllium	< 0.01	< 0.02	
Bismuth	0.016	< 0.1	
Cadmium	0.048	< 0.1	
Lead	0.09	< 1.0	
Mercury	0.36	< 1.1	
Platinum	< 0.003	< 0.005	
Thallium	< 0.001	< 0.01	
Thorium	< 0.001	< 0.005	
Uranium	0.025	< 0.06	
Nickel	0.29	< 0.4	
Silver	0.13	< 0.15	
Tin	0.06	< 0.3	
Titanium	0.81	< 1.0	
Total Toxic Representation			

ESSENTIAL AND OTHER ELEMENTS

ELEMENTS	RESULT µg/g	REFERENCE RANGE	PERCENTILE 2.5th 16th 50th 84th 97.5th
Calcium	1840	300- 1200	
Magnesium	150	35- 120	
Sodium	24	12- 90	
Potassium	3.0	8.0- 38	
Copper	37	12- 35	
Zinc	220	140- 220	
Manganese	0.1	0.15- 0.65	
Chromium	0.31	0.2- 0.4	
Vanadium	0.071	0.018- 0.065	
Molybdenum	0.069	0.028- 0.056	
Boron	0.68	0.3- 2.0	
Iodine	0.49	0.25- 1.3	
Lithium	0.013	0.007- 0.023	
Phosphorus	213	160- 250	
Selenium	1.3	0.95- 1.7	
Strontium	4.4	0.5- 7.6	
Sulfur	51000	44500- 52000	
Barium	1.4	0.26- 3.0	
Cobalt	0.032	0.013- 0.05	
Iron	8.5	5.4- 14	
Germanium	0.049	0.045- 0.065	
Rubidium	0.005	0.007- 0.096	
Zirconium	0.23	0.02- 0.42	

SPECIMEN DATA

COMMENTS:

Date Collected:
Date Received:
Date Completed: 2/10/2003

Methodology: ICP-MS

Sample Size:
Sample Type:
Hair Color:
Treatment:
Shampoo:

V06.99

RATIOS

ELEMENTS	RATIOS	EXPECTED RANGE
Ca/Mg	12.3	4- 30
Ca/P	8.64	1- 12
Na/K	8	0.5- 10
Zn/Cu	5.95	4- 20
Zn/Cd	> 999	> 800

This form is provided to Andrew Cutler to be used as an educational tool and does not constitute endorsement by *Doctor's Data.*

Hair Tests do improve with Chelation

As you saw in young woman's case above, her hair test changed around but didn't improve how it counted on assorted inteventions that don't help remove heavy mteals, but then improved once she started to chelate. This section provides several examples of before and after chelation tests.

The 3 year old boy whose hair tests are presented on pages 180 and 181 developed normally until 13 months of age at which point he reacted badly to his MMR vaccine, lost the words he had been using, and did not begin talking again at all until age 3. Following the MMR reaction, he had poor eye contact, did not point, wave or clap and did not respond to simple directions. He also had growth failure during his second year following the MMR reaction. He has a diagnosis of speech and language disorder.

The parents chelated this boy on the 3-4 hour protocol, first using DMSA only for four cycles, then using DMSA + ALA. The parents immediately noticed improved eye contact, more attempts at speech and some hyperactivity. With continuing chelation much improvement in all areas was seen. After 11 months of chelation a second hair test was performed. Note the normal mineral transport in the essential element section as well as the greatly reduced level of toxic elements. The hair tests count as follows:

	Test 1, page 183 before chelation	Test 2, page 184 after chelation
Counting rule #1: bars to right.	16	15
Counting rule #2: bars to left.	7	8
Counting rule #3: bars in red.	2	1
Counting rule #4: bars in middle	11	15
Counting rule #5: is not met by either test.		

The first test meets counting rule 4. The second test's essential elements section does not look uncommon.

Note that the first test had apparently very elevated antimony, arsenic, lead, silver and titanium. This is an effect of deranged mineral transport, his body did not actually contain vast amounts of all these metals. The antimony and lead may have been real – lead is not falsely elevated very often by deranged mineral transport (and chelation would remove some), while antimony goes up at least in part because it is retained during deranged mineral transport.

Titanium and zirconium go up together in deranged mineral transport as these have quite similar chemistry, thus the zirconium and titanium probably are not actually present in his body at vastly elevated levels. They are among the first elements to respond to mineral transport derangement so it is not uncommon to see them very elevated even with orderly normal mineral transport when someone is being detoxified.

HAIR ELEMENTS

DOCTOR'S DATA INC.

SEX: Male
AGE: 2

POTENTIALLY TOXIC ELEMENTS

TOXIC ELEMENTS	RESULT μg/g	REFERENCE RANGE	PERCENTILE 68th 95th
Aluminum	14	< 8.0	
Antimony	0.44	< 0.066	
Arsenic	0.26	< 0.08	
Beryllium	< 0.01	< 0.02	
Bismuth	0.19	< 0.13	
Cadmium	0.11	< 0.15	
Lead	4.9	< 1.0	
Mercury	0.31	< 0.4	
Platinum	< 0.003	< 0.005	
Thallium	< 0.001	< 0.01	
Thorium	< 0.001	< 0.005	
Uranium	0.022	< 0.06	
Nickel	0.34	< 0.4	
Silver	1.9	< 0.2	
Tin	0.58	< 0.3	
Titanium	3.9	< 1.0	
Total Toxic Representation			

ESSENTIAL AND OTHER ELEMENTS

ELEMENTS	RESULT μg/g	REFERENCE RANGE	PERCENTILE 2.5th 16th 50th 84th 97.5th
Calcium	218	125- 370	
Magnesium	9.0	12- 30	
Sodium	160	12- 90	
Potassium	220	12- 40	
Copper	17	8.0- 16	
Zinc	64	100- 190	
Manganese	0.12	0.2- 0.55	
Chromium	0.57	0.26- 0.5	
Vanadium	0.091	0.03- 0.1	
Molybdenum	0.16	0.05- 0.13	
Boron	5.6	0.6- 4.0	
Iodine	2.2	0.25- 1.3	
Lithium	0.017	0.007- 0.023	
Phosphorus	159	160- 250	
Selenium	1.1	0.95- 1.7	
Strontium	0.69	0.16- 1.0	
Sulfur	49200	45500- 53000	
Barium	0.22	0.16- 0.8	
Cobalt	0.02	0.013- 0.035	
Iron	11	8.0- 19	
Germanium	0.064	0.045- 0.065	
Rubidium	0.18	0.016- 0.18	
Zirconium	6.2	0.04- 1.0	

SPECIMEN DATA

COMMENTS:

Date Collected:
Date Received: 9/3/2002
Date Completed:

Methodology: ICP-MS

Sample Size:
Sample Type:
Hair Color:
Treatment:
Shampoo:

V06.99

RATIOS

ELEMENTS	RATIOS	EXPECTED RANGE
Ca/Mg	24.2	4- 30
Ca/P	1.37	0.8- 8
Na/K	0.727	0.5- 10
Zn/Cu	3.76	4- 20
Zn/Cd	582	> 800

This form is provided to Andrew Cutler to be used as an educational tool and does not constitute endorsement by *Doctor's Data*.

HAIR ELEMENTS

DOCTOR'S DATA INC

SEX: Male
AGE: 3

POTENTIALLY TOXIC ELEMENTS

TOXIC ELEMENTS	RESULT µg/g	REFERENCE RANGE	PERCENTILE 68th 95th
Aluminum	16	< 8.0	
Antimony	0.21	< 0.066	
Arsenic	0.076	< 0.08	
Beryllium	< 0.01	< 0.02	
Bismuth	0.23	< 0.13	
Cadmium	0.063	< 0.15	
Lead	2.3	< 1.0	
Mercury	0.3	< 0.4	
Platinum	< 0.003	< 0.005	
Thallium	< 0.001	< 0.01	
Thorium	< 0.001	< 0.005	
Uranium	0.043	< 0.06	
Nickel	0.3	< 0.4	
Silver	0.4	< 0.2	
Tin	0.54	< 0.3	
Titanium	1.7	< 1.0	
Total Toxic Representation			

ESSENTIAL AND OTHER ELEMENTS

ELEMENTS	RESULT µg/g	REFERENCE RANGE	PERCENTILE 2.5th 16th 50th 84th 97.5th
Calcium	236	125- 370	
Magnesium	15	12- 30	
Sodium	41	12- 90	
Potassium	54	12- 40	
Copper	13	8.0- 16	
Zinc	130	100- 190	
Manganese	0.14	0.2- 0.55	
Chromium	0.37	0.26- 0.5	
Vanadium	0.037	0.03- 0.1	
Molybdenum	0.1	0.05- 0.13	
Boron	3.1	0.6- 4.0	
Iodine	1.8	0.25- 1.3	
Lithium	0.012	0.007- 0.023	
Phosphorus	206	160- 250	
Selenium	0.75	0.95- 1.7	
Strontium	1.4	0.16- 1.0	
Sulfur	51900	45500- 53000	
Barium	0.32	0.16- 0.8	
Cobalt	0.01	0.013- 0.035	
Iron	8.8	8.0- 19	
Germanium	0.035	0.045- 0.065	
Rubidium	0.049	0.016- 0.18	
Zirconium	3.1	0.04- 1.0	

SPECIMEN DATA

COMMENTS:

Date Collected:
Date Received: 8/18/2003
Date Completed:
Methodology: ICP-MS

Sample Size:
Sample Type:
Hair Color:
Treatment:
Shampoo:

V06.99

RATIOS

ELEMENTS	RATIOS	EXPECTED RANGE
Ca/Mg	15.7	4- 30
Ca/P	1.15	0.8- 8
Na/K	0.759	0.5- 10
Zn/Cu	10	4- 20
Zn/Cd	> 999	> 800

This form is provided to Andrew Cutler to be used as an educational tool and does not constitute endorsement by *Doctor's Data*.

This boy was diagnosed with autism at 15 months of age and tested out at the six month level for social skills, expressive and receptive language at that point. He had received many mercury containing vaccines and the mother had received two Rhogam shots shortly before conception, three during pregnancy and one shortly after birth. She also received a mercury containing flu shot during pregnancy. The mother, who has a mouthful of mercury (silver amalgam) dental fillings breast fed her son.

The boy was born with severe congenital clubfeet which were operated on at 8 months of age. He also has congenital hydronephrosis which is being carefully watched but appears to be diminishing on its own.

The boy had no eye contact. He was fascinated by wheels and spinning objects and would watch them for long periods of time. He showed little interest in the world beyond his reach. He often made strange shrieking noises. He became hysterical when he heard certain sounds such as blenders and coffee grinders. He had low muscle tone, poor balance and was uncoordinated. He was very difficult to feed because he disliked many foods.

Immediately after diagnosis the boy was placed on a gluten free, casein free, soy free (GFCFSF) diet and foods identified by blood test as causing IgG mediated reactions were also removed. Testing showed many nutritional deficiencies and also that his thyroid hormone levels were low, as were his NK immune system cells. He was placed on a long list of nutritional supplements in addition to the dietary restrictions and this was quite helpful.

Hair testing was performed in order to check for heavy metal problems. The results of four tests are tabulated below, and the tests are shown on the following pages.

	First test page 186	Second test page 187	Third test page 188	Fourth test page 189
Counting rule #1: bars to right.	20	19	20	7
Counting rule #2: bars to left.	3	4	3	16
Counting rule #3: bars in red.	4	4	4	2
Counting rule #4: bars in middle	7	12	13	17
Counting rules met	2, 3, 4	2, 3	2, 3	none

There were actually five hair tests done. One of them appears to have been an error where some other little boy's hair was switched with this one's sample. The test that I left out would come between test 1 and test 2. The omitted test looked absolutely perfectly normal. Since the other four tests tell a consistent story and track what was happening with the boy the fifth test clearly was a mistake of some sort. Mistakes happen – often before the sample shows up at the laboratory with the wrong label on it. Because there is always the slight chance[99] of a mistake it is important to always consider whether a test result makes sense in light of everything else that is going on.

The first test was done using hair from his first baby haircut.

[99] While I believe the overall rate of mistakes involving misidentifying samples is less than 1%, I suggest using a 5% incidence because most of the possibilities for mixing up samples occurs before they are sent to the laboratory and probably comes from a few health care practitioners offices where adequate quality control measures to prevent this have not yet been implemented. Since there is no practical way to know if your sample was handled someplace where confusion is common unless you are the health care provider and start seeing this frequently, the best assumption is to use a high percentage error rate when making decisions. This may lead you to spend some money getting an extra test to be sure of things when a hair test comes back otherwise than you expect, but this is a small cost compared to months or years of avoidable suffering if the test comes back wrong and you believe it.

HAIR ELEMENTS

DOCTOR'S DATA INC.

SEX: Male
AGE: 1

POTENTIALLY TOXIC ELEMENTS

TOXIC ELEMENTS	RESULT µg/g	REFERENCE RANGE
Aluminum	45	< 8.0
Antimony	0.39	< 0.066
Arsenic	0.041	< 0.08
Beryllium	< 0.01	< 0.02
Bismuth	1.5	< 0.13
Cadmium	0.32	< 0.15
Lead	4.8	< 1.0
Mercury	1.1	< 0.4
Platinum	< 0.003	< 0.005
Thallium	< 0.001	< 0.01
Thorium	< 0.001	< 0.005
Uranium	0.45	< 0.06
Nickel	1.1	< 0.4
Silver	0.7	< 0.2
Tin	1.7	< 0.3
Titanium	3.1	< 1.0
Total Toxic Representation		

PERCENTILE: 68th, 95th

ESSENTIAL AND OTHER ELEMENTS

ELEMENTS	RESULT µg/g	REFERENCE RANGE
Calcium	626	125- 370
Magnesium	66	12- 30
Sodium	49	12- 90
Potassium	78	12- 40
Copper	23	8.0- 16
Zinc	170	100- 190
Manganese	0.66	0.2- 0.55
Chromium	0.53	0.26- 0.5
Vanadium	0.33	0.03- 0.1
Molybdenum	0.2	0.05- 0.13
Boron	12	0.6- 4.0
Iodine	31	0.25- 1.3
Lithium	0.022	0.007- 0.023
Phosphorus	140	160- 250
Selenium	1.2	0.95- 1.7
Strontium	5.7	0.16- 1.0
Sulfur	47300	45500- 53000
Barium	2.5	0.16- 0.8
Cobalt	0.058	0.013- 0.035
Iron	25	8.0- 19
Germanium	0.068	0.045- 0.065
Rubidium	0.085	0.016- 0.18
Zirconium	0.48	0.04- 1.0

PERCENTILE: 2.5th, 16th, 50th, 84th, 97.5th

SPECIMEN DATA

COMMENTS:

Date Collected: 4/25/2002
Date Received:
Date Completed:
Methodology: ICP-MS

Sample Size:
Sample Type:
Hair Color:
Treatment:
Shampoo:

V06 99

RATIOS

ELEMENTS	RATIOS	EXPECTED RANGE
Ca/Mg	9.48	4- 30
Ca/P	4.47	0.8- 8
Na/K	0.628	0.5- 10
Zn/Cu	7.39	4- 20
Zn/Cd	531	> 800

This form is provided to Andrew Cutler to be used as an educational tool and does not constitute endorsement by *Doctor's Data.*

HAIR ELEMENTS

DOCTOR'S DATA INC

SEX: Male
AGE: 2

POTENTIALLY TOXIC ELEMENTS

TOXIC ELEMENTS	RESULT µg/g	REFERENCE RANGE
Aluminum	17.0	< 8.0
Antimony	0.280	< 0.066
Arsenic	0.140	< 0.080
Beryllium	< 0.01	< 0.020
Bismuth	0.820	< 0.130
Cadmium	0.160	< 0.150
Lead	1.40	< 1.00
Mercury	0.28	< 0.40
Platinum	< 0.003	< 0.005
Thallium	< 0.001	< 0.010
Thorium	< 0.001	< 0.005
Uranium	0.085	< 0.060
Nickel	0.35	< 0.40
Silver	0.17	< 0.20
Tin	1.00	< 0.30
Titanium	1.50	< 1.00
Total Toxic Representation		

PERCENTILE: 68th, 95th

ESSENTIAL AND OTHER ELEMENTS

ELEMENTS	RESULT µg/g	REFERENCE RANGE
Calcium	235	125- 370
Magnesium	31	12- 30
Sodium	280	12- 90
Potassium	800	12- 40
Copper	10.0	8.0- 16.0
Zinc	61	100- 190
Manganese	0.35	0.20- 0.55
Chromium	0.39	0.26- 0.50
Vanadium	0.150	0.030- 0.100
Molybdenum	0.170	0.050- 0.130
Boron	15.00	0.60- 4.00
Iodine	3.70	0.25- 1.30
Lithium	0.028	0.007- 0.023
Phosphorus	228	160- 250
Selenium	1.00	0.95- 1.70
Strontium	0.53	0.16- 1.00
Sulfur	45100	45500- 53000
Barium	0.42	0.16- 0.80
Cobalt	0.032	0.013- 0.035
Iron	22.0	8.0- 19.0
Germanium	0.060	0.045- 0.065
Rubidium	1.400	0.016- 0.180
Zirconium	0.520	0.040- 1.000

PERCENTILE: 2.5th, 16th, 50th, 84th, 97.5th

SPECIMEN DATA

COMMENTS:

Date Collected: 10/18/2003
Date Received:
Date Completed:
Methodology: ICP-MS

Sample Size:
Sample Type:
Hair Color:
Treatment:
Shampoo:

V06.99

RATIOS

ELEMENTS	RATIOS	EXPECTED RANGE
Ca/Mg	7.58	4- 30
Ca/P	1.03	0.8- 8
Na/K	0.35	0.5- 10
Zn/Cu	6.1	4- 20
Zn/Cd	381	> 800

This form is provided to Andrew Cutler to be used as an educational tool and does not constitute endorsement by *Doctor's Data.*

HAIR ELEMENTS

DOCTOR'S DATA INC.

SEX: Male
AGE: 2

POTENTIALLY TOXIC ELEMENTS

TOXIC ELEMENTS	RESULT µg/g	REFERENCE RANGE	PERCENTILE (68th, 95th)
Aluminum	13	< 8.0	
Antimony	0.29	< 0.066	
Arsenic	0.13	< 0.080	
Beryllium	< 0.01	< 0.020	
Bismuth	0.21	< 0.13	
Cadmium	0.21	< 0.15	
Lead	1.1	< 1.0	
Mercury	0.14	< 0.40	
Platinum	< 0.003	< 0.005	
Thallium	0.002	< 0.010	
Thorium	< 0.001	< 0.005	
Uranium	0.041	< 0.060	
Nickel	0.40	< 0.40	
Silver	0.09	< 0.20	
Tin	0.68	< 0.30	
Titanium	1.5	< 1.0	
Total Toxic Representation			

ESSENTIAL AND OTHER ELEMENTS

ELEMENTS	RESULT µg/g	REFERENCE RANGE	PERCENTILE (2.5th, 16th, 50th, 84th, 97.5th)
Calcium	251	125- 370	
Magnesium	28	12- 30	
Sodium	350	12- 90	
Potassium	490	12- 40	
Copper	9.9	8.0- 16	
Zinc	120	100- 190	
Manganese	0.39	0.20- 0.55	
Chromium	0.37	0.26- 0.50	
Vanadium	0.15	0.030- 0.10	
Molybdenum	0.14	0.050- 0.13	
Boron	38	0.60- 4.0	
Iodine	1.4	0.25- 1.3	
Lithium	0.036	0.007- 0.023	
Phosphorus	295	160- 250	
Selenium	1.2	0.95- 1.7	
Strontium	0.70	0.16- 1.0	
Sulfur	51100	45500- 53000	
Barium	0.58	0.16- 0.80	
Cobalt	0.030	0.013- 0.035	
Iron	19	8.0- 19	
Germanium	0.067	0.045- 0.065	
Rubidium	0.70	0.016- 0.18	
Zirconium	0.32	0.040- 1.0	

SPECIMEN DATA

COMMENTS:

Date Collected: 1/21/2004
Date Received:
Date Completed:

Methodology: ICP-MS

Sample Size:
Sample Type:
Hair Color:
Treatment:
Shampoo:

V06.99

RATIOS

ELEMENTS	RATIOS	EXPECTED RANGE
Ca/Mg	8.96	4- 30
Ca/P	0.851	0.8- 8
Na/K	0.714	0.5- 10
Zn/Cu	12.1	4- 20
Zn/Cd	571	> 800

This form is provided to Andrew Cutler to be used as an educational tool and does not constitute endorsement by *Doctor's Data.*

HAIR ELEMENTS

DOCTOR'S DATA INC.

SEX: Male
AGE: 2

POTENTIALLY TOXIC ELEMENTS

TOXIC ELEMENTS	RESULT µg/g	REFERENCE RANGE	PERCENTILE (68th, 95th)
Aluminum	8.2	< 8.0	
Antimony	0.13	< 0.066	
Arsenic	0.11	< 0.080	
Beryllium	< 0.01	< 0.020	
Bismuth	0.54	< 0.13	
Cadmium	0.089	< 0.15	
Lead	0.77	< 1.0	
Mercury	0.23	< 0.40	
Platinum	< 0.003	< 0.005	
Thallium	< 0.001	< 0.010	
Thorium	< 0.001	< 0.005	
Uranium	0.038	< 0.060	
Nickel	0.20	< 0.40	
Silver	0.07	< 0.20	
Tin	0.54	< 0.30	
Titanium	1.9	< 1.0	
Total Toxic Representation			

ESSENTIAL AND OTHER ELEMENTS

ELEMENTS	RESULT µg/g	REFERENCE RANGE	PERCENTILE (2.5th, 16th, 50th, 84th, 97.5th)
Calcium	248	125- 370	
Magnesium	32	12- 30	
Sodium	200	12- 90	
Potassium	340	12- 40	
Copper	8.8	8.0- 16	
Zinc	140	100- 190	
Manganese	0.25	0.20- 0.55	
Chromium	0.40	0.26- 0.50	
Vanadium	0.12	0.030- 0.10	
Molybdenum	0.11	0.050- 0.13	
Boron	13	0.60- 4.0	
Iodine	1.3	0.25- 1.3	
Lithium	0.020	0.007- 0.023	
Phosphorus	193	160- 250	
Selenium	1.1	0.95- 1.7	
Strontium	0.58	0.16- 1.0	
Sulfur	46700	45500- 53000	
Barium	0.32	0.16- 0.80	
Cobalt	0.015	0.013- 0.035	
Iron	12	8.0- 19	
Germanium	0.053	0.045- 0.065	
Rubidium	0.37	0.016- 0.18	
Zirconium	0.30	0.040- 1.0	

SPECIMEN DATA

COMMENTS:

Date Collected: 4/28/2004
Date Received:
Date Completed:

Methodology: ICP-MS

Sample Size:
Sample Type:
Hair Color:
Treatment:
Shampoo:

V06.99

RATIOS

ELEMENTS	RATIOS	EXPECTED RANGE
Ca/Mg	7.75	4- 30
Ca/P	1.28	0.8- 8
Na/K	0.588	0.5- 10
Zn/Cu	15.9	4- 20
Zn/Cd	> 999	> 800

This form is provided to Andrew Cutler to be used as an educational tool and does not constitute endorsement by *Doctor's Data*.

The second test is just before chelation started, but after a number of other interventions including a GFCF diet and a long list of nutritional supplements were begun.

The third test was after two months of chelation.

The fourth test was after five months of chelation.

It is quite apparent from the hair tests that this boy has mercury poisoning, not autism. Or to put it another way, autism is a good description of the symptoms that mercury poisoning is causing the boy to exhibit.

Chelation was with DMPS and ALA on an every 3 hour protocol, with two initial cycles using DMSA instead of DMPS.

Two other therapies that caused immediate positive responses were frequent injections of methylcobalamin (the biologically active form of vitamin B-12) in large quantities, and high doses of vitamin A. The B-12 increased his babbling. On adding the vitamin A (with a loading dose of 4,000 IU per pound given once, followed by 400 IU per pound daily thereafter) his vocabulary went from 10 words to 150 words in a two week period.

Other things that can be seen on the tests are:

The second test has an abnormal sodium to potassium (Na/K) ratio, suggesting thyroid problems. This was about the time that blood testing independently determined that the boy's thyroid hormone levels were too low.

Note how the "all high" presentation of the first test is reflected in the great elevation of many toxic metals. Over tests 2, 3, and 4 as the essential elements become more normal the toxic elements also recede, until by the time of the fourth test both the toxic and essential element portions of the test look relatively normal.

This developmentally disordered (but not autistic) girl has a complicated history and situation. She is of very superior intelligence, has some deficit of social perception, has assorted mild "autism spectrum" symptoms that don't add up to a diagnosis, and had some congenital abnormalities of the lungs and heart at birth which were corrected by using a catheter to form blood clots in selected locations and letting some lung tissue die. She often walks on her toes and has tight heel cords[100]. She is the birth daughter of a mother with bipolar (manic depressive) disorder and was adopted at birth. The birth mother had active genital herpes at the time of birth and the daughter's condition improves when she is given antiviral medications. The adoptive mother pursued a very wide variety of interventions of all sorts for her daughter which led to great overall improvement. Hair testing was done at various times. The two tests presented here were before chelation and after several months of chelation. The tests are face to face on the next two pages and they count as follows:

	First test (pg 191)	Second test (pg 192)
Counting rule #1: bars to right.	5	12
Counting rule #2: bars to left.	18	11
Counting rule #3: bars in red.	2	0
Counting rule #4: bars in middle	15	16
Counting rule #5: is not met by either test.		

The first test meets counting rule #1. The second test counts out as an ordinary and average test. There was intensive chelation with DMSA and alpha lipoic acid on a 3 hour schedule between the two tests.

On the first test the ratio of calcium to magnesium (Ca/Mg) was abnormal, indicating problems metabolizing carbohydrates which were entirely apparent to her parents from her behavior after eating sugary treats.

It is also worth nothing that the birth mother's hair test showed deranged mineral transport suggesting that mercury poisoning may underlie her bipolar

[100] Toe walking is common in autistic children.

HAIR ELEMENTS

DOCTOR'S DATA INC.

SEX: Female
AGE: 8

POTENTIALLY TOXIC ELEMENTS

TOXIC ELEMENTS	RESULT µg/g	REFERENCE RANGE	PERCENTILE (68th, 95th)
Aluminum	9.5	< 8.0	
Antimony	0.032	< 0.066	
Arsenic	0.054	< 0.06	
Beryllium	< 0.01	< 0.02	
Bismuth	0.002	< 0.12	
Cadmium	0.012	< 0.1	
Lead	0.13	< 1.0	
Mercury	0.48	< 0.4	
Platinum	< 0.003	< 0.005	
Thallium	< 0.001	< 0.01	
Thorium	< 0.001	< 0.005	
Uranium	0.057	< 0.06	
Nickel	0.12	< 0.4	
Silver	0.11	< 0.16	
Tin	0.18	< 0.3	
Titanium	0.58	< 1.0	
Total Toxic Representation			

ESSENTIAL AND OTHER ELEMENTS

ELEMENTS	RESULT µg/g	REFERENCE RANGE	PERCENTILE (2.5th, 16th, 50th, 84th, 97.5th)
Calcium	288	250- 800	
Magnesium	360	25- 90	
Sodium	18	12- 90	
Potassium	31	7.0- 40	
Copper	39	12- 35	
Zinc	190	120- 220	
Manganese	0.07	0.18- 0.6	
Chromium	0.29	0.23- 0.45	
Vanadium	0.03	0.025- 0.1	
Molybdenum	0.032	0.037- 0.083	
Boron	0.3	0.3- 1.7	
Iodine	0.28	0.25- 1.3	
Lithium	0.008	0.007- 0.023	
Phosphorus	165	160- 250	
Selenium	0.91	0.95- 1.7	
Strontium	1.2	0.37- 3.6	
Sulfur	45300	45500- 53000	
Barium	0.34	0.21- 1.9	
Cobalt	0.01	0.013- 0.05	
Iron	7.7	6.0- 17	
Germanium	0.029	0.045- 0.065	
Rubidium	0.009	0.008- 0.08	
Zirconium	0.4	0.03- 0.4	

SPECIMEN DATA

COMMENTS:
Date Collected:
Date Received:
Date Completed: 10/15/2002
Methodology: ICP-MS

Sample Size:
Sample Type:
Hair Color:
Treatment:
Shampoo:

V06.99

RATIOS

ELEMENTS	RATIOS	EXPECTED RANGE
Ca/Mg	0.8	4- 30
Ca/P	1.75	1- 12
Na/K	0.581	0.5- 10
Zn/Cu	4.87	4- 20
Zn/Cd	> 999	> 800

This form is provided to Andrew Cutler to be used as an educational tool and does not constitute endorsement by *Doctor's Data.*

HAIR ELEMENTS

DOCTOR'S DATA INC.

SEX: Female
AGE: 8

POTENTIALLY TOXIC ELEMENTS

TOXIC ELEMENTS	RESULT µg/g	REFERENCE RANGE
Aluminum	5.0	< 8.0
Antimony	0.058	< 0.066
Arsenic	0.026	< 0.06
Beryllium	< 0.01	< 0.02
Bismuth	0.006	< 0.12
Cadmium	0.038	< 0.1
Lead	0.15	< 1.0
Mercury	0.17	< 0.4
Platinum	< 0.003	< 0.005
Thallium	< 0.001	< 0.01
Thorium	< 0.001	< 0.005
Uranium	0.083	< 0.06
Nickel	0.26	< 0.4
Silver	0.05	< 0.16
Tin	0.09	< 0.3
Titanium	0.53	< 1.0
Total Toxic Representation		

ESSENTIAL AND OTHER ELEMENTS

ELEMENTS	RESULT µg/g	REFERENCE RANGE
Calcium	1090	250- 800
Magnesium	86	25- 90
Sodium	16	12- 90
Potassium	7.0	7.0- 40
Copper	18	12- 35
Zinc	260	120- 220
Manganese	0.11	0.18- 0.6
Chromium	0.23	0.23- 0.45
Vanadium	0.023	0.025- 0.1
Molybdenum	0.037	0.037- 0.083
Boron	0.37	0.3- 1.7
Iodine	0.2	0.25- 1.3
Lithium	0.013	0.007- 0.023
Phosphorus	212	160- 250
Selenium	1.4	0.95- 1.7
Strontium	4.2	0.37- 3.6
Sulfur	50900	45500- 53000
Barium	0.96	0.21- 1.9
Cobalt	0.019	0.013- 0.05
Iron	11	6.0- 17
Germanium	0.067	0.045- 0.065
Rubidium	0.006	0.008- 0.08
Zirconium	0.17	0.03- 0.4

SPECIMEN DATA

COMMENTS:

Date Collected:
Date Received:
Date Completed: 5/17/2003
Methodology: ICP-MS

Sample Size:
Sample Type:
Hair Color:
Treatment:
Shampoo:

V06.99

RATIOS

ELEMENTS	RATIOS	EXPECTED RANGE
Ca/Mg	12.7	4- 30
Ca/P	5.14	1- 12
Na/K	2.29	0.5- 10
Zn/Cu	14.4	4- 20
Zn/Cd	> 999	> 800

This form is provided to Andrew Cutler to be used as an educational tool and does not constitute endorsement by *Doctor's Data.*

disorder[101] and that her daughter may have inherited a propensity to mineral transport derangement at low mercury levels.

Hair Testing and Shampoo

This gentleman has anxiety and a recurring depression characterized by hopeless, helpless despair. He also has a loss of short term memory, difficulty concentrating, low energy, he does not find sleep refreshing, he has chronic sinus problems and postnasal drip apparently due to allergy, frequent respiratory infections and constipation. In an attempt to understand the cause of these conditions in order to correct them he had a hair test done. It is on the next page and counts as follows:

Counting rule #1: 10 bars go right. This does not meet rule #1.

Counting rule #2: 13 bars go left. This does not meet counting rule #2.

Counting rule #3: 5 bars go into the red bands. This meets counting rule #3.

Counting rule #4: 11 bars are contained entirely in the middle green + white stripe. This meets counting rule #4.

Counting rule #5 is irrelevant since two other rules are met.

Now you are probably wondering about that typo where I counted 5 in the red since you count 6. One of the bars is vastly elevated selenium and this is usually a contaminant due to certain shampoo ingredients. This gentleman did fill in the test order form thoroughly and listed the shampoo he uses – Selsun Blue®. "Selsun" is a contraction of selenium sulfide, the active ingredient in this and some other dandruff shampoos. The vast elevation of selenium is due to external contamination from shampoo and can not be assumed to relate to body burden – not should it be counted as an abnormal for the counting rules.

The relation of sodium and potassium to calcium and magnesium suggests the possibility of adrenal problems. In fact this gentleman has substantial problems with anxiety, obsession, allergy and environmental reactivity which are consistent with adrenal problems, and he does much better when taking adrenal support supplements plus hydrocortisone and DHEA.

Note the relation of calcium to potassium. This suggests the possibility of thyroid problems.

The low ratio of calcium to phosphorus (Ca/P) suggests that taking digestive enzymes with meals would be helpful.

---

[101] Psychometric Evidence that Dental Amalgam Mercury may be an Etiological Factor in Manic Depression, Siblerud, Motl and Kienholz, *The Journal of Orthomolecular Medicine*, volume **13** pages 31-40, 1998.

HAIR ELEMENTS

DOCTOR'S DATA INC.

SEX: Male
AGE: 54

POTENTIALLY TOXIC ELEMENTS

TOXIC ELEMENTS	RESULT µg/g	REFERENCE RANGE
Aluminum	1.8	< 7.0
Antimony	0.074	< 0.066
Arsenic	0.025	< 0.08
Beryllium	< 0.01	< 0.02
Bismuth	0.049	< 0.06
Cadmium	0.029	< 0.15
Lead	0.39	< 2.0
Mercury	0.03	< 1.1
Platinum	< 0.003	< 0.005
Thallium	< 0.001	< 0.01
Thorium	< 0.001	< 0.005
Uranium	0.027	< 0.06
Nickel	0.15	< 0.4
Silver	0.04	< 0.12
Tin	0.69	< 0.3
Titanium	1.3	< 1.0
Total Toxic Representation		

ESSENTIAL AND OTHER ELEMENTS

ELEMENTS	RESULT µg/g	REFERENCE RANGE
Calcium	147	200- 750
Magnesium	9.0	25- 75
Sodium	88	12- 90
Potassium	74	9.0- 40
Copper	9.8	10- 28
Zinc	160	130- 200
Manganese	0.05	0.15- 0.65
Chromium	0.26	0.2- 0.4
Vanadium	0.027	0.018- 0.065
Molybdenum	0.051	0.025- 0.064
Boron	1.7	0.4- 3.0
Iodine	1.2	0.25- 1.3
Lithium	< 0.004	0.007- 0.023
Phosphorus	205	160- 250
Selenium	12	0.95- 1.7
Strontium	0.17	0.3- 3.5
Sulfur	48800	44500- 52000
Barium	0.11	0.16- 1.6
Cobalt	0.003	0.013- 0.035
Iron	6.2	5.4- 13
Germanium	0.019	0.045- 0.065
Rubidium	0.07	0.011- 0.12
Zirconium	0.54	0.02- 0.44

SPECIMEN DATA

COMMENTS:

Date Collected:
Date Received:
Date Completed:

Methodology: ICP-MS

Sample Size:
Sample Type:
Hair Color:
Treatment:
Shampoo: Selsun Blue

V06.99

RATIOS

ELEMENTS	RATIOS	EXPECTED RANGE
Ca/Mg	16.3	4- 30
Ca/P	0.717	0.8- 8
Na/K	1.19	0.5- 10
Zn/Cu	16.3	4- 20
Zn/Cd	> 999	> 800

This form is provided to Andrew Cutler to be used as an educational tool and does not constitute endorsement by *Doctor's Data.*

Ambiguous Hair Tests can be Confirmed by other Means

This is the case of a boy with a diagnosis of autism. His autism was severe, he was non-verbal and self abusive. He had very little receptive language, no expressive language, and many sensory problems. The parents got him a hair test to try to decide if he really had mercury poisoning instead of autism. The test is on the next page and it counts as follows:

Counting rule #1: 13 bars go right. This does not meet rule #1.

Counting rule #2: 10 bars go left. This does not meet counting rule #2.

Counting rule #3: 1 bar goes into the red band. This does not meet counting rule #3.

Counting rule #4: 16 bars are contained entirely in the middle green + white stripe. This does not meet counting rule #4.

Counting rule #5: is not met.

For bars that reach the red – rule #3 – one definitely goes in. Two more are so close it is hard to tell. And another two are REALLY close. But to be a statistically valid approach the counting rules have to be applied rigidly, either the bar makes it in or not. Close doesn't count, even so close you can't be sure if it is in or not. This is a situation where common sense is more important than robot like adherence to procedure. The test counts normal but does not look normal.

The vastly elevated antimony could be the problem, or it could be there because this child has a lot of mercury and that makes his body hold on to antimony.

Even though the test does not meet any counting rule and in fact by count looks ordinary and average, it very clearly does NOT look ordinary and average to the eye. The counting rules need to be applied properly (it is either in or it isn't, close doesn't matter) to be statistically valid, but they should not be applied thoughtlessly.

The parents and doctor chose to do a urine test to help remove the ambiguity from the hair test. The urine test is on page 197. As you can see, mercury came back very elevated while antimony came back normal. This does confirm that the boy has too much mercury, which is causing deranged mineral transport and making his body hold on to antimony. Also note that some children with deranged mineral transport could have low urine mercury results – in this kind of situation it is sometimes necessary to do several different tests in a row before getting a clear cut answer. For most people a hair test is enough. For this boy it took two tests, a hair test then a urine test.

Note that the hair test does show very low calcium and magnesium compared to sodium and potassium, indicating excessive levels of stress. This information alone strongly suggests that whatever is causing the boy's developmental problems is a disease process that is hard on his body rather than some psychological problem or neurological issue due to improper development of the brain. This information alone is very helpful in clarifying that further medical testing is both appropriate and necessary to identify the underlying problem so it can be corrected.

Based on the results of the hair and urine tests, the parents tried chelation with alpha lipoic acid (ALA) and DMSA on a 3 hour schedule. After 60 weekend rounds of chelation his autism is now moderate rather than severe, his receptive language has exploded and he does have some very limited speech. He still bites himself when angry or frustrated but self injurious behavior is greatly reduced. All of his sensory problems are gone except for some sound sensitivity. His tantrums are almost gone. He is now toilet trained and stays dry at night, which was not something his doctors thought would ever be possible when he was first diagnosed.

HAIR ELEMENTS

06/99

DOCTOR'S DATA INC.

SEX: Male
AGE: 3

POTENTIALLY TOXIC ELEMENTS

TOXIC ELEMENTS	RESULT µg/g	REFERENCE RANGE	PERCENTILE 68th 95th
Aluminum	10	< 8.0	
Antimony	0.92	< 0.066	
Arsenic	0.12	< 0.08	
Beryllium	< 0.01	< 0.02	
Bismuth	0.14	< 0.13	
Cadmium	0.059	< 0.15	
Lead	0.33	< 1.0	
Mercury	< 0.03	< 0.4	
Platinum	< 0.003	< 0.005	
Thallium	< 0.001	< 0.01	
Thorium	< 0.001	< 0.005	
Uranium	0.007	< 0.06	
Nickel	0.12	< 0.4	
Silver	0.07	< 0.2	
Tin	0.46	< 0.3	
Titanium	1.3	< 1.0	
Total Toxic Representation			

ESSENTIAL AND OTHER ELEMENTS

ELEMENTS	RESULT µg/g	REFERENCE RANGE	PERCENTILE 2.5th 16th 50th 84th 97.5th
Calcium	83.0	125- 370	
Magnesium	6.0	12- 30	
Sodium	37	12- 90	
Potassium	31	12- 40	
Copper	11	8.0- 16	
Zinc	130	100- 190	
Manganese	0.1	0.2- 0.55	
Chromium	0.33	0.26- 0.5	
Vanadium	0.029	0.03- 0.1	
Molybdenum	0.082	0.05- 0.13	
Boron	1.9	0.6- 4.0	
Iodine	0.83	0.25- 1.3	
Lithium	0.011	0.007- 0.023	
Phosphorus	213	160- 250	
Selenium	1.6	0.95- 1.7	
Strontium	0.07	0.16- 1.0	
Sulfur	50000	45500- 53000	
Barium	0.08	0.16- 0.8	
Cobalt	0.011	0.013- 0.035	
Iron	12	8.0- 19	
Germanium	0.061	0.045- 0.065	
Rubidium	0.046	0.016- 0.18	
Zirconium	2.6	0.04- 1.0	

SPECIMEN DATA

COMMENTS:

Date Collected:
Date Received:
Date Completed:

Methodology: ICP-MS

Sample Size:
Sample Type:
Hair Color:
Treatment:
Shampoo:

RATIOS

ELEMENTS	RATIOS	EXPECTED RANGE
Ca/Mg	13.6	4- 30
Ca/P	0.389	0.8- 8
Na/K	1.17	0.5- 10
Zn/Cu	12.3	4- 20
Zn/Cd	> 999	> 800

This form is provided to Andrew Cutler to be used as an educational tool and does not constitute endorsement by *Doctor's Data.*

URINE TOXIC METALS

10/00

DOCTOR'S DATA INC

SEX: Male
AGE: 3

POTENTIALLY TOXIC METALS

TOXIC METALS	RESULT µg/g CREAT	REFERENCE RANGE	WITHIN REFERENCE RANGE	ELEVATED	VERY ELEVATED
Aluminum	< dl	< 35			
Antimony	< dl	< 5			
Arsenic	20	< 100			
Beryllium	< dl	< 0.5			
Bismuth	46	< 30			
Cadmium	< dl	< 2			
Lead	4.6	< 15			
Mercury	6.4	< 3			
Nickel	4.6	< 12			
Platinum	< dl	< 2			
Thallium	0.6	< 14			
Thorium	< dl	< 12			
Tin	2.9	< 6			
Tungsten	< dl	< 23			
Uranium	< dl	< 1			

CREATININE

	RESULT mg/dL	REFERENCE RANGE	2SD LOW	1SD LOW	MEAN	1SD HIGH	2SD HIGH
Creatinine	17	21- 76					

SPECIMEN DATA

Comments:
Date Collected:
Date Received:
Date Completed:

Method: ICP-MS
<dl: less than detection limit

Collection Period: timed: 6 hrs
Volume: 360 ml
Provocation:

Toxic metals are reported as µg/g creatinine to account for urine dilution variations. Reference ranges are representative of a healthy population under non-challenge or non-provoked conditions. No safe reference levels for toxic metals have been established.

This form is provided to Andrew Cutler to be used as an educational tool and does not constitute endorsement by *Doctor's Data.*

Toxic People can have Normal Hair Tests

This 51 year old gentleman had been a general dentist for 24 years. He had his amalgam fillings replaced with gold at age 38 due their condition, the poorer durability of composite and awareness of Huggins' work. At age 49 he had to cease practicing due to poor health. He had multiple stress related symptoms including severe depression with anxiety, difficulty concentrating, indecision, stomach ulcers, as well as shoulder and neck pain. Four months later he tested extremely high in mercury. He chelated with 250 mg DMSA by mouth 3 times a week for six weeks and his depression became markedly worse. Then he chelated for 3 months with intravenous EDTA weekly and intramuscular DMPS injections every other week.

Presently he has moderate fatigue, depression and anxiety, difficulty concentrating, indecision, memory problems, moodiness, generally low stress tolerance, and some persisting stomach problems and shoulder and neck pain.

As you can see from the hair element profiles he has very normal mineral transport and a relatively low mercury level. All of this chelation has successfully cleared this gentleman's bloodstream and some other tissues (like hair) of mercury. Unfortunately he did not use a chelation protocol or chelating agents that was able to clear the brain or really get inside the cells of organs that hold on to mercury tightly so he continues to be quite sick – even though the numbers on the piece of paper that comes from the laboratory are the picture of health.

In order to ensure the accuracy of the first test, he split his hair sample in two, carefully mixing it, and sent part to *Doctor's Data* and part to *Great Smokies Diagnostic Laboratories.* Both hair tests came back with very similar values for the elements that both labs determine. This hair test accurately reflects the health of his hair. It is the other parts of his body that it is not a completely accurate reflection of.

Subsequent to the first test he began chelating on a proper protocol using 250 mg of alpha lipoic acid[102] every 3 hours so as to clear his brain and internal organs of mercury. After 55 chelation days he had another hair test done. This second hair test shows zinc wasting (elevated zinc) and increased stress as indicated by the movement of sodium and potassium away from calcium and magnesium as compared to the first test. Hair mercury levels were also reduced quite substantially from the first to the second test. The first hair test is on the facing page and the second is on page 200.

	First test, page 199	Second test, page 200
Counting rule #1: bars to right.	12	12
Counting rule #2: bars to left.	11	11
Counting rule #3: bars in red.	0	1
Counting rule #4: bars in middle	16	18

Neither test meets counting rule number 5.

In fact, both tests are really very average and ordinary. They look exactly the way you would expect them to despite the fact that this gentleman is still severely poisoned to the point of disability.

[102] This is a very high dose that many people would not be able to tolerate.

HAIR ELEMENTS

DOCTOR'S DATA INC.

SEX: Male
AGE: 51

POTENTIALLY TOXIC ELEMENTS

TOXIC ELEMENTS	RESULT µg/g	REFERENCE RANGE	PERCENTILE 68th 95th
Aluminum	1.7	< 7.0	
Antimony	< 0.01	< 0.066	
Arsenic	0.026	< 0.08	
Beryllium	< 0.01	< 0.02	
Bismuth	0.008	< 0.06	
Cadmium	0.082	< 0.15	
Lead	0.58	< 2.0	
Mercury	1.4	< 1.1	
Platinum	< 0.003	< 0.005	
Thallium	< 0.001	< 0.01	
Thorium	< 0.001	< 0.005	
Uranium	0.025	< 0.06	
Nickel	0.14	< 0.4	
Silver	0.02	< 0.12	
Tin	0.07	< 0.3	
Titanium	0.56	< 1.0	
Total Toxic Representation			

ESSENTIAL AND OTHER ELEMENTS

ELEMENTS	RESULT µg/g	REFERENCE RANGE	PERCENTILE 2.5th 16th 50th 84th 97.5th
Calcium	489	200- 750	
Magnesium	56	25- 75	
Sodium	7.0	12- 90	
Potassium	8.0	9.0- 40	
Copper	27	10- 28	
Zinc	240	130- 200	
Manganese	0.24	0.15- 0.65	
Chromium	0.32	0.2- 0.4	
Vanadium	0.042	0.018- 0.065	
Molybdenum	0.025	0.025- 0.064	
Boron	1.2	0.4- 3.0	
Iodine	0.97	0.25- 1.3	
Lithium	0.007	0.007- 0.023	
Phosphorus	185	160- 250	
Selenium	1.5	0.95- 1.7	
Strontium	1.1	0.3- 3.5	
Sulfur	49200	44500- 52000	
Barium	0.13	0.16- 1.6	
Cobalt	0.009	0.013- 0.035	
Iron	5.7	5.4- 13	
Germanium	0.056	0.045- 0.065	
Rubidium	0.008	0.011- 0.12	
Zirconium	0.013	0.02- 0.44	

SPECIMEN DATA

COMMENTS:

Date Collected:	Sample Size:
Date Received:	Sample Type:
Date Completed: 8/12/2003	Hair Color:
	Treatment:
Methodology: ICP-MS	Shampoo:

V06.99

RATIOS

ELEMENTS	RATIOS	EXPECTED RANGE
Ca/Mg	8.73	4- 30
Ca/P	2.64	0.8- 8
Na/K	0.875	0.5- 10
Zn/Cu	8.89	4- 20
Zn/Cd	> 999	> 800

This form is provided to Andrew Cutler to be used as an educational tool and does not constitute endorsement by *Doctor's Data.*

HAIR ELEMENTS

DOCTOR'S DATA INC.

SEX: Male
AGE: 51

POTENTIALLY TOXIC ELEMENTS

TOXIC ELEMENTS	RESULT µg/g	REFERENCE RANGE
Aluminum	2.9	< 7.0
Antimony	< 0.01	< 0.066
Arsenic	0.030	< 0.080
Beryllium	< 0.01	< 0.020
Bismuth	0.10	< 0.060
Cadmium	0.10	< 0.15
Lead	0.46	< 2.0
Mercury	0.13	< 1.1
Platinum	< 0.003	< 0.005
Thallium	0.005	< 0.010
Thorium	< 0.001	< 0.005
Uranium	0.019	< 0.060
Nickel	0.36	< 0.40
Silver	0.03	< 0.12
Tin	0.20	< 0.30
Titanium	1.2	< 1.0
Total Toxic Representation		

ESSENTIAL AND OTHER ELEMENTS

ELEMENTS	RESULT µg/g	REFERENCE RANGE
Calcium	742	200- 750
Magnesium	71	25- 75
Sodium	9	12- 90
Potassium	8	9- 40
Copper	15	10- 28
Zinc	300	130- 200
Manganese	0.21	0.15- 0.65
Chromium	0.28	0.20- 0.40
Vanadium	0.055	0.018- 0.065
Molybdenum	0.039	0.025- 0.064
Boron	0.48	0.40- 3.0
Iodine	0.18	0.25- 1.3
Lithium	0.017	0.007- 0.023
Phosphorus	215	160- 250
Selenium	1.7	0.95- 1.7
Strontium	1.3	0.30- 3.5
Sulfur	54500	44500- 52000
Barium	0.21	0.16- 1.6
Cobalt	0.017	0.013- 0.035
Iron	7.7	5.4- 13
Germanium	0.057	0.045- 0.065
Rubidium	0.016	0.011- 0.12
Zirconium	0.18	0.020- 0.44

SPECIMEN DATA

COMMENTS:

Date Collected:
Date Received:
Date Completed: 1/28/2004
Methodology: ICP-MS

Sample Size:
Sample Type:
Hair Color:
Treatment:
Shampoo:

V06.99

RATIOS

ELEMENTS	RATIOS	EXPECTED RANGE
Ca/Mg	10.5	4- 30
Ca/P	3.45	0.8- 8
Na/K	1.13	0.5- 10
Zn/Cu	20	4- 20
Zn/Cd	> 999	> 800

This form is provided to Andrew Cutler to be used as an educational tool and does not constitute endorsement by *Doctor's Data*.

Autoimmune Diseases like Multiple Sclerosis and Lupus Erythrematosus can be due to Mercury

This 38 year old man has a long history of complex autoimmunity and multiple chemical sensitivity of 15 years duration. A diagnosis of multiple sclerosis was offered at age 24 after optic neuritis. MS like symptoms of bladder problems, weakness, fatigue and numbness were present in his early 20's. Sleep disorder, itchiness and allergies plus asthma have been ongoing since the beginning of the illness. He has also had reactive arthritis episodes after colds and flus. Laboratory abnormalities included anti-nuclear antibodies, anti-myelin antibodies, anti-parietal cell antibodies, anti-mitochondrial antibodies, anti-smooth muscle antibodies and anti-brush border antibodies as well as a low CD8 suppresser T cell level (thus elevating the CD4/CD8 ratio). Remission of much of this was achieved 3 months after amalgam filling removal at age 24. Minor to moderate problems in addition to the long-standing chemical sensitivities waxed and waned from age 24 to age 38.

In the fall of 2002 he took alpha lipoic acid 300-600 mg 2 or 3 times a day and suffered a serious relapse and exaggeration of MS symptoms as well as greatly increased sensitivity to chemicals and foods. Following 3-4 months of improper ALA use peripheral neuropathies were at times incapacitating with burning pain, numbness and weakness in all limbs.

Following proper chelation with ALA on a 3 hour protocol, there was much symptomatic improvement. He was also able to stop taking thyroid medication and stop using a steroid inhaler.

This gentleman performed hair tests every 3 months (it takes 9-12 months for adults to have improved hair tests). Three tests taken starting just before he began alpha lipoic acid detoxification are presented here. As you can see, the tests do come out very similarly, with all the significant changes from test to test being accounted for by specific interventions.

This person also has elevated plasma cysteine, yet you will note that his sulfur on the hair test is almost exactly average. Hair sulfur levels are not relevant to body levels.

Note the very elevated uranium. This is common for people who live in the southwestern United States as this gentleman does. The uranium declined substantially from test #2 to test #3 following the addition of supplementary iron to this gentleman's regimen, with significant improvement in energy.

Also note the adrenal problem sign in the calcium and magnesium being much different (lower) than the sodium and potassium on the first test, but not on later tests. One of the therapies he is taking is hydrocortisone and it is quite helpful. He began it after the first test.

Also note the somewhat elevated arsenic. It is unclear how significant this is due to the derangement of mineral transport, however the combination of arsenic plus mercury often causes chemical sensitivity, sometimes severe.

How these tests count:

	First test, page 202	Second test, page 203	Third test, page 204
Counting rule #1: bars to right.	13	15	16
Counting rule #2: bars to left.	10	8	7
Counting rule #3: bars in red.	3	3	4
Counting rule #4: bars in middle	12	12	13

Counting rule #5: The first and second test meet this with 3 reds and 12 in the middle band. The third test meets counting rule #3.

HAIR ELEMENTS

DOCTOR'S DATA INC.

SEX: Male
AGE: 38

POTENTIALLY TOXIC ELEMENTS

TOXIC ELEMENTS	RESULT µg/g	REFERENCE RANGE
Aluminum	1.7	< 7.0
Antimony	0.013	< 0.066
Arsenic	0.17	< 0.08
Beryllium	< 0.01	< 0.02
Bismuth	0.002	< 0.06
Cadmium	0.058	< 0.15
Lead	0.35	< 2.0
Mercury	1.8	< 1.1
Platinum	< 0.003	< 0.005
Thallium	< 0.001	< 0.01
Thorium	0.001	< 0.005
Uranium	0.5	< 0.06
Nickel	0.06	< 0.4
Silver	0.04	< 0.12
Tin	0.08	< 0.3
Titanium	0.79	< 1.0
Total Toxic Representation		

ESSENTIAL AND OTHER ELEMENTS

ELEMENTS	RESULT µg/g	REFERENCE RANGE
Calcium	244	200- 750
Magnesium	28	25- 75
Sodium	150	12- 90
Potassium	110	9.0- 40
Copper	13	10- 28
Zinc	180	130- 200
Manganese	0.09	0.15- 0.65
Chromium	0.48	0.2- 0.4
Vanadium	0.1	0.018- 0.065
Molybdenum	0.14	0.025- 0.064
Boron	38	0.4- 3.0
Iodine	0.22	0.25- 1.3
Lithium	0.027	0.007- 0.023
Phosphorus	209	160- 250
Selenium	2.4	0.95- 1.7
Strontium	0.56	0.3- 3.5
Sulfur	48500	44500- 52000
Barium	0.21	0.16- 1.6
Cobalt	0.007	0.013- 0.035
Iron	6.6	5.4- 13
Germanium	0.055	0.045- 0.065
Rubidium	0.087	0.011- 0.12
Zirconium	0.046	0.02- 0.44

SPECIMEN DATA

COMMENTS:

Date Collected:
Date Received:
Date Completed: 8/27/2003
Methodology: ICP-MS

Sample Size:
Sample Type:
Hair Color:
Treatment:
Shampoo:

V06.99

RATIOS

ELEMENTS	RATIOS	EXPECTED RANGE
Ca/Mg	8.71	4- 30
Ca/P	1.17	0.8- 8
Na/K	1.36	0.5- 10
Zn/Cu	13.8	4- 20
Zn/Cd	> 999	> 800

This form is provided to Andrew Cutler to be used as an educational tool and does not constitute endorsement by *Doctor's Data.*

HAIR ELEMENTS

DOCTOR'S DATA INC

SEX: Male
AGE: 39

POTENTIALLY TOXIC ELEMENTS

TOXIC ELEMENTS	RESULT µg/g	REFERENCE RANGE
Aluminum	2.0	< 7.0
Antimony	0.021	< 0.066
Arsenic	0.13	< 0.080
Beryllium	< 0.01	< 0.020
Bismuth	0.015	< 0.060
Cadmium	0.020	< 0.15
Lead	0.36	< 2.0
Mercury	2.9	< 1.1
Platinum	< 0.003	< 0.005
Thallium	< 0.001	< 0.010
Thorium	< 0.001	< 0.005
Uranium	0.12	< 0.060
Nickel	0.08	< 0.40
Silver	0.04	< 0.12
Tin	0.10	< 0.30
Titanium	0.65	< 1.0
Total Toxic Representation		

PERCENTILE: 68th, 95th

ESSENTIAL AND OTHER ELEMENTS

ELEMENTS	RESULT µg/g	REFERENCE RANGE
Calcium	325	200- 750
Magnesium	55	25- 75
Sodium	62	12- 90
Potassium	44	9- 40
Copper	16	10- 28
Zinc	250	130- 200
Manganese	0.10	0.15- 0.65
Chromium	0.33	0.20- 0.40
Vanadium	0.081	0.018- 0.065
Molybdenum	0.15	0.025- 0.064
Boron	12	0.40- 3.0
Iodine	0.36	0.25- 1.3
Lithium	0.028	0.007- 0.023
Phosphorus	256	160- 250
Selenium	2.3	0.95- 1.7
Strontium	0.76	0.30- 3.5
Sulfur	48900	44500- 52000
Barium	0.22	0.16- 1.6
Cobalt	0.004	0.013- 0.035
Iron	5.4	5.4- 13
Germanium	0.082	0.045- 0.065
Rubidium	0.040	0.011- 0.12
Zirconium	0.054	0.020- 0.44

PERCENTILE: 2.5th, 16th, 50th, 84th, 97.5th

SPECIMEN DATA

COMMENTS:

Date Collected:
Date Received:
Date Completed: 11/21/2003
Methodology: ICP-MS

Sample Size:
Sample Type:
Hair Color:
Treatment:
Shampoo:

V06.99

RATIOS

ELEMENTS	RATIOS	EXPECTED RANGE
Ca/Mg	5.91	4- 30
Ca/P	1.27	0.8- 8
Na/K	1.41	0.5- 10
Zn/Cu	15.6	4- 20
Zn/Cd	> 999	> 800

This form is provided to Andrew Cutler to be used as an educational tool and does not constitute endorsement by *Doctor's Data*.

HAIR ELEMENTS

DOCTOR'S DATA INC

SEX: Male
AGE: 39

POTENTIALLY TOXIC ELEMENTS

TOXIC ELEMENTS	RESULT µg/g	REFERENCE RANGE	PERCENTILE 68th 95th
Aluminum	2.4	< 7.0	
Antimony	0.011	< 0.066	
Arsenic	0.15	< 0.080	
Beryllium	0.012	< 0.020	
Bismuth	0.012	< 0.060	
Cadmium	0.015	< 0.15	
Lead	0.27	< 2.0	
Mercury	3.0	< 1.1	
Platinum	< 0.003	< 0.005	
Thallium	< 0.001	< 0.010	
Thorium	< 0.001	< 0.005	
Uranium	0.13	< 0.060	
Nickel	0.16	< 0.40	
Silver	0.02	< 0.12	
Tin	0.06	< 0.30	
Titanium	0.85	< 1.0	
Total Toxic Representation			

ESSENTIAL AND OTHER ELEMENTS

ELEMENTS	RESULT µg/g	REFERENCE RANGE	PERCENTILE 2.5th 16th 50th 84th 97.5th
Calcium	422	200- 750	
Magnesium	53	25- 75	
Sodium	93	12- 90	
Potassium	67	9- 40	
Copper	16	10- 28	
Zinc	260	130- 200	
Manganese	0.07	0.15- 0.65	
Chromium	0.46	0.20- 0.40	
Vanadium	0.054	0.018- 0.065	
Molybdenum	0.070	0.025- 0.064	
Boron	17	0.40- 3.0	
Iodine	0.30	0.25- 1.3	
Lithium	0.036	0.007- 0.023	
Phosphorus	276	160- 250	
Selenium	3.1	0.95- 1.7	
Strontium	0.79	0.30- 3.5	
Sulfur	51600	44500- 52000	
Barium	0.23	0.16- 1.6	
Cobalt	0.005	0.013- 0.035	
Iron	7.4	5.4- 13	
Germanium	0.053	0.045- 0.065	
Rubidium	0.058	0.011- 0.12	
Zirconium	0.11	0.020- 0.44	

SPECIMEN DATA

COMMENTS:

Date Collected:
Date Received:
Date Completed: 2/5/2004

Methodology: ICP-MS

Sample Size:
Sample Type:
Hair Color:
Treatment:
Shampoo:

V06.99

RATIOS

ELEMENTS	RATIOS	EXPECTED RANGE
Ca/Mg	7.96	4- 30
Ca/P	1.53	0.8- 8
Na/K	1.39	0.5- 10
Zn/Cu	16.3	4- 20
Zn/Cd	> 999	> 800

This form is provided to Andrew Cutler to be used as an educational tool and does not constitute endorsement by *Doctor's Data.*

Now we come to a case of a woman diagnosed with Lupus Erythromatosus. She has complained of severe fatigue for 3 years and currently complains of multiple symptoms: fatigue, numbness, slurring words, sinus problems, nose bleeds, skin markings due to blood vessel problems[103], excessive numbers of moles, weakness, migraines, dizziness, ear/hearing problems, ringing in the ears and hoarseness. She has had shingles which is unusual for a woman this young. She has also had three miscarriages.

She had a thyroid nodule that was making excessive thyroid hormones. This was treated with radioactive iodine, which resulted in her going from having too much thyroid hormone to having too little[104].

Other laboratory test results include positive antinuclear antibodies (a sign of lupus), elevated thyroid peroxidase antibodies (a sign of autoimmune disease attacking the thyroid gland), and low ferritin (a marker of body iron stores) and vitamin B-12.

Her health care practitioner also believes she has chronic yeast infections[105].

Her sister died at age 51 from breast cancer which first occurred at age 35.

This woman's hair test is presented on the next page. It counts as follows:

Counting rule #1: 3 bars go right. This meets rule #1.

Counting rule #2: 20 bars go left. This does not meet counting rule #2.

Counting rule #3: 1 bar goes into the red band. This does not meet counting rule #3.

Counting rule #4: 14 bars are contained entirely in the middle green + white stripe. This does not meet counting rule #4.

Counting rule #5 is irrelevant since another counting rule was met.

Note the relation of calcium and magnesium to sodium and potassium. This suggests the possibility of adrenal problems.

Note the low ratio of calcium to phosphorus (Ca/P). This suggests that taking digestive enzymes with meals would be helpful for this woman.

[103] the proper medical term for the blood vessel problem is telangiectasia.

[104] in medical terms, going from being hyperthyroid to being hypothyroid.

[105] the proper medical term is candidiasis.

HAIR ELEMENTS

DOCTOR'S DATA INC

SEX: Female
AGE: 33

POTENTIALLY TOXIC ELEMENTS

TOXIC ELEMENTS	RESULT µg/g	REFERENCE RANGE	PERCENTILE 68th 95th
Aluminum	2.0	< 12	
Antimony	0.025	< 0.060	
Arsenic	0.094	< 0.090	
Beryllium	< 0.01	< 0.020	
Bismuth	0.024	< 0.13	
Cadmium	0.016	< 0.10	
Lead	0.18	< 1.0	
Mercury	0.69	< 1.1	
Platinum	< 0.003	< 0.005	
Thallium	< 0.001	< 0.010	
Thorium	< 0.001	< 0.005	
Uranium	0.011	< 0.060	
Nickel	0.09	< 0.40	
Silver	0.29	< 0.10	
Tin	0.10	< 0.30	
Titanium	0.65	< 1.0	
Total Toxic Representation			

ESSENTIAL AND OTHER ELEMENTS

ELEMENTS	RESULT µg/g	REFERENCE RANGE	PERCENTILE 2.5th 16th 50th 84th 97.5th
Calcium	227	475- 1500	
Magnesium	30	45- 180	
Sodium	200	24- 180	
Potassium	89	20- 80	
Copper	7.2	9.0- 24	
Zinc	140	130- 200	
Manganese	0.12	0.20- 0.65	
Chromium	0.25	0.20- 0.45	
Vanadium	0.025	0.018- 0.065	
Molybdenum	0.051	0.040- 0.10	
Boron	1.3	0.70- 4.0	
Iodine	0.52	0.25- 1.3	
Lithium	0.015	0.008- 0.040	
Phosphorus	239	350- 600	
Selenium	1.2	0.95- 1.7	
Strontium	0.42	1.0- 8.0	
Sulfur	46200	43000- 50000	
Barium	0.17	0.50- 5.0	
Cobalt	0.009	0.013- 0.050	
Iron	8.0	5.8- 14	
Germanium	0.038	0.045- 0.065	
Rubidium	0.087	0.030- 0.25	
Zirconium	0.072	0.040- 1.0	

SPECIMEN DATA

COMMENTS:

Date Collected:
Date Received:
Date Completed:

Methodology: ICP-MS

Sample Size:
Sample Type:
Hair Color:
Treatment:
Shampoo:

V06.99

RATIOS

ELEMENTS	RATIOS	EXPECTED RANGE
Ca/Mg	7.57	4- 30
Ca/P	0.95	1- 12
Na/K	2.25	0.5- 10
Zn/Cu	19.4	4- 20
Zn/Cd	> 999	> 800

This form is provided to Andrew Cutler to be used as an educational tool and does not constitute endorsement by *Doctor's Data.*

Many Autistic Children are Toxic

This autistic boy had his hair tested, was found to have deranged mineral transport suggestive of mercury toxicity, was chelated with improvement, and later the parents submitted hair from his first baby hair cut for analysis. As you can see from the tests facing each other on pages 208 and 209, these results are consistent with literature reports[106] and the uniform experience of health care practitioners in this field that most mercury poisoned children misdiagnosed as autistic have low hair mercury levels.

	Baby hair (page 208)	Current hair (page 209)
Counting rule #1: bars to right.	17	16
Counting rule #2: bars to left.	6	7
Counting rule #3: bars in red.	1	2
Counting rule #4: bars in middle	15	11
Counting rule #5: is not met by either test.		

The first test is suspicious but meets no counting rule. The second test meets counting rule #4.

Aluminum, antimony, and bismuth are vastly elevated on the baby hair test, while arsenic and titanium are very elevated. This is relatively typical of children with deranged mineral transport.

The second test, which does meet the counting rules, shows antimony, bismuth, tin and titanium as very elevated. This is quite characteristic of children with deranged mineral transport.

The abnormal calcium to magnesium (Ca/Mg) ratio on the second test suggests poor regulation of blood sugar. The low calcium to phosphorus (Ca/P) ratio on the second test suggests that supplementary digestive enzymes would be helpful.

The abnormal ratio of sodium to potassium (Na/K) on the first test suggests thyroid function may have been deranged at that time. This is not uncommon in autistic children and as normal, healthy levels of thyroid hormone are necessary for proper brain development may contribute to the development of autism and associated neurological problems.

---

[106] Reduced levels of mercury in first baby haircuts of autistic children, by Holmes, Blaxill and Haley, *International Journal of Toxicology*, volume **22**, issue 4, pages 277-85, July-August (2003).

HAIR ELEMENTS

06/99

DOCTOR'S DATA INC.

SEX: Male
AGE:

POTENTIALLY TOXIC ELEMENTS

TOXIC ELEMENTS	RESULT µg/g	REFERENCE RANGE
Aluminum	57	< 8.0
Antimony	0.27	< 0.066
Arsenic	0.58	< 0.08
Beryllium	0.012	< 0.02
Bismuth	21	< 0.13
Cadmium	0.2	< 0.15
Lead	1.9	< 1.0
Mercury	0.25	< 0.4
Platinum	< 0.003	< 0.005
Thallium	< 0.001	< 0.01
Thorium	< 0.001	< 0.005
Uranium	0.041	< 0.06
Nickel	0.36	< 0.4
Silver	0.09	< 0.2
Tin	0.58	< 0.3
Titanium	3.4	< 1.0
Total Toxic Representation		

PERCENTILE: 68th, 95th

ESSENTIAL AND OTHER ELEMENTS

ELEMENTS	RESULT µg/g	REFERENCE RANGE
Calcium	238	125- 370
Magnesium	20	12- 30
Sodium	6.0	12- 90
Potassium	16	12- 40
Copper	21	8.0- 16
Zinc	120	100- 190
Manganese	0.81	0.2- 0.55
Chromium	0.5	0.26- 0.5
Vanadium	0.055	0.03- 0.1
Molybdenum	0.11	0.05- 0.13
Boron	9.3	0.6- 4.0
Iodine	3.1	0.25- 1.3
Lithium	0.049	0.007- 0.023
Phosphorus	189	160- 250
Selenium	1.6	0.95- 1.7
Strontium	0.4	0.16- 1.0
Sulfur	50600	45500- 53000
Barium	0.27	0.16- 0.8
Cobalt	0.021	0.013- 0.035
Iron	22	8.0- 19
Germanium	0.059	0.045- 0.065
Rubidium	0.029	0.016- 0.18
Zirconium	5.0	0.04- 1.0

PERCENTILE: 2.5th, 16th, 50th, 84th, 97.5th

SPECIMEN DATA

COMMENTS: insufficient sample to recheck results

Date Collected: 8/12/1997
Date Received:
Date Completed:
Methodology: ICP-MS

Sample Size:
Sample Type:
Hair Color:
Treatment:
Shampoo:

RATIOS

ELEMENTS	RATIOS	EXPECTED RANGE
Ca/Mg	11.6	4- 30
Ca/P	1.26	0.8- 8
Na/K	0.369	0.5- 10
Zn/Cu	5.95	4- 20
Zn/Cd	623	> 800

This form is provided to Andrew Cutler to be used as an educational tool and does not constitute endorsement by *Doctor's Data*.

HAIR ELEMENTS

06/99

DOCTOR'S DATA INC

SEX: Male
AGE: 4

POTENTIALLY TOXIC ELEMENTS

TOXIC ELEMENTS	RESULT µg/g	REFERENCE RANGE	PERCENTILE 68th 95th
Aluminum	12	< 8.0	
Antimony	0.44	< 0.066	
Arsenic	0.14	< 0.08	
Beryllium	< 0.01	< 0.02	
Bismuth	0.9	< 0.13	
Cadmium	0.21	< 0.15	
Lead	1.0	< 1.0	
Mercury	0.21	< 0.4	
Platinum	< 0.003	< 0.005	
Thallium	0.003	< 0.01	
Thorium	< 0.001	< 0.005	
Uranium	0.091	< 0.06	
Nickel	0.34	< 0.4	
Silver	0.68	< 0.2	
Tin	1.5	< 0.3	
Titanium	2.2	< 1.0	
Total Toxic Representation			

ESSENTIAL AND OTHER ELEMENTS

ELEMENTS	RESULT µg/g	REFERENCE RANGE	PERCENTILE 2.5th 16th 50th 84th 97.5th
Calcium	88.0	125- 370	
Magnesium	33	12- 30	
Sodium	69	12- 90	
Potassium	100	12- 40	
Copper	15	8.0- 16	
Zinc	35	100- 190	
Manganese	0.13	0.2- 0.55	
Chromium	1.4	0.26- 0.5	
Vanadium	0.034	0.03- 0.1	
Molybdenum	0.15	0.05- 0.13	
Boron	5.4	0.6- 4.0	
Iodine	0.87	0.25- 1.3	
Lithium	0.027	0.007- 0.023	
Phosphorus	200	160- 250	
Selenium	0.71	0.95- 1.7	
Strontium	0.38	0.16- 1.0	
Sulfur	47800	45500- 53000	
Barium	0.23	0.16- 0.8	
Cobalt	0.025	0.013- 0.035	
Iron	16	8.0- 19	
Germanium	0.078	0.045- 0.065	
Rubidium	0.11	0.016- 0.18	
Zirconium	1.7	0.04- 1.0	

SPECIMEN DATA

COMMENTS:

Date Collected: 2/7/2002
Date Received:
Date Completed:
Methodology: ICP-MS

Sample Size:
Sample Type:
Hair Color:
Treatment:
Shampoo:

RATIOS

ELEMENTS	RATIOS	EXPECTED RANGE
Ca/Mg	2.69	4- 30
Ca/P	0.441	0.8- 8
Na/K	0.664	0.5- 10
Zn/Cu	2.36	4- 20
Zn/Cd	168	> 800

This form is provided to Andrew Cutler to be used as an educational tool and does not constitute endorsement by *Doctor's Data.*

The young boy in the following case had a normal birth. He was breast fed for 19 months, exclusively for the first six. He was constipated from the beginning which was resolved by the mother removing dairy products from her diet. There was growth failure at age 1. At age 19 months there was delayed speech so blood lead testing was performed with the following results:

19 months: finger stick, 57
19 months, blood lead 28
22 months: blood lead 23
25 months: blood lead 19
37 months: blood lead 5

The toxic lead level is 10 or more The only intervention performed was the mother giving the boy vitamin C and calcium. There were no medical interventions to reduce blood lead levels prescribed by the pediatricians based on these tests. The source of the lead is believed to have been repair and remodeling of an older construction home at age 18-19 months.

Currently the boy has delayed, slurred, slushy speech despite 6 months of speech therapy, and he has ADHD[106]. The mother decided to chelate him at 36 months, starting with 10 mg DMSA every 4 hours.

A hair test was performed at 38 months and is shown on the facing page. It counts as follows:

Counting rule #1: 9 bars go right. This does not meet rule #1.

Counting rule #2: 14 bars go left. This does not meet counting rule #2.

Counting rule #3: 1 bar goes into the red band. This does not meet counting rule #3.

Counting rule #4: 11 bars are contained entirely in the middle green + white stripe. This meets counting rule #4.

Counting rule #5 is irrelevant since another rule is met.

While the test meets rule #4 for excessive scatter, the essential element bars do not mostly make it into the red. Thus the red range elevated bismuth may be significant and may reflect elevated body levels of bismuth. This could be explored further with a blood test for bismuth[108] but it isn't necessary to do so since chelation for mercury which is presumed to be a problem from the hair test results, and for lead which is known to be a problem from prior blood testing will clear out whatever bismuth is around.

Note especially that the hair lead levels are not elevated despite the history of elevated blood lead. It is common for hair lead not to be elevated even when body burden is high if the exposure was ended many months before the hair test. In this case the hair test was about 19 months after exposure and was done well after blood lead levels were in the normal range. Also note that with deranged mineral transport the levels of the toxic elements are not particularly reliable anyway.

Note the relation of calcium and magnesium to sodium and potassium. This suggests the possibility of quite significant adrenal problems, with treatment for these reasonably expected to lead to significant overall improvement.

Also note that neither lead nor bismuth causes derangement of mineral transport on hair tests. These results represent a mixed metal toxicity, lead plus mercury with the additional possibility of bismuth.

---

[107] ADHD is Attention Deficit Hyperactivity Disorder.

[108] Erythrocyte elements as determined by *Great Smokies Diagnostic Laboratories* measures bismuth as well as other relevant elements.

HAIR ELEMENTS

DOCTOR'S DATA INC

SEX: Male
AGE: 3

POTENTIALLY TOXIC ELEMENTS

TOXIC ELEMENTS	RESULT µg/g	REFERENCE RANGE
Aluminum	6.8	< 8.0
Antimony	0.031	< 0.066
Arsenic	0.036	< 0.08
Beryllium	< 0.01	< 0.02
Bismuth	0.8	< 0.13
Cadmium	< 0.009	< 0.15
Lead	0.68	< 1.0
Mercury	0.11	< 0.4
Platinum	< 0.003	< 0.005
Thallium	< 0.001	< 0.01
Thorium	0.002	< 0.005
Uranium	0.005	< 0.06
Nickel	0.02	< 0.4
Silver	0.03	< 0.2
Tin	0.42	< 0.3
Titanium	1.0	< 1.0
Total Toxic Representation		

PERCENTILE: 68th, 95th

ESSENTIAL AND OTHER ELEMENTS

ELEMENTS	RESULT µg/g	REFERENCE RANGE
Calcium	120	125- 370
Magnesium	9.0	12- 30
Sodium	170	12- 90
Potassium	120	12- 40
Copper	8.1	8.0- 16
Zinc	220	100- 190
Manganese	0.09	0.2- 0.55
Chromium	0.32	0.26- 0.5
Vanadium	0.049	0.03- 0.1
Molybdenum	0.034	0.05- 0.13
Boron	1.1	0.6- 4.0
Iodine	0.44	0.25- 1.3
Lithium	0.012	0.007- 0.023
Phosphorus	162	160- 250
Selenium	1.4	0.95- 1.7
Strontium	0.09	0.16- 1.0
Sulfur	51500	45500- 53000
Barium	0.08	0.16- 0.8
Cobalt	0.009	0.013- 0.035
Iron	15	8.0- 19
Germanium	0.038	0.045- 0.065
Rubidium	0.093	0.016- 0.18
Zirconium	1.6	0.04- 1.0

PERCENTILE: 2.5th, 16th, 50th, 84th, 97.5th

SPECIMEN DATA

COMMENTS:

Date Collected:
Date Received:
Date Completed:

Methodology: ICP-MS

Sample Size:
Sample Type:
Hair Color:
Treatment:
Shampoo:

V06.99

RATIOS

ELEMENTS	RATIOS	EXPECTED RANGE
Ca/Mg	13.3	4- 30
Ca/P	0.741	0.8- 8
Na/K	1.42	0.5- 10
Zn/Cu	27.2	4- 20
Zn/Cd	> 999	> 800

This form is provided to Andrew Cutler to be used as an educational tool and does not constitute endorsement by *Doctor's Data.*

Here is the case of an autistic boy who was delivered normally at 42 weeks of gestation. At birth there was evidence of septic spots so oral antibiotics were prescribed. These were vomited up so the antibiotics were given intravenously for 5 days. He had apparently normal development until seven months at which time the parents noticed various odd behaviors. He was eventually diagnosed with autism after a large number of doctor visits. He does not appear to have other health problems. He was breast fed for 6 months. His mother put him on a gluten free, casein free (GFCF) diet for 10 months at age 4 but discontinued it due to lack of any effect, good or bad.

His hair test is presented on the facing page and counts as follows:

Counting rule #1: 10 bars go right. This does not meet rule #1.

Counting rule #2: 13 bars go left. This does not meet counting rule #2.

Counting rule #3: 4 bars go into the red bands. This meets counting rule #3.

Counting rule #4: 14 bars are contained entirely in the middle green + white stripe. This does not meet counting rule #4.

Counting rule #5: is irrelevant since counting rule #3 was met.

Note that with the "excessive scatter" presentation of the essential elements the toxic elements become extremely hard to interpret since there is no particular reason to expect their level in hair to bear much relation to body burden. The elevation of antimony and titanium is of unclear significance. Since deranged mineral transport often causes children to accumulate antimony, and since children naturally absorb antimony more readily than adults, it would make sense to remove any possible sources of exposure like flame retardant bedding and pajamas that have been treated with antimony.

Note the relation of calcium and magnesium to sodium and potassium. This suggests the possibility of adrenal problems.

Note the relation of calcium to potassium and the very high absolute potassium level. This suggests the possibility of thyroid problems.

HAIR ELEMENTS

06/99

DOCTOR'S DATA INC

SEX: Male
AGE: 5

POTENTIALLY TOXIC ELEMENTS

TOXIC ELEMENTS	RESULT µg/g	REFERENCE RANGE
Aluminum	13	< 8.0
Antimony	0.71	< 0.066
Arsenic	0.014	< 0.08
Beryllium	< 0.01	< 0.02
Bismuth	0.033	< 0.13
Cadmium	< 0.009	< 0.15
Lead	0.29	< 1.0
Mercury	< 0.03	< 0.4
Platinum	< 0.003	< 0.005
Thallium	< 0.001	< 0.01
Thorium	< 0.001	< 0.005
Uranium	< 0.001	< 0.06
Nickel	0.73	< 0.4
Silver	0.09	< 0.2
Tin	0.28	< 0.3
Titanium	3.0	< 1.0
Total Toxic Representation		

PERCENTILE: 68th, 95th

ESSENTIAL AND OTHER ELEMENTS

ELEMENTS	RESULT µg/g	REFERENCE RANGE
Calcium	177	125- 370
Magnesium	19	12- 30
Sodium	420	12- 90
Potassium	490	12- 40
Copper	9.2	8.0- 16
Zinc	150	100- 190
Manganese	0.13	0.2- 0.55
Chromium	0.28	0.26- 0.5
Vanadium	0.019	0.03- 0.1
Molybdenum	0.031	0.05- 0.13
Boron	1.5	0.6- 4.0
Iodine	0.46	0.25- 1.3
Lithium	0.011	0.007- 0.023
Phosphorus	279	160- 250
Selenium	1.4	0.95- 1.7
Strontium	0.19	0.16- 1.0
Sulfur	50300	45500- 53000
Barium	< 0.04	0.16- 0.8
Cobalt	0.011	0.013- 0.035
Iron	9.5	8.0- 19
Germanium	0.038	0.045- 0.065
Rubidium	0.11	0.016- 0.18
Zirconium	0.82	0.04- 1.0

PERCENTILE: 2.5th, 16th, 50th, 84th, 97.5th

SPECIMEN DATA

COMMENTS:

Date Collected: Sample Size:
Date Received: Sample Type:
Date Completed: Hair Color:
Treatment:
Methodology: ICP-MS Shampoo:

RATIOS

ELEMENTS	RATIOS	EXPECTED RANGE
Ca/Mg	9.41	4- 30
Ca/P	0.635	0.8- 8
Na/K	0.868	0.5- 10
Zn/Cu	16.1	4- 20
Zn/Cd	> 999	> 800

This form is provided to Andrew Cutler to be used as an educational tool and does not constitute endorsement by *Doctor's Data.*

The next few pages present the case of an entire family with toxicity issues that manifest as very different diagnoses.

This family of four includes a father with Parkinson's disease (his case was discussed on pages 34-36), a mother, and two sons, both on the autistic spectrum, the older one also considered mentally retarded and the younger with severe vision problems diagnosed as ametropic amblyopia, with vision of 20/800 in both eyes. The first son had regression into autism at 18 months following normal development until an adverse reaction to MMR and other vaccines, the second son was autistic from birth with no eye contact, no speech he didn't stop screaming for the first year due to colic, as well as constant diarrhea.

Two interesting "family traits" are much improved on chelation. Everyone in the family used to be fairly obsessive. Also every family member had what the father describes as "angry bugged out eyes" of the sort that go along with thyroid disease. These have resolved for all family members subsequent to chelation.

The younger son's hair test is presented on the facing page and counts as follows:

Counting rule #1: 13 bars go right. This does not meet rule #1.

Counting rule #2: 10 bars go left. This does not meet counting rule #2.

Counting rule #3: 4 bars go into the red bands. This meets counting rule #3.

Counting rule #4: 15 bars are contained entirely in the middle green + white stripe. This does not meet counting rule #4.

Counting rule #5 is irrelevant since another counting rule was met.

This test meets a counting rule, suggesting deranged mineral transport due to mercury toxicity. This child improved dramatically on chelation, confirming that he in fact had mercury poisoning.

Note the very elevated zinc, suggesting zinc wasting and deficiency.

Note the vastly elevated antimony and the very elevated tin as well as the almost very elevated bismuth. These often rise with deranged mineral transport as the body accumulates antimony and tin when mercury impairs their excretion. The significance of the bismuth is unclear but since the test both shows deranged mineral transport and also shows four red zone essential elements on the high side it most likely is an artifact of mineral transport derangement and has no special significance regarding body burden or exposure to bismuth.

The abnormal sodium to potassium (Na/K) ratio suggests the possibility of abnormal thyroid function. Since adequate thyroid hormone levels are essential for brain development in children testing free T3 and free T4 and ensuring free T3 is at least half way up the range and free T4 at least 1/3 of the way up the range would be reasonable follow up.

Chelation on a 3-4 hour protocol using DMSA and alpha lipoic acid resulted in great improvement over time. In addition, certain interventions are worth mentioning since their effects were prompt and dramatic: vinpocetine turned speech on like a switch, and bowel problems were resolved promptly with digestive enzymes (the boy was already on a gluten free, casein free diet).

It is also worth nothing that glasses to correct his farsightedness greatly improved his behavior and development.

Special interventions provided by charity and public agencies were eventually withdrawn since the boy went from the 5th percentile to the 75th percentile on the *Vineland Extended Inventory of Adaptive Skills.*

HAIR ELEMENTS

DOCTOR'S DATA INC

SEX: Male
AGE: 1

POTENTIALLY TOXIC ELEMENTS

TOXIC ELEMENTS	RESULT µg/g	REFERENCE RANGE	PERCENTILE 68th 95th
Aluminum	15	< 8.0	
Antimony	4.0	< 0.066	
Arsenic	0.13	< 0.08	
Beryllium	< 0.01	< 0.02	
Bismuth	0.6	< 0.13	
Cadmium	0.11	< 0.15	
Lead	1.5	< 1.0	
Mercury	1.1	< 0.4	
Platinum	< 0.003	< 0.005	
Thallium	< 0.001	< 0.01	
Thorium	< 0.001	< 0.005	
Uranium	0.006	< 0.06	
Nickel	0.27	< 0.4	
Silver	0.75	< 0.2	
Tin	1.4	< 0.3	
Titanium	1.2	< 1.0	
Total Toxic Representation			

ESSENTIAL AND OTHER ELEMENTS

ELEMENTS	RESULT µg/g	REFERENCE RANGE	PERCENTILE 2.5th 16th 50th 84th 97.5th
Calcium	157	125- 370	
Magnesium	16	12- 30	
Sodium	13	12- 90	
Potassium	45	12- 40	
Copper	12	8.0- 16	
Zinc	300	100- 190	
Manganese	0.19	0.2- 0.55	
Chromium	1.2	0.26- 0.5	
Vanadium	3.5	0.03- 0.1	
Molybdenum	0.13	0.05- 0.13	
Boron	1.7	0.6- 4.0	
Iodine	5.1	0.25- 1.3	
Lithium	0.011	0.007- 0.023	
Phosphorus	163	160- 250	
Selenium	0.79	0.95- 1.7	
Strontium	0.26	0.16- 1.0	
Sulfur	48000	45500- 53000	
Barium	0.76	0.16- 0.8	
Cobalt	0.014	0.013- 0.035	
Iron	16	8.0- 19	
Germanium	0.039	0.045- 0.065	
Rubidium	0.059	0.016- 0.18	
Zirconium	0.76	0.04- 1.0	

SPECIMEN DATA

COMMENTS:

Date Collected:	Sample Size:
Date Received:	Sample Type:
Date Completed:	Hair Color:
	Treatment:
Methodology: ICP-MS	Shampoo:

V06.99

RATIOS

ELEMENTS	RATIOS	EXPECTED RANGE
Ca/Mg	9.81	4- 30
Ca/P	0.963	0.8- 8
Na/K	0.289	0.5- 10
Zn/Cu	25	4- 20
Zn/Cd	> 999	> 800

This form is provided to Andrew Cutler to be used as an educational tool and does not constitute endorsement by *Doctor's Data.*

The older son was small and listless, but otherwise normal. He received all his vaccinations and regressed promptly and dramatically due to his 18 month vaccinations. He suddenly started crying continuously, lost language and eye contact, and began having chronic diarrhea. He was then diagnosed with pervasive developmental disorder, not otherwise specified (PDD-NOS), which is an "autism spectrum" disorder for children who don't quite fit the standard diagnostic checklist for autism but clearly are impaired.

This was devastating to the parents. They arranged a wide variety of services and interventions for him and were then frustrated when little apparent progress was being made. The parents felt that all of the service providers reported in glowing terms about what great jobs they were doing in order to prevent complaints or demands from the parents. This motivated them to seek further information from a variety of sources, one being the internet.

The parents found out about a variety of helpful interventions for autistic children such as the GFCF diet. Their son's hyperactivity and opposition behavior were eliminated by appropriate diet modification. They also found accurate information regarding the link between autism and mercury from vaccines and other sources. This motivated them to do a hair test on their son.

The older son's hair test is presented on the facing page. It counts as follows:

Counting rule #1: 12 bars go right. This does not meet rule #1.

Counting rule #2: 11 bars go left. This does not meet counting rule #2.

Counting rule #3: 2 bars go into the red bands. This does not meet counting rule #3.

Counting rule #4: 14 bars are contained entirely in the middle green + white stripe. This does not meet counting rule #4.

Counting rule #5 is not met.

The counting rules suggest normal orderly mineral transport.

The very elevated hair mercury suggests excessive body burden of mercury.

This child improved dramatically on chelation, confirming that he in fact had mercury poisoning.

The vastly elevated vanadium is of unclear significance.

The low calcium to phosphorus (Ca/P) ratio suggests he would benefit from the use of digestive enzymes, and in fact he did benefit from these when the parents tried them for other reasons.

Since mineral transport is orderly and normal, the ratio of zinc to copper is meaningful and in this case it indicates these two elements are in balance.

Prior to chelation this boy's IQ was measured at 84 using standard tests administered by a Ph.D. psychologist skilled in their use. After more than a year of chelation his IQ was measured a second time and it was 132.

HAIR ELEMENTS

DOCTOR'S DATA INC

SEX: Male
AGE: 4

POTENTIALLY TOXIC ELEMENTS

TOXIC ELEMENTS	RESULT µg/g	REFERENCE RANGE	PERCENTILE (68th, 95th)
Aluminum	11	< 8.0	
Antimony	0.063	< 0.066	
Arsenic	0.14	< 0.08	
Beryllium	< 0.01	< 0.02	
Bismuth	0.15	< 0.13	
Cadmium	0.082	< 0.15	
Lead	0.73	< 1.0	
Mercury	2.6	< 0.4	
Platinum	< 0.003	< 0.005	
Thallium	< 0.001	< 0.01	
Thorium	< 0.001	< 0.005	
Uranium	0.003	< 0.06	
Nickel	0.19	< 0.4	
Silver	0.18	< 0.2	
Tin	0.34	< 0.3	
Titanium	1.1	< 1.0	
Total Toxic Representation			

ESSENTIAL AND OTHER ELEMENTS

ELEMENTS	RESULT µg/g	REFERENCE RANGE	PERCENTILE (2.5th, 16th, 50th, 84th, 97.5th)
Calcium	159	125- 370	
Magnesium	29	12- 30	
Sodium	110	12- 90	
Potassium	29	12- 40	
Copper	17	8.0- 16	
Zinc	220	100- 190	
Manganese	0.15	0.2- 0.55	
Chromium	0.41	0.26- 0.5	
Vanadium	2.5	0.03- 0.1	
Molybdenum	0.077	0.05- 0.13	
Boron	0.6	0.6- 4.0	
Iodine	1.5	0.25- 1.3	
Lithium	< 0.004	0.007- 0.023	
Phosphorus	220	160- 250	
Selenium	1.0	0.95- 1.7	
Strontium	0.48	0.16- 1.0	
Sulfur	49300	45500- 53000	
Barium	0.14	0.16- 0.8	
Cobalt	0.016	0.013- 0.035	
Iron	13	8.0- 19	
Germanium	0.035	0.045- 0.065	
Rubidium	0.028	0.016- 0.18	
Zirconium	0.35	0.04- 1.0	

SPECIMEN DATA

COMMENTS:

Date Collected:
Date Received:
Date Completed:
Methodology: ICP-MS

Sample Size:
Sample Type:
Hair Color:
Treatment:
Shampoo:

V06.99

RATIOS

ELEMENTS	RATIOS	EXPECTED RANGE
Ca/Mg	5.48	4- 30
Ca/P	0.723	0.8- 8
Na/K	3.79	0.5- 10
Zn/Cu	12.9	4- 20
Zn/Cd	> 999	> 800

This form is provided to Andrew Cutler to be used as an educational tool and does not constitute endorsement by *Doctor's Data.*

The mother was unable to get pregnant for ten years. She conceived some time after having her mercury (silver amalgam) dental fillings removed. She used to experience significant heart pain frequently and this was eliminated by chelation. She was a dance major in college but could not work out much and had poor wind and endurance. She is emotionally volatile and has quite a temper. She is obsessive and compulsive in the sense that she has to do routines meticulously and can't go on until they are done, but she is not stereotypic about this and the routines are reasonable parts of day to day activities, not unusual or aberrant behaviors.

The mother's hair test is presented on the facing page and counts as follows:

Counting rule #1: 10 bars go right. This does not meet rule #1.

Counting rule #2: 13 bars go left. This does not meet counting rule #2.

Counting rule #3: 5 bars go into the red bands. This meets counting rule #3.

Counting rule #4: 12 bars are contained entirely in the middle green + white stripe. This does not meet counting rule #4.

Counting rule #5 is irrelevant since another counting rule was met.

Note that that this woman lives in a geographical area where soil and hence food lithium is low, so one could rationally consider the test to have only four bars in the red – still enough to have a high probability of deranged mineral transport due to mercury problems.

Note the vastly elevated bismuth. Bismuth is an ingredient in many women's cosmetic products so this finding has to be interpreted cautiously – especially in the presence of deranged mineral transport. One likely explanation of this result is that the mother uses bismuth containing make up and is affectionate with her sons, one of whom had some bismuth contamination in his hair at the time of testing just as the mother did.

It is not really necessary to resolve the issue of whether bismuth is significant or not since interventions to reduce body inventories of mercury will also reduce body inventories of bismuth.

Note the relation of calcium and magnesium to sodium and potassium. This suggests the possibility of mild adrenal problems.

HAIR ELEMENTS

DOCTOR'S DATA INC.

SEX: Female
AGE: 35

POTENTIALLY TOXIC ELEMENTS

TOXIC ELEMENTS	RESULT µg/g	REFERENCE RANGE	PERCENTILE (68th, 95th)
Aluminum	1.7	< 7.0	
Antimony	0.034	< 0.05	
Arsenic	0.13	< 0.06	
Beryllium	< 0.01	< 0.02	
Bismuth	1.0	< 0.1	
Cadmium	0.01	< 0.1	
Lead	0.08	< 1.0	
Mercury	2.6	< 1.1	
Platinum	< 0.003	< 0.005	
Thallium	< 0.001	< 0.01	
Thorium	< 0.001	< 0.005	
Uranium	0.027	< 0.06	
Nickel	0.02	< 0.4	
Silver	0.06	< 0.15	
Tin	0.19	< 0.3	
Titanium	0.98	< 1.0	
Total Toxic Representation			

ESSENTIAL AND OTHER ELEMENTS

ELEMENTS	RESULT µg/g	REFERENCE RANGE	PERCENTILE (2.5th, 16th, 50th, 84th, 97.5th)
Calcium	227	300- 1200	
Magnesium	25	35- 120	
Sodium	54	12- 90	
Potassium	27	8.0- 38	
Copper	10	12- 35	
Zinc	200	140- 220	
Manganese	0.05	0.15- 0.65	
Chromium	0.38	0.2- 0.4	
Vanadium	0.25	0.018- 0.065	
Molybdenum	0.027	0.028- 0.056	
Boron	1.1	0.3- 2.0	
Iodine	0.55	0.25- 1.3	
Lithium	< 0.004	0.007- 0.023	
Phosphorus	207	160- 250	
Selenium	1.1	0.95- 1.7	
Strontium	0.33	0.5- 7.6	
Sulfur	48000	44500- 52000	
Barium	0.07	0.26- 3.0	
Cobalt	0.004	0.013- 0.05	
Iron	5.7	5.4- 14	
Germanium	0.028	0.045- 0.065	
Rubidium	0.032	0.007- 0.096	
Zirconium	0.12	0.02- 0.42	

SPECIMEN DATA

COMMENTS:

Date Collected:
Date Received:
Date Completed:

Methodology: ICP-MS

Sample Size:
Sample Type:
Hair Color:
Treatment:
Shampoo:

V06.99

RATIOS

ELEMENTS	RATIOS	EXPECTED RANGE
Ca/Mg	9.08	4- 30
Ca/P	1.1	1- 12
Na/K	2	0.5- 10
Zn/Cu	20	4- 20
Zn/Cd	> 999	> 800

This form is provided to Andrew Cutler to be used as an educational tool and does not constitute endorsement by *Doctor's Data*.

Mercury and Silver impair Vision

This eight year old boy has normal social skills and intellect, no apparent developmental issues, and good verbal expressive and receptive skills. However he has serious reading problems that come on after he has been reading for a while, and he is also unable to hit a pitched ball despite otherwise being agile.

The mother considered it key that, despite passing the school vision screening, her son would consistently read a page or two well and then his comprehension and decoding ability would decline. Examination by a behavioral optometrist identified that there were visual issues but they were not clearly delineated by the initial examination though there were clearly substantial deficits in eye tracking ability. Vision therapy was instituted with almost immediate improvement in the boy's reading ability.

The boy also had eczema which was resolved by using digestive enzymes with food. The eczema returns if the enyzmes are discontinued.

The boy's older brother has significant learning and other issues. The older brother isn't bad enough off to legitimately be diagnosed as "high functioning autistic," which has led to a lot of diagnostic gyrations as to exactly what to label him as.

His older sister is very sensitive to sounds.

Neither older brother nor older sister have apparent visual problems.

A maternal uncle has a language specific learning disability, with a much lower verbal than nonverbal IQ. A maternal aunt is believed to have a similar but much less severe problem. There is a first cousin who is autistic and a second cousin with very low functioning autism. All mothers of affected children have night vision problems and early menopause.

Many genetic issues were ruled out by medical testing. The family history does suggest the possibility of an X linked genetic disorder that possibly does not cause problems unless there is a heavy metal exposure.

The older brother is being chelated and there is notable improvement.

The hair test is presented on the facing page, and it counts as follows:

Counting rule #1: 19 bars go right. This does not meet rule #1.

Counting rule #2: 4 bars go left. This meets counting rule #2.

Counting rule #3: 0 bars go into the red bands. This does not meet counting rule #3.

Counting rule #4: 21 bars are contained entirely in the middle green + white stripe. This does not meet counting rule #4.

Counting rule #5: is irrelevant since another rule is met.

In addition there is wildly, vastly elevated silver while the essential elements are not elevated very much. The silver most likely is a true reflection of body burden rather than an artifact of deranged mineral transport.

HAIR ELEMENTS

DOCTOR'S DATA INC

SEX: Male
AGE: 8

POTENTIALLY TOXIC ELEMENTS

TOXIC ELEMENTS	RESULT µg/g	REFERENCE RANGE	PERCENTILE 68th 95th
Aluminum	16	< 8.0	
Antimony	0.085	< 0.066	
Arsenic	0.031	< 0.08	
Beryllium	< 0.01	< 0.02	
Bismuth	0.057	< 0.12	
Cadmium	0.32	< 0.15	
Lead	3.1	< 1.0	
Mercury	0.1	< 0.4	
Platinum	< 0.003	< 0.005	
Thallium	< 0.001	< 0.01	
Thorium	< 0.001	< 0.005	
Uranium	0.024	< 0.06	
Nickel	0.31	< 0.4	
Silver	2.2	< 0.13	
Tin	0.21	< 0.3	
Titanium	0.94	< 1.0	
Total Toxic Representation			

ESSENTIAL AND OTHER ELEMENTS

ELEMENTS	RESULT µg/g	REFERENCE RANGE	PERCENTILE 2.5th 16th 50th 84th 97.5th
Calcium	679	160- 500	
Magnesium	46	12- 50	
Sodium	23	12- 90	
Potassium	31	10- 40	
Copper	30	9.0- 30	
Zinc	170	110- 190	
Manganese	0.18	0.18- 0.6	
Chromium	0.33	0.23- 0.5	
Vanadium	0.044	0.025- 0.1	
Molybdenum	0.079	0.04- 0.089	
Boron	1.8	0.5- 3.5	
Iodine	0.74	0.25- 1.3	
Lithium	0.005	0.007- 0.023	
Phosphorus	239	160- 250	
Selenium	1.6	0.95- 1.7	
Strontium	2.0	0.21- 2.1	
Sulfur	49900	45500- 53000	
Barium	1.0	0.19- 1.6	
Cobalt	0.037	0.013- 0.035	
Iron	9.5	6.0- 17	
Germanium	0.066	0.045- 0.065	
Rubidium	0.035	0.008- 0.08	
Zirconium	0.36	0.06- 0.7	

SPECIMEN DATA

COMMENTS:

Date Collected:
Date Received:
Date Completed:

Methodology: ICP-MS

Sample Size:
Sample Type:
Hair Color:
Treatment:
Shampoo:

V06.99

RATIOS

ELEMENTS	RATIOS	EXPECTED RANGE
Ca/Mg	14.8	4- 30
Ca/P	2.84	0.8- 8
Na/K	0.742	0.5- 10
Zn/Cu	5.67	4- 20
Zn/Cd	531	> 800

This form is provided to Andrew Cutler to be used as an educational tool and does not constitute endorsement by *Doctor's Data*.

Chronic Fatigue and Fibromyalgia are usually due to Heavy Metals

This lady is a pleasant, 65 year old retired nurse who has rheumatoid arthritis, chronic fatigue syndrome, fibromyalgia, GERD[109], irritable bowel syndrome, cataracts, and a chronic intermittent itchy skin rash. She has a total left hip replacement with chronic pain along the titanium implant. She complains of fatigue, hair loss, fuzzy thinking and itchy pain.

She does not show significant neurological problems, no attention deficit, depression, etc., and has no apparent mental, psychiatric or personality issues.

Highlights of her laboratory testing are:

- C reactive protein, very elevated
- Plasma cysteine, very elevated
- Plasma sulfate, midrange normal
- Fibrinogen, elevated
- Estradiol, elevated

Her test, presented on the facing page, counts as follows:

Counting rule #1: 11 bars go right. This does not meet rule #1.

Counting rule #2: 12 bars go left. This does not meet counting rule #2.

Counting rule #3: 3 bars go into the red bands. This does not meet counting rule #3.

Counting rule #4: 12 bars are contained entirely in the middle green + white stripe. This does not meet counting rule #4.

Counting rule #5: 3 bars enter the red zone and 12 bars are entirely within the middle band. This meets counting rule #5, that the test counts one short of positive on any two rules.

Since sodium is not too near potassium, and calcium is not too near magnesium, but they are also not far apart compared to the separation of sodium and potassium from calcium and magnesium this is ambiguous whether there are adrenal problems. Adrenal steroids were prescribed based on the woman's symptoms and they were quite helpful, without negative side effects.

Note the relation of calcium to potassium. This suggests the possibility of thyroid problems. This woman's relevant symptoms were much improved by using thyroid hormone at less than a full replacement dose.

The significance of the vastly elevated silver is unclear. This woman did work in an occupation[110] where substantial exposure to silver was possible. Her symptoms and other tests do not clearly differentiate between mercury poisoning versus a mixed mercury-silver poisoning.

Treatment for H. Pylorii cured her GERD.

Dietary limitation of thiol/sulfur foods and of items excluded on the Feingold diet was implemented and was quite helpful in symptom control.

Nattokinase was prescribed to treat the elevated fibrinogen.

Progesterone cream was prescribed for hormone balancing.

A variety of "supportive" nutritional supplements were prescribed.

Removal of dental mercury amalgam ("silver") fillings followed by chelation with oral DMPS every 8 hours, then with oral DMPS + ALA every 3 hours led to great improvement.

---

[109] GERD is gastroesophageal reflux disease.

[110] She worked as a nurse for most of her adult life and at various times many silver compounds were routinely used for medical purposes.

HAIR ELEMENTS

DOCTOR'S DATA INC.

SEX: Female
AGE: 65

POTENTIALLY TOXIC ELEMENTS

TOXIC ELEMENTS	RESULT µg/g	REFERENCE RANGE
Aluminum	1.3	< 7.0
Antimony	0.091	< 0.05
Arsenic	0.069	< 0.06
Beryllium	< 0.01	< 0.02
Bismuth	0.025	< 0.1
Cadmium	0.21	< 0.1
Lead	0.62	< 1.0
Mercury	0.2	< 1.1
Platinum	< 0.003	< 0.005
Thallium	< 0.001	< 0.01
Thorium	< 0.001	< 0.005
Uranium	0.007	< 0.06
Nickel	0.24	< 0.4
Silver	14	< 0.15
Tin	0.14	< 0.3
Titanium	1.6	< 1.0
Total Toxic Representation		

PERCENTILE: 68th, 95th

ESSENTIAL AND OTHER ELEMENTS

ELEMENTS	RESULT µg/g	REFERENCE RANGE
Calcium	279	300- 1200
Magnesium	17	35- 120
Sodium	100	12- 90
Potassium	120	8.0- 38
Copper	43	12- 35
Zinc	170	140- 220
Manganese	0.23	0.15- 0.65
Chromium	0.38	0.2- 0.4
Vanadium	0.043	0.018- 0.065
Molybdenum	0.031	0.028- 0.056
Boron	1.2	0.3- 2.0
Iodine	2.4	0.25- 1.3
Lithium	0.007	0.007- 0.023
Phosphorus	163	160- 250
Selenium	220	0.95- 1.7
Strontium	0.59	0.5- 7.6
Sulfur	55600	44500- 52000
Barium	0.21	0.26- 3.0
Cobalt	0.005	0.013- 0.05
Iron	5.7	5.4- 14
Germanium	0.065	0.045- 0.065
Rubidium	0.15	0.007- 0.096
Zirconium	0.061	0.02- 0.42

PERCENTILE: 2.5th, 16th, 50th, 84th, 97.5th

SPECIMEN DATA

COMMENTS:
Date Collected: Sample Size:
Date Received: Sample Type:
Date Completed: Hair Color:
Treatment:
Methodology: ICP-MS Shampoo:

V06.99

RATIOS

ELEMENTS	RATIOS	EXPECTED RANGE
Ca/Mg	16.4	4- 30
Ca/P	1.71	1- 12
Na/K	0.833	0.5- 10
Zn/Cu	3.95	4- 20
Zn/Cd	810	> 800

This form is provided to Andrew Cutler to be used as an educational tool and does not constitute endorsement by *Doctor's Data*.

Trisomy 18 Victim and Her Family

The significance of heavy metals in chromosomal abberations is not known. However, the presumption that these things "just happen" with a few uncontrollable risk factors like the mother's age seems unlikely to be the complete story. Also, the families of victims of chromosomal disorders are quite likely to seek alternative care at some point and end up with a hair test since mainstream medicine offers little assistance for these unfortunate people.

The following four tests are hair tests for a family where one of the children has trisomy 18[111], where there are three number 18 chromosomes instead of the usual two. This is an unusual case because trisomy 18 babies rarely survive the first year and this girl has made it to age 11 so far. She is severely mentally retarded and has no language, is fed through a feeding tube surgically placed in her stomach, needs to wear diapers, is extremely constipated, has a red raised rash on her arms and leg which she does not scratch, has a ventral septal heart defect, pulmonary hypertension, has had all immunizations, has had pneumonia in the past, has osteoporosis, bad scoliosis, poor muscle tone, dry eyes, has to lie down all the time, and weighs only 40 pounds. She had been learning to eat with a spoon but began having seizures about 2 year ago and regressed, losing that skill.

The seizures were treated with Valium, Kepra and Lamictil. Later GABA and phosphatidylserine were added to try to reduce the antiseizure medications. Other supplements and dietary modifications were also implemented to try to improve her quality of life. Her hair test is on the facing page. It counts out as shown below:

Counting rule #1: 8 bars go right. This does not meet rule #1.

Counting rule #2: 15 bars go left. This does not meet counting rule #2.

Counting rule #3: 3 bars go into the red bands. This does not meet counting rule #3.

Counting rule #4: 11 bars are contained entirely in the middle green + white stripe. This meets counting rule #4.

Counting rule #5 is irrelevant since another rule is met.

Since this girl has a chromosomal abnormality it is a very open question whether she should be expected to have the same kind of mineral transport as a chromosomally typical person would.

One possible interpretation of this test is that it meets the counting rules and has deranged mineral transport due to mercury, as supported by the high tin levels, and the arsenic may or may not mean anything.

Another possible interpretation of is that the observed derangement of mineral transport is due to the effect of trisomy 18 on mineral metabolism, has nothing to do with mercury, and the tin and arsenic levels may be completely normal for someone with trisomy 18.

The interpretation of too much mercury is supported by the finding that her mother and sister both have deranged mineral transport on their hair tests, on the following pages. The possibility that the arsenic measurement truly represents a high level of arsenic in her body is somewhat supported by the significant amount of arsenic in her father's hair.

[111] Edward's syndrome.

HAIR ELEMENTS

DOCTOR'S DATA INC.

SEX: Female
AGE: 11

POTENTIALLY TOXIC ELEMENTS

TOXIC ELEMENTS	RESULT µg/g	REFERENCE RANGE	PERCENTILE 68th 95th
Aluminum	5.5	< 8.0	
Antimony	0.084	< 0.066	
Arsenic	0.26	< 0.06	
Beryllium	< 0.01	< 0.02	
Bismuth	0.08	< 0.12	
Cadmium	0.052	< 0.1	
Lead	0.1	< 1.0	
Mercury	0.07	< 0.4	
Platinum	< 0.003	< 0.005	
Thallium	< 0.001	< 0.01	
Thorium	< 0.001	< 0.005	
Uranium	< 0.001	< 0.06	
Nickel	0.41	< 0.4	
Silver	0.03	< 0.16	
Tin	1.1	< 0.3	
Titanium	0.86	< 1.0	
Total Toxic Representation			

ESSENTIAL AND OTHER ELEMENTS

ELEMENTS	RESULT µg/g	REFERENCE RANGE	PERCENTILE 2.5th 16th 50th 84th 97.5th
Calcium	333	250- 800	
Magnesium	20	25- 90	
Sodium	29	12- 90	
Potassium	41	7.0- 40	
Copper	10	12- 35	
Zinc	200	120- 220	
Manganese	0.07	0.18- 0.6	
Chromium	0.3	0.23- 0.45	
Vanadium	0.024	0.025- 0.1	
Molybdenum	0.038	0.037- 0.083	
Boron	6.2	0.3- 1.7	
Iodine	1.8	0.25- 1.3	
Lithium	0.004	0.007- 0.023	
Phosphorus	188	160- 250	
Selenium	0.89	0.95- 1.7	
Strontium	0.34	0.37- 3.6	
Sulfur	49100	45500- 53000	
Barium	0.17	0.21- 1.9	
Cobalt	0.012	0.013- 0.05	
Iron	7.9	6.0- 17	
Germanium	0.063	0.045- 0.065	
Rubidium	0.032	0.008- 0.08	
Zirconium	0.57	0.03- 0.4	

SPECIMEN DATA

COMMENTS:

Date Collected:	Sample Size:
Date Received:	Sample Type:
Date Completed:	Hair Color:
	Treatment:
Methodology: ICP-MS	Shampoo:

V06.99

RATIOS

ELEMENTS	RATIOS	EXPECTED RANGE
Ca/Mg	16.7	4- 30
Ca/P	1.77	1- 12
Na/K	0.707	0.5- 10
Zn/Cu	20	4- 20
Zn/Cd	> 999	> 800

This form is provided to Andrew Cutler to be used as an educational tool and does not constitute endorsement by *Doctor's Data*.

The ten year old sister of the girl with trisomy 18 described on the previous page is having a really hard time in school and probably has some kind of learning disability. Her hair test is presented on the facing page. It counts as shown here:

Counting rule #1: 18 bars go right. This does not meet rule #1.

Counting rule #2: 5 bars go left. This does meet counting rule #2.

Counting rule #3: 2 bars go into the red bands. This does not meet counting rule #3.

Counting rule #4: 19 bars are contained entirely in the middle green + white stripe. This does not meet counting rule #4.

Counting rule #5 is irrelevant since another rule has already been met.

Note the vast elevation of copper. This represents a very elevated body burden of copper, well into the toxic region. Since most copper and the majority of mercury are excreted by the liver in bile, anything that reduces bile flow or impairs the transfer of metals into bile can lead to copper + mercury toxicity as is seen here.

The elevated silver is of unclear significance. It needs to be considered in light of what is going on with her, what her personality is like, and whether there is some known source of either contamination or exposure.

HAIR ELEMENTS

DOCTOR'S DATA INC.

SEX: Female
AGE: 10

POTENTIALLY TOXIC ELEMENTS

TOXIC ELEMENTS	RESULT µg/g	REFERENCE RANGE	PERCENTILE 68th 95th
Aluminum	15	< 8.0	
Antimony	0.060	< 0.066	
Arsenic	0.023	< 0.060	
Beryllium	< 0.01	< 0.020	
Bismuth	0.12	< 0.12	
Cadmium	0.24	< 0.10	
Lead	2.4	< 1.0	
Mercury	0.09	< 0.40	
Platinum	< 0.003	< 0.005	
Thallium	< 0.001	< 0.010	
Thorium	< 0.001	< 0.005	
Uranium	0.029	< 0.060	
Nickel	0.76	< 0.40	
Silver	2.1	< 0.16	
Tin	0.56	< 0.30	
Titanium	0.77	< 1.0	
Total Toxic Representation			

ESSENTIAL AND OTHER ELEMENTS

ELEMENTS	RESULT µg/g	REFERENCE RANGE	PERCENTILE 2.5th 16th 50th 84th 97.5th
Calcium	1430	250- 800	
Magnesium	39	25- 90	
Sodium	17	12- 90	
Potassium	8	7- 40	
Copper	140	12- 35	
Zinc	210	120- 220	
Manganese	0.53	0.18- 0.60	
Chromium	0.32	0.23- 0.45	
Vanadium	0.041	0.025- 0.10	
Molybdenum	0.062	0.037- 0.083	
Boron	1.1	0.30- 1.7	
Iodine	0.35	0.25- 1.3	
Lithium	0.077	0.007- 0.023	
Phosphorus	211	160- 250	
Selenium	1.3	0.95- 1.7	
Strontium	3.2	0.37- 3.6	
Sulfur	50800	45500- 53000	
Barium	0.98	0.21- 1.9	
Cobalt	0.050	0.013- 0.050	
Iron	12	6.0- 17	
Germanium	0.078	0.045- 0.065	
Rubidium	0.028	0.008- 0.080	
Zirconium	0.25	0.030- 0.40	

SPECIMEN DATA

COMMENTS:

Date Collected:	Sample Size:
Date Received:	Sample Type:
Date Completed:	Hair Color:
	Treatment:
Methodology: ICP-MS	Shampoo:

V06.99

RATIOS

ELEMENTS	RATIOS	EXPECTED RANGE
Ca/Mg	36.7	4- 30
Ca/P	6.78	1- 12
Na/K	2.13	0.5- 10
Zn/Cu	1.5	4- 20
Zn/Cd	875	> 800

This form is provided to Andrew Cutler to be used as an educational tool and does not constitute endorsement by *Doctor's Data*.

The mother of the two girls discussed in the last few pages has severe hearing loss believed to be due to poor nerve function. She is overweight, has mercury (silver amalgam) dental fillings and root canals, high cholesterol, poor regulation of blood sugar, is somewhat obsessive and has other cognitive and behavioral issues that suggest the possibility of an underlying metabolic problem of the sort that heavy metals can cause.

Her hair test is presented on the facing page and counts as follows:

Counting rule #1: 6 bars go right. This does not meet rule #1.

Counting rule #2: 17 bars go left. This does not meet counting rule #2.

Counting rule #3: 4 bars go into the red bands. This meets counting rule #3.

Counting rule #4: 12 bars are contained entirely in the middle green + white stripe. This does not meet counting rule #4.

Counting rule #5 is irrelevant since another counting rule was already met.

Note the substantial difference between potassium and calcium, suggesting the possibility of thyroid problems.

The Ca/Mg ratio is almost out of range, suggesting blood sugar regulation problems which are apparent from other clinical laboratory testing and from what is going on with her. The numbers at the end of the “expected range” or “normal range” are not magic, some people who score just inside the range have major problems while a few who score just outside the range are fine.

HAIR ELEMENTS

DOCTOR'S DATA INC.

SEX: Female
AGE: 49

POTENTIALLY TOXIC ELEMENTS

TOXIC ELEMENTS	RESULT µg/g	REFERENCE RANGE	PERCENTILE 68th 95th
Aluminum	7.1	< 7.0	
Antimony	0.025	< 0.050	
Arsenic	0.023	< 0.060	
Beryllium	< 0.01	< 0.020	
Bismuth	0.085	< 0.10	
Cadmium	0.073	< 0.10	
Lead	0.19	< 1.0	
Mercury	0.15	< 1.1	
Platinum	< 0.003	< 0.005	
Thallium	< 0.001	< 0.010	
Thorium	< 0.001	< 0.005	
Uranium	0.004	< 0.060	
Nickel	0.07	< 0.40	
Silver	0.01	< 0.15	
Tin	0.36	< 0.30	
Titanium	0.76	< 1.0	
Total Toxic Representation			

ESSENTIAL AND OTHER ELEMENTS

ELEMENTS	RESULT µg/g	REFERENCE RANGE	PERCENTILE 2.5th 16th 50th 84th 97.5th
Calcium	562	300- 1200	
Magnesium	19	35- 120	
Sodium	27	12- 90	
Potassium	< 3	8- 38	
Copper	14	12- 35	
Zinc	200	140- 220	
Manganese	0.12	0.15- 0.65	
Chromium	0.19	0.20- 0.40	
Vanadium	0.015	0.018- 0.065	
Molybdenum	0.014	0.028- 0.056	
Boron	0.38	0.30- 2.0	
Iodine	0.63	0.25- 1.3	
Lithium	< 0.004	0.007- 0.023	
Phosphorus	225	160- 250	
Selenium	1.0	0.95- 1.7	
Strontium	1.3	0.50- 7.6	
Sulfur	57500	44500- 52000	
Barium	0.37	0.26- 3.0	
Cobalt	0.010	0.013- 0.050	
Iron	5.6	5.4- 14	
Germanium	0.067	0.045- 0.065	
Rubidium	0.005	0.007- 0.096	
Zirconium	0.31	0.020- 0.42	

SPECIMEN DATA

COMMENTS:

Date Collected: | Sample Size:
Date Received: | Sample Type:
Date Completed: | Hair Color:
Treatment:
Methodology: ICP-MS | Shampoo:

V06.99

RATIOS

ELEMENTS	RATIOS	EXPECTED RANGE
Ca/Mg	29.6	4- 30
Ca/P	2.5	1- 12
Na/K	9	0.5- 10
Zn/Cu	14.3	4- 20
Zn/Cd	> 999	> 800

This form is provided to Andrew Cutler to be used as an educational tool and does not constitute endorsement by *Doctor's Data.*

The father of the two girls discussed above is a self employed handyman. He denies having any significant health problems. His health care provider was concerned this might not be so but he declined to do testing or fill in questionnaires to help identify it. His wife filled out some for him which suggested the possibility of significant neurological or cognitive problems with attention deficit and slow cognition being the most apparent but not the only issues. He does not smoke and is somewhat overweight. He was quite obsessed with his 11 year old daughter's high hair arsenic, possibly due to concern that he might have somehow caused it through working with treated lumber.

The father's hair test is presented on the facing page. It counts as follows:

Counting rule #1: 13 bars go right. This does not meet rule #1.

Counting rule #2: 12 bars go left. This does not meet counting rule #2.

Counting rule #3: 1 bar goes into the red bands. This does not meet counting rule #3.

Counting rule #4: 18 bars are contained entirely in the middle green + white stripe. This does not meet counting rule #4.

Counting rule #5 is not met.

This is a very ordinary, average, normal looking test result. The arsenic and antimony are not elevated enough to be expected to have a significant impact on health and there is a source of occupational exposure to account for them being moderately elevated (arsenic preserved lumber and antimony in carpets and paint).

HAIR ELEMENTS

DOCTOR'S DATA INC

SEX: Male
AGE: 47

POTENTIALLY TOXIC ELEMENTS

TOXIC ELEMENTS	RESULT µg/g	REFERENCE RANGE	PERCENTILE 68th 95th
Aluminum	12	< 7.0	
Antimony	0.13	< 0.066	
Arsenic	0.19	< 0.080	
Beryllium	< 0.01	< 0.020	
Bismuth	0.026	< 0.060	
Cadmium	0.37	< 0.15	
Lead	2.0	< 2.0	
Mercury	0.17	< 1.1	
Platinum	< 0.003	< 0.005	
Thallium	< 0.001	< 0.010	
Thorium	< 0.001	< 0.005	
Uranium	0.019	< 0.060	
Nickel	0.28	< 0.40	
Silver	0.09	< 0.12	
Tin	0.20	< 0.30	
Titanium	0.61	< 1.0	
Total Toxic Representation			

ESSENTIAL AND OTHER ELEMENTS

ELEMENTS	RESULT µg/g	REFERENCE RANGE	PERCENTILE 2.5th 16th 50th 84th 97.5th
Calcium	275	200- 750	
Magnesium	13	25- 75	
Sodium	32	12- 90	
Potassium	20	9- 40	
Copper	15	10- 28	
Zinc	180	130- 200	
Manganese	0.23	0.15- 0.65	
Chromium	0.49	0.20- 0.40	
Vanadium	0.095	0.018- 0.065	
Molybdenum	0.048	0.025- 0.064	
Boron	1.3	0.40- 3.0	
Iodine	1.3	0.25- 1.3	
Lithium	0.010	0.007- 0.023	
Phosphorus	182	160- 250	
Selenium	1.2	0.95- 1.7	
Strontium	0.67	0.30- 3.5	
Sulfur	51800	44500- 52000	
Barium	0.35	0.16- 1.6	
Cobalt	0.010	0.013- 0.035	
Iron	9.5	5.4- 13	
Germanium	0.089	0.045- 0.065	
Rubidium	0.024	0.011- 0.12	
Zirconium	0.27	0.020- 0.44	

SPECIMEN DATA

COMMENTS:

Date Collected:	Sample Size:
Date Received:	Sample Type:
Date Completed:	Hair Color:
	Treatment:
Methodology: ICP-MS	Shampoo:

V06.99

RATIOS

ELEMENTS	RATIOS	EXPECTED RANGE
Ca/Mg	21.2	4- 30
Ca/P	1.51	0.8- 8
Na/K	1.6	0.5- 10
Zn/Cu	12	4- 20
Zn/Cd	486	> 800

This form is provided to Andrew Cutler to be used as an educational tool and does not constitute endorsement by *Doctor's Data.*

How Lactation affects a Hair Test

The mother of the autistic boy whose case was just discussed on pages 216-218 had hair tests done while she was breast feeding, and several months after she stopped. Here are the two hair tests.

While these tests look relatively normal, and the mineral transport in the other tests shown in this book for a lactating woman (pages 166-167) also appeared relatively normal, I have also seen lactating women's hair tests that met counting rule #1, the "all low" presentation. Some of these women were toxic but some apparently were in good health. Some of ones in good health were on an exclusively vegetarian diet and this may have accounted for the appearance of their test results despite the fact that vegetarianism does not otherwise appear to change the results for normal healthy people.

This particular woman's hair tests count as follows:

	While nursing (page 233)	After weaning (page 234)
Counting rule #1: bars to right.	10	16
Counting rule #2: bars to left.	13	7
Counting rule #3: bars in red.	2	0
Counting rule #4: bars in middle	15	16
Counting rule #5: on short on two othe rules	no	no

Neither test meets any counting rule. The first test is not uncommon statistically, while the second one is unusual but neither suspicious nor abnormal due to the seven bars going left.

This woman did live in a geographical region where very high hair uranium levels are common. Typically those over 0.5 are associated with fatigue and other nonspecific symptoms which improve (and hair uranium falls) on increasing dietary iron intake.

The sudden apparent elevation of bismuth on test #2 seems more likely due to contamination from cosmetic use than due to an actual sudden increase in body burden.

The significance of the aluminum is unclear. Aluminum can become elevated due to increased retention caused by mercury induced derangement of mineral transport. Aluminum is also the active ingredient in many antiperspirants and is in many food products, antacids and some medicines.

HAIR ELEMENTS

06/99

DOCTOR'S DATA INC

SEX: Female
AGE: 38

POTENTIALLY TOXIC ELEMENTS

TOXIC ELEMENTS	RESULT µg/g	REFERENCE RANGE	PERCENTILE (68th, 95th)
Aluminum	17	< 7.0	
Antimony	0.02	< 0.05	
Arsenic	0.018	< 0.06	
Beryllium	< 0.01	< 0.02	
Bismuth	0.068	< 0.1	
Cadmium	0.022	< 0.1	
Lead	0.17	< 1.0	
Mercury	0.26	< 1.1	
Platinum	< 0.003	< 0.005	
Thallium	< 0.001	< 0.01	
Thorium	< 0.001	< 0.005	
Uranium	0.42	< 0.06	
Nickel	0.25	< 0.4	
Silver	0.05	< 0.15	
Tin	0.25	< 0.3	
Titanium	1.6	< 1.0	
Total Toxic Representation			

ESSENTIAL AND OTHER ELEMENTS

ELEMENTS	RESULT µg/g	REFERENCE RANGE	PERCENTILE (2.5th, 16th, 50th, 84th, 97.5th)
Calcium	876	300- 1200	
Magnesium	140	35- 120	
Sodium	29	12- 90	
Potassium	14	8.0- 38	
Copper	12	12- 35	
Zinc	140	140- 220	
Manganese	0.1	0.15- 0.65	
Chromium	0.22	0.2- 0.4	
Vanadium	0.058	0.018- 0.065	
Molybdenum	0.085	0.028- 0.056	
Boron	< 0.07	0.3- 2.0	
Iodine	6.3	0.25- 1.3	
Lithium	0.019	0.007- 0.023	
Phosphorus	194	160- 250	
Selenium	0.89	0.95- 1.7	
Strontium	8.6	0.5- 7.6	
Sulfur	46600	44500- 52000	
Barium	2.1	0.26- 3.0	
Cobalt	0.048	0.013- 0.05	
Iron	7.9	5.4- 14	
Germanium	0.034	0.045- 0.065	
Rubidium	0.021	0.007- 0.096	
Zirconium	0.37	0.02- 0.42	

SPECIMEN DATA

COMMENTS:

Date Collected: 10/26/2001
Date Received:
Date Completed:
Methodology: ICP-MS

Sample Size:
Sample Type:
Hair Color:
Treatment:
Shampoo:

RATIOS

ELEMENTS	RATIOS	EXPECTED RANGE
Ca/Mg	6.22	4- 30
Ca/P	4.51	1- 12
Na/K	2.07	0.5- 10
Zn/Cu	12.1	4- 20
Zn/Cd	> 999	> 800

This form is provided to Andrew Cutler to be used as an educational tool and does not constitute endorsement by *Doctor's Data.*

HAIR ELEMENTS

DOCTOR'S DATA INC

SEX: Female
AGE: 40

POTENTIALLY TOXIC ELEMENTS

TOXIC ELEMENTS	RESULT µg/g	REFERENCE RANGE	PERCENTILE 68th 95th
Aluminum	23	< 7.0	
Antimony	0.018	< 0.050	
Arsenic	0.010	< 0.060	
Beryllium	< 0.01	< 0.020	
Bismuth	1.0	< 0.10	
Cadmium	0.023	< 0.10	
Lead	0.12	< 1.0	
Mercury	< 0.03	< 1.1	
Platinum	< 0.003	< 0.005	
Thallium	< 0.001	< 0.010	
Thorium	< 0.001	< 0.005	
Uranium	0.70	< 0.060	
Nickel	0.30	< 0.40	
Silver	0.04	< 0.15	
Tin	0.26	< 0.30	
Titanium	1.1	< 1.0	
Total Toxic Representation			

ESSENTIAL AND OTHER ELEMENTS

ELEMENTS	RESULT µg/g	REFERENCE RANGE	PERCENTILE 2.5th 16th 50th 84th 97.5th
Calcium	1900	300- 1200	
Magnesium	300	35- 120	
Sodium	58	12- 90	
Potassium	7	8- 38	
Copper	16	12- 35	
Zinc	220	140- 220	
Manganese	0.10	0.15- 0.65	
Chromium	0.23	0.20- 0.40	
Vanadium	0.081	0.018- 0.065	
Molybdenum	0.053	0.028- 0.056	
Boron	0.32	0.30- 2.0	
Iodine	0.65	0.25- 1.3	
Lithium	0.041	0.007- 0.023	
Phosphorus	209	160- 250	
Selenium	1.1	0.95- 1.7	
Strontium	13	0.50- 7.6	
Sulfur	49600	44500- 52000	
Barium	1.9	0.26- 3.0	
Cobalt	0.035	0.013- 0.050	
Iron	7.5	5.4- 14	
Germanium	0.062	0.045- 0.065	
Rubidium	0.013	0.007- 0.096	
Zirconium	0.38	0.020- 0.42	

SPECIMEN DATA

COMMENTS:

Date Collected: 8/10/2003
Date Received:
Date Completed:
Methodology: ICP-MS

Sample Size:
Sample Type:
Hair Color:
Treatment:
Shampoo:

V06.99

RATIOS

ELEMENTS	RATIOS	EXPECTED RANGE
Ca/Mg	6.33	4- 30
Ca/P	9.09	1- 12
Na/K	8.29	0.5- 10
Zn/Cu	13.8	4- 20
Zn/Cd	> 999	> 800

This form is provided to Andrew Cutler to be used as an educational tool and does not constitute endorsement by *Doctor's Data.*

What to do about Heavy Metals and other Hair Test Results

Important Notes and Caveats

This section is NOT intended as a complete guide to treatment. It is intended as an outline of a few important factors. More complete discussions may be found in *Amalgam Illness: Diagnosis and Treatment*, *The Biochemical Option*, and a variety of other sources.

Remember that people whose bodies manage to accumulate some or other element in excess when their neighbors don't often have uniquely individual biochemistries and all kinds of interesting and unexpected reactions to supposedly benign medicines and nutritional supplements. Also remember that this book is based on probabilities and statistics, but most of its use is going to be finding people who do not fit the statistical rule, are far from average, and thus may not respond like most other people to whatever intervention is tried.

Also please note that I list a lot of treatments that involve taking prescription drugs or nutritional supplements in the same manner as drugs are used. Part of why I do this is that it is a natural way to think of things since I was trained as a chemist and have a lot of experience doing things this way. Partly it is because some people are quite toxic and not much else seems to work for them. However this is not meant to imply that other (non-drug) methods may not be adequate for people with less severe toxicity.

Besides, being a chemist, I understand how you can use chemicals to affect the basic chemical processes of metabolism. Some of these chemicals are considered food in a legal and practical sense, some are considered nutritional supplements in a legal sense, and some are considered prescription or over the counter medication in a legal sense.

These legal distinctions are not firmly based in chemical reality. Some foods and nutritional supplements are unnatural synthetic materials, some prescription medications are natural herbal extracts (e. g. morphine, digitalis). Some supplements are quite powerful, some medications have mild effects. While it is true that there is a correlation between something requiring a prescription and being able to really mess people up with severe side effects, this does not mean that only prescription agents are powerful (and dangerous), or that only over the counter agents are safe.

Prescription drugs do have a bad reputation among many health care practitioners and members of the public because in certain circumstances very powerful and dangerous medications are routinely used in a cavalier and indiscriminate fashion, resulting in enormous human suffering. The real problem here is that some prescribers are irresponsible, not that prescription drugs are of necessity all unpleasant and dangerous.

Since I find little chemical sense in the existing legal distinctions, I just talk about the chemicals and what you can use them for. I do this because I know what the chemicals do to people. This is not to say that chemicals are the only way to help sick people – far from it! There are many and varied approaches that to some extent overlap and are interchangeable, and to some extent have their own unique benefits. For example, it is seldom possible to replace manual medicine such as chiropractic adjustment, osteopathic manipulation, or various forms of therapeutic massage or occupational therapy with a handful of pills.

It is very important to interpret unusual hair test results in light of other information, rather than using them in isolation. The most important information to consider is what is going on with the person the hair is from. Review of other materials is crucial to relating the numbers and bars on the hair test to the signs and symptoms of the person whose hair was analyzed. One source is the discussion earlier in this book of how the different elements affect people. If there is any uncertainty or need for additional information, further research is appropriate.

Health care professionals as well as readers who are managing their own care are strongly suggested to review some quality mainstream sources such as *Merritt's Textbook of Neurology*, *Neurology* (by Goetz), *Hunter's Diseases of Occupations* (as published prior to 1970), and 1940-1960 editions of standard internal medicine and toxicology books.

The homeopathic literature is extremely helpful in figuring out whether a given person has a problem with a particular element. It doesn't matter if you believe in homeopathy or not, a homeopathic materia medica is a collection of notes physicians took on observing people poisoned with the relevant substances. One must keep in mind that homeopathic writings are organized in a very different way than allopathic writings and that many of them are quite old, dating to times when concepts of health and intoxication were somewhat different than they are now. With careful thought and reading, however, the homeopathic literature is the most detailed and complete description available of the effects of most toxins.

Please also note that some laboratories and practitioners suggest adrenal cortex extract and vitamin E for essentially everyone, and in fact these are helpful for almost everyone who has a health condition for which they end up getting a hair test.

General Commentary on managing Toxic People

See *Amalgam Illness: Diagnosis and Treatment* for adequately detailed information to actually manage a case. A book forthcoming in 2005 entitled *The Biochemical Option: Effective treatments for autism, ADHD, developmental disorder and learning disabilities* will focus on children. Other books of interest are *Biological Treatment of Autism and PDD* as well as *Children with Starving Brains*[112]. The material below is the barest smattering of information so as to avoid going off in the wrong direction.

Select the appropriate detoxification protocol.

Select at least some basic antioxidant nutritional supplements to be taken several times a day. Without these chelation is unnecessarily stressful and unpleasant.

[112] For information on these and other books, or to obtain them, see www.noamalgam.com.

In the long term, detoxification is the most powerful intervention by far. In the short term, dietary modification is the most powerful intervention for symptom relief, the use of nutritional supplements in large quantities is next best, and the use of prescription medications is usually not too helpful except for specific situations where the medication is an exact match for a problem the person has.

Dietary modification is dramatically harder for people to do than is taking supplements. For practical reasons most people start with nutritional supplements and later decide if dietary modification is justified by their need for symptom relief. It is relatively common for dietary control to be used for impaired children and less common for it to be used for adults.

After a hair test suggests toxicity, or it is suspected for some other reason, many people choose to do some medical laboratory testing to understand what is going on in their bodies as a consequence of the heavy metal problem. There are many, many possible tests that might be relevant to a given person's situation. There are a few tests that are relevant often enough that I will mention them.

- Plasma cysteine
- Plasma sulfate
- Blood count (CBC with differential)
- Free T3, free T4, T3, T4, TSH
- Hemoglobin A1c

Of the thyroid tests, free T3 and free T4 are far more valuable than the others, and TSH is of almost no value unless some of the other parameters have also been measured[113].

With people in whom the status of mineral transport is unclear, or who have unusual values of an essential element, red blood cell or whole blood elements can be useful.

The natural history of mercury detoxification is quite complex. It is discussed on page 52 of *Amalgam Illness: Diagnosis and Treatment.*[114] For most other elements, slow improvement with ongoing detoxification is the rule. For essential elements improvement is often complete in months, while for toxic elements it may proceed over a period of years.

Generally, heavy metal detoxification involves an exponential decay in symptoms. For example, half the problems might be resolved in the first 6 months, half the remaining problems (a quarter of the original ones) resolved in the next 6 months, etc. Mercury is the exception to this, with a few months of improvement, several months of worsening, and then slow improvement over many more months.

General comments on chelation

I will discuss the chelating agents alpha lipoic acid (ALA), dimercaptosuccinic acid (DMSA), dimercaptopropane sulfonate sodium (DMPS), and ethylenediamine tetraacetic acid (EDTA). Here is some general discussion of their use and properties, applicable to all mentions of them later in this section.

Ultimately curative therapy for people who have a heavy metal problem involves use of chelating agents that effectively remove the toxic metal from all relevant parts of their bodies.

ALA crosses cell membranes and crosses the blood-brain barrier. DMPS, DMSA and EDTA do not.

Glutathione, methionine, chlorella, MSM, NAC, TTFD, allithiamine and high thiol diets are not chelating agents in the chemical sense and do not remove most heavy metals.

ALA is over the counter and sold as a nutritional supplement.

[113] See pages 115-118 of *Amalgam Illness: Diagnosis and Treatment* for a discussion of thyroid issues. This book may be obtained at www.noamalgam.com.

[114] To obtain this book see www.noamalgam.com.

DMPS is prescription only, available only from compounding pharmacies. Prices vary A LOT from pharmacy to pharmacy, I suggest calling around for quotes.

DMSA and EDTA are available both by prescription and as over the counter nutritional supplements.

It is very seldom necessary to inject these, and in fact doing so can be quite dangerous.

It is necessary to give chelating agents (and most other pharmacologic agents) often enough to maintain reasonably steady blood levels. Doing this gently removes toxin from the body, giving these agents less often than necessary causes concentration of the toxin into the target organ and often makes the person worse rather than better.

EDTA and DMPS have to be given every 8 hours or more often.

DMSA has to be given every 4 hours or more often.

ALA has to be given every 3 hours or more often. Four hours is OK at night.

Generally it is necessary to chelate for 3 or more days in a row, then skip some and repeat. For some agents this requires taking them at night. People who do this find out it is not at all difficult to either roll over, gulp down a pill and go back to sleep, or to pop a spoonful of something soft with the chelator in it into their sleeping child's mouth and then go back to bed. Skipping the nighttime doses can lead to disastrous worsening of symptoms and it is very important to stop the cycle if one is missed.

Some people can chelate for 3 days a week, others need longer breaks. This is somewhat individual and needs to be determined by trial and error.

For chronically poisoned people it often takes many many cycles of chelation to get them well.

For DMPS, DMSA, EDTA and ALA, start with about 1/4 mg per pound of person per dose. For DMPS, DMSA and ALA, adjust up or down a bit depending on side effects, but don't use more than 1 mg per pound. Once it is adjusted, stick with it. For EDTA double the dose every couple of days until you reach about 5-10 mg per pound of person. If you start too high with the EDTA it causes a lot of nausea, by working up tolerance develops and nausea is avoided.

All the chelators cause oxidative stress, you have to give lots of antioxidants frequently before, during and after chelation cycles. Vitamins C, E, zinc and magnesium are very important supplements. Using chelators with no antioxidants can be truly miserable.

People who find a chelation protocol "rough" often do much better if the interval between doses is DECREASED, and find it gets rougher if the interval is INCREASED. While obvious from pharmacokinetic considerations, this is apparently counterintuitive for many health care providers.

Another counterintuitive factor is that metal removal is NOT proportional to dosage. It goes as a bit less than the square root. People often make themselves miserable trying to struggle through taking the highest tolerable dose and are generally much better off halving it, only slowing down extraction 25-30%, and being enough more comfortable they really can chelate on a routine schedule. Routine and frequent chelation is essential. Being miserable is not at all a necessary part of chelation.

When I refer to alkalinizing the urine, I mean to give sodium bicarbonate (baking soda) or citrate or malate form supplements until the pH of the urine is 8 or more. Your doctor may be able to come up with pH test strips or you can try to get something appropriate at a swimming pool supply store. The thiol based chelators will drop off some of the metals in the kidneys if the urine is too acidic. Alkalinizing the urine can make some people less able to tolerate protein in their diets (by making it harder to get rid of ammonia).

Expected side effects are an increase in symptoms or appearance of new symptoms during the chelation cycle and for up to one day afterwards. People are often tired and perhaps cranky or depressed the day after chelation. Prolonged exacerbation of symptoms or appearance of new symptoms indicate something is wrong – there is no "healing crisis" with these substances.

When using ALA with DMPS or DMSA, leaving the ALA out of the last dose or two of the cycle often reduces the day after effects substantially.

There is no appropriate tracking test to decide when chelation is over. All laboratory tests will be normal before the toxin is adequately cleared from the brain and internal organs. The way to decide when chelation is over is when there are no side effects on chelation and no further progress is obtained.

Based on the high reported adverse reaction rate, the innaccurate summaries of some journal papers given in the sales literature, and the lack of consideration for heat transfer and volatile control in most designs it is my opinion infrared saunas are never appropriate for detoxification.

There are many nutritional supplements advertized for chelation or detoxification. The agents above work as stated, when taken as stated. The fundamental chemical and kinetic properties of these agents are not changed by advertizing claims.

Heavy Metal Hair Test Results

Aluminum (Al)

The first thing to be sure of if there is high hair aluminum is that it is not a sign of a mercury problem since mercury often causes aluminum retention.

Regardless of the source, aluminum levels can be lowered by reducing exposure. This is done by excluding processed cheese, foods using aluminum containing baking powders, and foods cooked in aluminum cookware from the diet (coated aluminum, e.g. Teflon, is OK for this purpose). Also ensure that no antacids are being used that contain aluminum and that any antiperspirant product does not contain it. The "natural crystal" antiperspirants are in fact aluminum salts.

It is not generally appropriate to chelate aluminum. In cases of acute and very high level toxicity, the injectable agent deferoxamine can be used. This must be used in frequent doses, and it is best if it is administered over some time in an iv drip rather than over a shorter period by syringe. It should if possible be given every 6 to 8 hours. Less frequent use is well known to cause problems and this is even discussed in recent editions of many mainstream medical textbooks.

Antimony (Sb)

Antimony can be chelated with DMSA or DMPS and alpha lipoic acid (ALA). The use of methylating supplements such as TMG, vitamin B-12 and folic acid may improve antimony clearance.

Supportive therapy specific to antimony includes a high intake of magnesium and of potassium, and heart support supplements such as coenzyme Q10, carnitine, and hawthorne extract. A high vitamin content skin cream[115] can be used for skin problems. A high protein

[115] for example, Rejuvenex® from the Life Extension Foundation, www.lef.org or any similar product.

restricted carbohydrate diet will make many antimony victims feel better.

Antimony toxicity will sometimes impair the handling of magnesium and potassium and have consequent effects on the heart. In this case serum potassium should be kept over 4.0 meq/dl and serum magnesium over 1.8 mg/dl if there is QT prolongation on the EKG or any symptoms of heart problems. Red blood cell potassium and magnesium are more useful measures than serum levels. RBC levels should be kept in the green zone (*Doctor's Data* test), or in the normal range (*Metametrix* or *Great Smokies* test) or at least 1/3 of the way up from the lower limit of the normal range on a "health plan" laboratory report. Doing this may require taking magnesium and potassium supplements with every meal.

A potassium chloride containing salt substitute may be used liberally on food as a more convenient source of potassium than pills. Potassium should not be taken by itself without food as it can cause dangerous irregularities in heart function. Taste should be relied on as a guide to preventing too great a potassium intake – that is, only salt the food to taste with potassium containing salt and do not eat it if it tastes bad. Some modest metallic taste is to be expected but should not be unpleasant. In any case, intakes of potassium exceeding 1/2 tsp. of potassium chloride based salt substitute spread throughout the day and mixed with food require a physician's supervision and relevant testing.

Laboratory tests to consider are plasma cysteine, blood or urine pyruvate, fasting insulin, heart disease markers and a liver function panel.

Arsenic (As)

If hair arsenic is very elevated in the presence of normal orderly mineral transport then arsenic toxicity should be suspected and treated by chelation. Since arsenic is a well known carcinogen the presence of clear and unambiguous symptoms of arsenic poisoning are not necessary in order to justify chelation to reduce body levels of arsenic.

Alpha lipoic acid is the best agent for arsenic chelation. Optionally DMPS can be used with it as an accessory chelator. DMSA is less effective in this accessory chelator role than DMPS is.

The body naturally clears some arsenic by methylation. Vitamin B-12, folic acid and trimethylgycline (TMG) should be used to the extent tolerated. Typically several thousand micrograms (several milligrams) of B-12 one or more times a day, 400-1,200 mcg of folic acid with meals and at bedtime, and 500-2000 mg of TMG earlier in the day are used. Some people become angry, agitated aggressive and unmotivated if they take too much in the way of methylating supplements.

Of there are skin eruptions some high vitamin content skin cream[116] will improve them.

Arsenic poisoned people often do not handle carbohydrates well. A trial of a low carbohydrate or even Atkins style diet may be worthwhile. Elevated pyruvate in blood or urine is a sign of these carbohydrate handling problems. People with inadequate adrenal function need some carbohydrate in their diet. Signs of inadequate adrenal function are low blood pressure (less than 100/67), frequent nausea, general weakness, or stress intolerance.

People with arsenic and particularly those with arsenic and mercury often become chemically sensitive due to deranged liver metabolism. These people will often say they are "allergic" to chemicals to fragrances like perfumes. They may have anxiety or panic attacks when exposed to automobile exhaust. There are many means of addressing this

[116] for example, Rejuvenex® from the Life Extension Foundation, www.lef.org or any similar product.

problem such as avoiding carotenes, eating grapefruit products and taking niacinamide 500 mg 3 or 4 times a day are quite helpful.

People with arsenic toxicity often have associated vision problems, that require examination and treatment by specialized eye doctors called Behavioral Optometrists. Such doctors can be found at www.covd.org, www.oep.org, www.nora.cc, and www.optometrist.org (search under vision therapy on the last site). Other eye doctors will be unfamiliar with how to identify the relevant vision problems and will almost invariably miss them and think everything is fine. Children who became toxic at a young age are very likely to need treatment, adults who had normal vision before becoming toxic often regain it spontaneously after detoxification.

Arsenic toxicity will sometimes impair the handling of magnesium and potassium and have consequent effects on the heart. In this case serum potassium should be kept over 4.0 meq/dl and serum magnesium over 1.8 mg/dl if there is QT prolongation on the EKG or any symptoms of heart problems. Red blood cell potassium and magnesium are more useful measures than serum levels. RBC levels should be kept in the green zone (*Doctor's Data* test), or in the normal range (*Metametrix* or *Great Smokies* test) or at least 1/3 of the way up from the lower limit of the normal range on a "health plan" laboratory report. Doing this may require taking magnesium and potassium supplements with every meal.

A potassium chloride containing salt substitute may be used liberally on food as a more convenient source of potassium than pills. Potassium should not be taken by itself without food as it can cause dangerous irregularities in heart function. Taste should be relied on as a guide to preventing too great a potassium intake – that is, only salt the food to taste with potassium containing salt and do not eat it if it tastes bad. Some modest metallic taste is to be expected but should not be unpleasant. In any case, intakes of potassium exceeding 1/2 tsp. of potassium chloride based salt substitute spread throughout the day and mixed with food require a physician's supervision and relevant testing.

Relevant laboratory tests for an arsenic toxic person to decide what supportive therapies they need are plasma cysteine, pyruvate, fasting insulin, and markers for heart disease and liver metabolism.

Beryllium (Be)

If hair beryllium is in the yellow or red range the source of exposure should be identified. If the exposure is not necessary for employment or some hobby it should be ended.

Anyone with sarcoidosis or related forms of autoimmunity and yellow or red range hair beryllium should be presumed to be reacting to the beryllium and be removed from the source of exposure until further testing (a beryllium lymphocyte proliferation test – done at special labs, the doctor may have to call occupational health authorities to find out how to get it done) can be performed to decide whether the beryllium is the culprit or not.

The typical treatment for beryllium induced autoimmune conditions is high amounts of glucocorticoid steroids like prednisone. Side effects of these may be reduced by giving large amounts of DHEA (an over the counter nutritional supplement) with them. EDTA can be used to chelate beryllium but chelation is seldom that helpful. Removal from exposure and prevention of future exposure is the most relevant and appropriate intervention.

Bismuth (Bi)

The first thing to do is end exposure. Since it can be in women's make up, some question may remain as to whether the bismuth was from contamination or from actual absorption of the make-up. Also it is in many medicinal products for digestive tract conditions, and in some diaper rash balms.

Bismuth can be chelated using DMSA every 4 hours or DMPS every 8 hours for several days in a row, every couple of weeks for a few months. If there are still some neurological, intellectual or psychiatric issues then chelate using ALA along with the DMPS or DMSA as described for mercury.

Alkalinize the urine to pH 7 or greater while chelating for bismuth for at least the first several months.

Cadmium (Cd)

Very elevated hair cadmium in the presence of normal orderly mineral transport suggests that the possibility of cadmium toxicity be investigated. If mineral transport is deranged only hair cadmium levels vastly higher than any other toxic or essential element are cause for concern.

The possibility of contamination needs to be considered (but not blindly accepted) whenever artists' pigments are the suspected source of exposure. For example, oil painters who point their brush on their tongue (common) are likely to be toxic, those who point it on their palette, wear gloves, and do not show any coloration on their hands from the pigments may well have simply gotten a very small speck of pigment on the hair that was tested.

If in doubt, repeat hair testing of pubic hair or of newly grown hair from the area the first sample was taken may be useful. Urine and blood tests are less reliable but may also be worth checking.

Cadmium is very difficult to clear from the body. Extensive chelation with DMSA possibly along with ALA is appropriate. The urine must be alkalinized to a pH of 7 or greater when doing this or what chelation results in is transfer of much cadmium from parts of the body where it isn't causing too much trouble into the kidney where it can be a problem.

There is some rather ambiguous and possibly misleading or incorrect medical research literature suggesting that thiamin given with DMSA may improve cadmium chelation results. Since thiamin is inexpensive and benign I do suggest giving 10-50 mg of thiamin with every DMSA dose.

Since cadmium is difficult to remove and is a synergistic toxin with most other heavy metals, people who definitely have high enough levels to impair their health would be well advised to avoid other heavy metal exposures scrupulously and undertake a lifelong program of nutritional supplementation in addition to attempting to get as detoxified as they can with DMSA and ALA.

Relevant supplements to oppose the toxic effects of cadmium are vitamin B1 (thiamin), potassium, vitamin D, vitamins C and E (for their antioxidant effects), zinc, copper, calcium, magnesium, carnitine or acetyl-l-carnitine, coenzyme Q-10, hawthorne extract, tyrosine, milk thistle extract and avoidance of low carbohydrate diets. The water soluble supplements are best taken several times a day, with meals and at bedtime. Vitamins D and E and coenzyme Q10 can be taken once a day. Red meat or iron supplements should be used unless clearly not necessary (blood ferritin over 100 as well as hemoglobin and hematocrit on a blood count being well above the lower limit of the normal range). Tyrosine may be taken once or twice a day and it can be energizing so most people take it in the morning.

Zinc should be used at a moderate level of 25-50 mg per day. Copper should be supplemented at 2-4 mg per day (a little less than 10% of zinc levels) unless it is elevated or very elevated in the hair test, in which case lower levels (elevated, 5% of zinc levels) or none (very elevated) should be used. Hair zinc levels are not relevant as they are often high when body zinc is inadequate, and additionally the zinc is being used for the pharmacologic purpose of increasing the body's production of proteins that bind and passivate cadmium rather than simply to ensure normal healthy levels of body zinc.

A potassium chloride containing salt substitute may be used liberally on food as a more convenient source of potassium than pills. Potassium should not be taken by itself without food as it can cause dangerous irregularities in heart function. Taste should be relied on as a guide to preventing too great a potassium intake – that is, only salt the food to taste with potassium containing salt and do not eat it if it tastes bad. Some modest metallic taste is to be expected but should not be unpleasant. In any case, intakes of potassium exceeding 1/2 tsp. of potassium chloride based salt substitute spread throughout the day and mixed with food require a physician's supervision and relevant testing.

Vitamin B-1 (thiamin) may be obtained as an individual supplement or as "B-50" or "B-100" tablets readily available at any grocery store or pharmacy. Taken with meals and at bedtime this promotes normal healthy levels of energy and subjective sensations of well being as well as supplying needed thiamin.

Gadolinium (Gd)

If hair gadolinium levels are very elevated and there are liver problems or anemia, then it is worth considering the possibility of gadolinium toxicity.

Gadolinium is most well chelated by DTPA[117]. EDTA is a reasonable gadolinium complexing agent and is readily available. It is best used by mouth rather than by injection.

No further use of gadolinium injections for MRI contrast should be permitted except in life threatening situations. Any other (very uncommon) exposures to gadolinium should be avoided.

Symptomatic support for gadolinium toxicity would include milk thistle extract, calcium, magnesium and vitamin C with meals and at bedtime, as well as vitamin K and vitamin E. The calcium and magnesium should be in the citrate, malate, aspartate or amino acid chelate forms. Calcium as hydroxyapatite is only useful if the person has adequate stomach acid production.

Appropriate interventions should also be implemented for any anemia suggested by blood count results that are near or outside of the normal ranges.

Gold (Au)

Very elevated hair gold levels in the presence of problems suggesting gold toxicity merit treatment to remove gold and improve symptoms.

As with any toxicity, the first thing to do is to identify the source of exposure and remove it.

[117] In the process of FDA approval at the time this book was written.

Chelation with DMPS or DMSA may be helpful. I suggest alkalinizing the urine though this is not known to be essential.

Supportive treatments depend on the problems. Skin rashes may respond to hydrocortisone and antihistamine creams as well as high vitamin content skin creams[118]. Antihistamines from skin creams are absorbed into the body to a limited extent, and some of them can be quite sedating and pro-depressant. Low white blood cells may sometimes be improved using thymus extract and related nutritional supplements, or if profound using injectable growth stimulating factors often used to control this problem in cancer chemotherapy. Liver function may be supported with vitamin C, magnesium and milk thistle extract with meals and at bedtime plus vitamin E and lecithin or phosphatidlcholine daily.

Lead (Pb)

Lead is best chelated with DMSA. If DMSA isn't tolerated, EDTA works well. DMPS is not effective. ALA is of unknown effectiveness but is known to improve brain biochemistry in lead poisoning and might possibly remove lead from the brain.

Lead is stored in the bones and also does not leave the brain and internal organs very fast. Chelation primarily clears the blood and some organs of lead. Then more is released from the bones and other sources and blood levels come back up. In order to effectively detoxify someone who has chronic lead poisoning it is necessary to chelate them over a period of several years. Two or three years is most appropriate. However, chelation need not be done intensively. Once intensive chelation has been done for the first month or two it is appropriate to chelate for a few days every month or two for the rest of the period so as to remove lead as it slowly enters circulation. Urine or blood lead tests during this period can be helpful in deciding how often to chelate.

It is very important to check all possible sources of exposure. Lead check swabs available at drugstores are indispensable in checking paint, plaster, old furniture, and anything else of interest. These can also be used to verify that copper plumbing supposedly installed with "lead free" solder wasn't actually put together with the far cheaper lead containing variety. Also check "lead crystal" decanters, glasses, etc. And any beverages stored in them. If in doubt, a lab analysis of drinking water can be done. The local water department may provide this service, or it can be gotten through commercial laboratories[119].

As soon as lead poisoning is suspected, calcium and magnesium intake should be maintained at high levels through supplements or diet, and vitamin D intake should be kept to a moderate level and not supplemented unless clearly necessary. Also if it is feasible, fluoridated water should be avoided. These steps reduce lead absorption.

Important things to check in lead poisoned people are:

Lab tests for the thyroid hormone T3 and in children this should be given unless they clearly have adequate levels (which does NOT mean results within normal limits, it means in the upper half of the age appropriate range, which is higher than the adult range). Slight deficiency of T3 can cause permanent irreversible intellectual impairment in children. Selenium supplementation or supplementation with "thyroid support herbs" may help with this.

[118] For example, Rejuvenex® from the Life Extension Foundation, www.lef.org or any similar product.

[119] *Doctor's Data* offers such a test, no doctor's signature needed, details on their website at www.doctorsdata.com.

In men, check levels of testosterone, LH and FSH[120].

In adults, check uric acid levels (an indicator of gout) and if it is over 8.0, restrict dietary purine sources, don't use the Atkins or similar diets for weight loss and avoid alcohol consumption.

Have an eye exam by a behavioral optometrist[121]. If amblyopia is found using an eye patch is inappropriate and harmful therapy, vision therapy should instead be used.

Mercury (Hg)

Very elevated hair mercury means mercury problems regardless of whether mineral transport is orderly or deranged, and deranged mineral transport also means mercury problems regardless of hair mercury levels.

Mercury detoxification through chelation is actually quite straightforward. DMSA or DMPS are used along with ALA (ALA can be used by itself if there is a reason to or the DMSA and DMSA are not available) and antioxidants are also given. Adequately frequent dosing is used to maintain reasonably constant blood levels of chelator. If exposure was just ended, there is a three month waiting period before ALA is introduced.

There are many possible supportive therapies depending on exactly how mercury is affecting the victim. These are discussed in detail in *Amalgam Illness: Diagnosis and Treatment* and reference should be made to that book[122]. A common error in treating toxic people is to try to optimize a complex suite of interventions to control symptoms before beginning chelation. Symptom control is difficult and complex, chelation is simple and straightforward. Symptoms are very difficult to control until chelation is under way. Delaying chelation for more than a month or two after exposure is ended is not appropriate.

Oxidative stress is a part of the mechanism of toxicity of mercury and all other heavy metals. Antioxidant supplements oppose this and are essential since chelation necessarily increases oxidative stress temporarily during the mobilization of the toxic metal.

Mercury is the one metal that interferes with the formation of cyclic AMP at the concentrations found in toxic people. The nutritional supplement forskolin increases formation of cyclic AMP and can be very helpful for people with mercury problems.

There are a large number of chelation and detoxification protocols and agents held out as helpful for mercury poisoning. Most are not, some are ineffective, many are harmful, some are quite dangerous. I base this general opinion on information available to me from a variety of sources including the manufacturer's own literature, adverse reaction reports on the web or transmitted privately to me, extensive discussions with both health care providers and patients, and my understanding of the relevant chemistry, pharmacology and kinetics.

It is worth special mention that many lab reports held out on the internet as "proving" effectiveness of this or that agent appear to me to be irreproducible reports that are unlikely to provide real evidence of toxic metal removal. I can absolutely attest from personal experience

[120] LH and FSH are Luteinizing Hormone and Follicle Stimulating Hormone, which the pituitary releases – these go up in lead intoxication and down in mercury intoxication.

[121] A behavioral optometrist can be found on www.oep.org, www.covd.org, www.opptometrist.org or www.nora.cc. Call their office and verify that they perform vision therapy before making an appointment.

[122] *Amalgam Illness: Diagnosis and Treatment* may be obtained at www.noamalgam.com.

(and years of frustration) as a chemist it is very easy to obtain misleading, wrong or uninterpretable data in trying to do experimental research.

Agents which should never be used with mercury toxic people under any circumstances include chlorella, TTFD, and EDTA.

NAC, whey protein, glutathione, and high thiol (often referred to as "sulfur") foods are not chelators, do not assist in detoxification, and should only be used if a plasma cysteine test comes back low.

Glutathione should not be administered by intravenous injection except in cases of neurological disease where disability or death loom and the potential benefits of temporary improvement outweigh the risks of severe permanent worsening.

DMPS should never be administered by injection[123]. DMPS injections and intravenous glutathione are the two interventions where dramatic negative reactions frequently occur without warning. Do note that many people tolerate these therapies for a while and then suddenly one time have severe reactions. Other agents and protocols that can cause problems usually, but not always, do lead to a worsening of symptoms that is reversible if the intervention is promptly discontinued before anything serious or permanent happens.

R-ALA and cilantro should only be used if repeated trials of low dose ALA chelation prove intolerable and the risks inherent in turning yourself into a guinea pig in a questionable medical research project are justified. In the rare cases where this is so, first try repeated trials of ALA at no more than 5 mg per dose on an every 2 hour schedule or more often, and if that is not tolerable, then try R-ALA at 5 mg per dose on an every 2-3 hour schedule, and if that is not tolerated, then try one drop of cilantro extract or 1 tsp. of chopped cilantro every 2-3 hours.

Much prompt symptomatic relief for mercury toxic people is often had by modifying diet based on plasma cysteine and sulfate test results and by treating functionally inadequate adrenal performance (in medical terms this is subclinical adrenal insufficiency) if they are present. Other endocrine problems are also very common, are easy to treat, and lead to much preventable misery if nothing is done.

The derangement of mineral transport mercury causes can affect heart function, and mercury may also have a direct toxic effect on the heart in certain instances. In cases where this seems to be happening, using adequate supplementary magnesium and potassium to suppress relevant symptoms or keep levels in the upper half of the normal range on a red blood cell element test helps, as does heart support supplements like carnitine or acetyl-l-carnitine, and coenzyme Q-10, and hawthorne extract or other relevant herbs.

A potassium chloride containing salt substitute may be used liberally on food as a more convenient source of potassium than pills. Potassium should not be taken by itself without food as it can cause dangerous irregularities in heart function. Taste should be relied on as a guide to preventing too great a potassium intake – that is, only salt the food to taste with potassium containing salt and do not eat it if it tastes bad. Some modest metallic taste is to be expected but should not be unpleasant. In any case, intakes of potassium exceeding 1/2 tsp. of potassium chloride based salt substitute spread throughout the day and mixed with food require a physician's supervision and relevant testing.

Mercury toxic people need a high protein diet, usually with digestive enzymes and stomach acid to aid in its digestion. Control or restriction of carbohydrate intake is helpful for some but not all victims.

[123] See www.dmpsbackfire.com for typical stories of the results.

People with mercury toxicity often have associated vision problems that require examination and treatment by specialized eye doctors called behavioral optometrists[124]. Other eye doctors will be unfamiliar with how to identify the relevant vision problems and will almost invariably miss them and think everything is fine. Children who became toxic at a young age are very likely to need treatment, adults who had normal vision before becoming toxic often regain it spontaneously after detoxification.

The two most common errors in treating mercury toxicity are failing to stick with proper chelation for the long term and thus not getting adequately detoxified to return to normal health, and trying other means of detoxification (e. g. chlorella, DMPS injections, TTFD, 10 mg/kg DMSA every 8 hours) and experiencing serious long term or permanent worsening as a result. This is usually done on the advice of a physician who insists they prescribe it all the time, it is perfectly safe, and they have never seen a problem. Regardless of the physician's lack of knowledge of the protocol, these methods should never be used.

For people who had deranged mineral transport with mercury toxicity despite having no obvious history of very high mercury exposures, it is appropriate to get hair tests on family members since the propensity to accumulate mercury this way is genetic. Also since mercury causes subtle emotional changes, people with mercury problems tend to self associate strongly and it is worth checking spouses despite the fact that they are not genetically related. I have found about a 95% rate of spouses both having toxicity issues in cases I have received testing from.

[124] Behavioral optometrists can be found at www.covd.org, www.oep.org, www.nora.cc, and www.optometrist.org (search under vision therapy on the last site). Call their office and make sure they offer vision therapy before making an appointment.

Nickel (Ni)

Elevated hair nickel in the presence of clearly normal mineral transport may suggest nickel toxicity. Since nickel is so often vastly elevated early on in the process of mineral transport derangement much suspicion of an underlying mercury problem is appropriate.

If there is a nickel problem the source of exposure should be identified and eliminated. Much jewelry contains nickel and this should be avoided to a reasonable extent, but since nickel is not well absorbed from jewelry unless it is corroded, such jewelry can be worn for special occasions if there is no skin reaction to it.

Nickel is well excreted in sweat and means to induce sweating are very helpful in reducing the body burden of nickel. Regular saunas are fine, see cautions elsewhere regarding infrared saunas. When inducing sweating by any means, ensure lots of magnesium is being taken frequently and use diaprhoretic teas such as elder flower, yarrow, and peppermint as well as eat hot spicy foods if tolerated. These all increase sweating.

The anti-alcoholism drug Antabuse should be avoided if at all possible as it redistributes nickel to the brain.

DMSA and DMPS chelation may be helpful.

Nickel is a known carcinogen and reasonable anticancer interventions such as lifelong supplementation with antioxidant vitamins are appropriate.

Palladium (Pd)

With elevated hair palladium the source of exposure should be identified and removed.

One source of exposure is a nutritional supplement used as an alternative cancer therapy. In this case the risks of palladium toxicity must be weighed against the risks of the underlying condition, cancer, and a reasonable decision made.

Palladium should be chelated similarly to mercury, with ALA and optionally DMPS or DMSA, and with a 3 month waiting period before introducing ALA if exposure was just ended.

Platinum (Pt)

If hair platinum is elevated then evaluation for platinum toxicity is appropriate except in the unusual cases where many toxic and essential elements are vastly elevated due to mineral transport derangement.

The source of platinum exposure should be identified and eliminated.

Platinum may be chelated similarly to mercury, with ALA plus optionally DMPS or DMSA.

The most important supportive therapy for platinum problems is magnesium with meals and at bedtime. It should be in the citrate, malate, aspartate or amino acid chelate form rather than oxide, hydroxide or carbonate.

Other supportive therapies include antioxidants (with the water soluble ones like vitamin C and B complex being given with meals and at bedtime) and if there are hearing problems then vinpocetine 5-10 mg 3-4 times a day and 5 mg manganese daily unless there is reason to believe there is already adequate or excessive manganese in the body.

Some test like a red blood cell or erythrocyte element test is appropriate to decide what other minerals may need to be supplemented.

Silver (Ag)

Very elevated hair silver in the presence of normal orderly mineral transport or hair silver vastly elevated compared to other elements in the presence of disordered mineral transport should be considered significant if there are problems of the sort silver toxicity can cause.

Any possible exposure to absorbable silver should be ended[125]. Sources include colloidal silver, silver nitrate sticks or silver nitrate for other medicinal use, silver protein, and photographic chemicals.

The impulsivity and emotional chaos that silver causes can be reduced by rigorously excluding hydrogenated fats from the diet, and by the use of carnosine and mood stabilizing medications. Sometimes a ketogenic diet such as the Atkins diet also brings relief. The use of the nutritional supplements carnitine or acetyl-l-carnitine may make an Atkins style diet more comfortable and well tolerated.

People with silver toxicity often have associated vision problems that require examination and treatment by specialized eye doctors called behavioral optometrists[126]. Other eye doctors will be unfamiliar with how to identify the relevant vision problems and will almost invariably miss them and think everything is fine.

Selenium and vitamin E are specifically known to be protective against silver toxicity. Selenium should be used at 200-400 mcg per day in an organic form such as selenomethionine,

---

[125] I think exposure to absorbable silver is unwise even in a normal healthy person.

[126] Behavioral optometrists can be found at www.covd.org, www.oep.org, www.nora.cc, and www.optometrist.org (search under vision therapy on the last site). Call their office and make sure they offer vision therapy before making an appointment

selenium yeast or S-selenocysteine. It should not be used in the form of selenite or selenium citrate[127].

The effectiveness of chelation for removing silver is unknown but it can be tried using DMSA or DMPS with alkalinization of the urine, along with ALA. The ALA can be added after 3-6 months of initial chelation if desired.

Thallium (Tl)

If hair thallium is elevated then careful investigation of whether thallium exposure and problems exist is appropriate except in the unusual cases where many toxic and essential elements are vastly elevated (in which case reasonable elevations of thallium are likely to be an artifact rather than reflecting body burden).

For detoxification, activated charcoal (and in a medical situation where it may reasonably be prescribed, Prussian blue) are given to prevent thallium from being reabsorbed in the intestines. DMSA or DMPS are given to reduce body levels and ALA is added after 2-3 months as is done in mercury toxicity. If DMPS and DMSA are not available or if it is known that exposure was not recent then ALA can be used immediately.

A high intake of potassium increases the rate of thallium clearance and also partially corrects some of the biochemical imbalance thallium causes. The easy way to arrange this is to use a potassium chloride containing salt substitute liberally on food. Potassium should not be taken by itself without food as it can cause dangerous irregularities in heart function. Taste should be relied on as a guide to preventing too great a potassium intake – that is, only salt the food to taste with potassium containing salt and do not eat it if it tastes bad. Some modest metallic taste is to be expected but should not be unpleasant. In any case, intakes of potassium exceeding 1/2 tsp. of potassium chloride based salt substitute spread throughout the day and mixed with food require a physician's supervision and relevant testing.

Due to the colorless, tasteless nature of thallium salts some diligence is needed in ensuring any possible source of exposure has been identified and eliminated.

Other supportive therapies that oppose the toxic effects of thallium are biotin 5-10 mg with meals and at bedtime, riboflavin (most easily obtained as "B-50" or "B-100" tablets at any grocery store or pharmacy) and magnesium (in the citrate, malate, aspartate or amino acid chelate forms) with meals and at bedtime, plus heart supporting supplements like carnitine, hawthorne extract or other relevant herbals, and coenzyme Q-10.

Despite the fact that blood tests may show low thallium levels in the presence of toxicity they are sometimes informative. At the time of writing this book, the Great Smokies Laboratories erythrocyte element (red blood cells) test checks for thallium which other labs offering similar test panels do not. In addition it checks for potassium and magnesium levels which are relevant in thallium toxicity.

Thorium (Th)

If elevated thorium is seen on a hair test an effort should be made to identify and end exposure, and provide supportive treatments, unless it is one of the tests with deranged mineral transport where lots of toxic and essential elements become vastly elevated.

It is not clear if it is possible to chelate out thorium. Supportive treatment is liver support and antioxidant nutritional

127 Selenite is used in laboratory experiments to cause cataracts in rats and is quite toxic.

supplements. Information on the risks of thorium exposure can easily be found by looking up "thorotrast" in the medical literature. The cancer risk can then be minimized by using complimentary and alternative treatments to control them[128].

Tin (Sn)

Elevations of hair tin due to mercury induced derangement of mineral transport are common and may happen before derangement in the body's handling of the essential elements is apparent on a hair test. Thus very elevated hair tin should always lead to suspicion regarding mercury issues, especially if some of aluminum, antimony, nickel and titanium are elevated.

If tin itself does appear to be present in excess, then the source of tin exposure needs to be identified and eliminated. If there clearly is not a source of exposure then mercury is the likely cause.

If tin exposure is ended and tin does not come down on its own over 6 months or so, then some mercury chelation is in order if there are no sources of mercury exposure present ("silver" amalgam dental fillings being the most common).

Supportive treatment for tin toxicity is supplementation with copper (1-2 mg a day), iron (4-8 mg a day or eat a lot of red meat along with digestive acid and enzymes), and antioxidants such as vitamin C, B complex, magnesium, and milk thistle extract with meals and at bedtime as well as vitamins A, E, and omega 3 fatty acid containing flax or fish oil daily. Any other supportive measures to relieve whatever symptoms the person may have should be used as well.

[128] and mainstream treatments where available, but mainstream medicine is not yet very advanced in this area.

Titanium (Ti)

As titanium is generally considered not to be significantly toxic and is also often vastly elevated very early in the derangement of mineral transport I would suggest not being too concerned about elevated hair titanium in and of itself, but only as a possible marker for mineral transport derangement.

Do note that anyone with a surgical implant made of titanium will have vastly elevated hair titanium and this is not significant.

Tungsten (W)

If hair tungsten is vastly elevated in the presence of normal orderly mineral transport (or dramatically more elevated than other elements in deranged mineral transport) then tungsten toxicity should be considered.

In tungsten toxicity blood uric acid and sulfate levels should be reduced. This may also occur in mercury toxicity.

The most important thing to do, as with any toxicity, is to identify and remove the source of exposure.

Administering high amounts of molybdenum both helps accelerate clearance of tungsten and also opposes its toxic effect (which is to interfere with the body's use of molybdenum).

Supportive therapy is to avoid high purine foods, sulfites and alcohol, and to limit the intake of foods containing thiols and their precursors.

Tungsten is believed to clear on its own without retention in any specific organ, so chelation is seldom appropriate. If chelation seems necessary, EDTA may be used (DMSA, DMPS and ALA most likely also work). If EDTA chelation is tried it should be by mouth, and care should be taken to avoid using EDTA products that also contain "sulfur" materials such as garlic, NAC, MSM, etc.

Uranium (U)

Greatly elevated hair uranium (over 0.5, at least half way across the red zone) is significant but lower levels may not be.

Uranium uptake can be reduced by increasing dietary iron or using iron supplements.

Uranium intake can be reduced by using distilled or bottled water and avoiding contact of food with any yellow colored ceramic items.

Essential and Other Element Hair Test Results - Major Minerals

Calcium (Ca)

If calcium is either high or very low, take calcium supplementation, vitamin D, magnesium and boron. There are many combination products containing all of these. Try to get the calcium in the citrate, malate or amino acid chelate form, not as oxide or carbonate. The hydroxyapatite form of calcium works well for people who aren't too unhealthy, but it is hard to digest and very sick people may not absorb it.

If there is evidence of substantial calcium wasting (red low calcium with normal orderly mineral transport, or a dramatic positive response to supplementation, or surprisingly bad osteoporosis) then plasma sulfate should be checked as this can cause calcium wasting. People with elevated plasma sulfate often feel great subjective improvement when using the Feingold diet restrictions on phenolic and salicylate compounds.

Magnesium (Mg)

What to do about hair magnesium is somewhat complicated. I have tabulated what to do based on the results below for clarity.

	What to do depending on mineral transport status	
Hair magnesium level	Mineral transport normal and orderly	Mineral transport deranged
Red low	supplement magnesium and taurine	supplement magnesium and taurine
Yellow low	supplement magnesium and taurine	No conclusion can be drawn
Yellow high and counting rule 1 met	N/A	supplement magnesium and taurine
Red high	Check red blood cell levels and supplement if low	supplement magnesium and taurine
Magnesium yellow low and (zinc yellow or red high or Ca/P high or Na/Mg low or blood alkaline phosphatase below 50)	supplement magnesium and taurine and vitamin B-6	No conclusion can be drawn
Magnesium red low and (zinc red high or Ca/P high or Na/Mg low or blood alkaline phosphatase below 50)	supplement magnesium and taurine and vitamin B-6	supplement magnesium and taurine and vitamin B-6

Please note that most people with disordered mineral transport do well on supplementary magnesium regardless of hair levels, and also that magnesium supplementation is very unlikely to cause problems.

If no counting rule is met but mineral transport does not appear particularly normal and orderly a red blood cell magnesium test can be used to decide what to do. Note that if magnesium is in the red low range mineral transport status doesn't affect the recommendation.

Magnesium supplements in the citrate, malate, aspartate, or amino acid chelate form are best. Those in the oxide, hydroxide or carbonate form should be avoided as they are poorly absorbed and use up stomach acid. Magnesium supplementation is done by taking 100-200 mg magnesium (adult basis) with each meal and possibly at bedtime. The amount used is adjusted to the maximum level that is not laxative. Calcium should also be given so that the ratio of calcium to magnesium intake from all sources (food and pills) is somewhere between 1 to 2 and 2 to 1.

Taurine is supplemented 500-1000 mg with meals (adult basis) as it helps retain magnesium in the body. People who are bothered by the chlorine smell of bleach and swimming pools are low in taurine and will find this especially helpful.

People who do not absorb adequate magnesium by eating it can get more in them by taking epsom salt baths where somewhere between a cup and a whole 4 pound carton is added to a hot tub. A solution of half a cup epsom salts can be made in 2-3 cups water and sponged on the skin where it is left to dry (and the white flaky powder washed off afterwards if desired), or an epsom salt foot bath can be used.

B-6 supplementation should be undertaken with some care as high levels do not agree with everyone. For an adult sized person, 100-250 mg of B-6 with meals and at bedtime is usually adequate. Excessive B-6 can cause tingling and numbness in the hands and feet which may occur in some adults at levels of supplementation greater than 500 mg 4 times a day. It is claimed that this is occasionally seen at levels as low as 250 mg in a day, though these claims are from sources that categorically state nobody should ever take any nutritional supplement and thus must be viewed with caution. If numbness or tingling in the hands and feet[129] occur discontinuation of the B-6 leads to its cessation in a few days. The relevant amounts for children need to be adjusted for weight based on an average adult weight of 135#. Generally if there is some noticeable response at a reasonable dose (e. g. 100-250 mg 3-4 times a day for an adult) then it is worth experimenting with amounts. If absolutely nothing happens, it is unlikely that jacking the dosage up and up and up will result in benefits and it will surely result in the numbness and tingling at some adequately high level.

Phosphorus (P)

If hair phosphorus is low in the presence of normal orderly mineral transport, investigate dietary protein digestion and adequacy, use betaine hydrochloride or some other source of acid with meals to aid digestion, and use supplementary digestive enzymes. Also if relevant evaluate liver health but do note that standard mainstream medical laboratory tests are seldom informative for this purpose. . In addition to liver support supplements such as milk thistle extract, the limitation or exclusion of fructose[130] from the diet is helpful.

[129] Referred to as peripheral neuropathy in medical terminology.

[130] A form of sugar found in fruit, table sugar, and high fructose corn syrup which is often used to sweeten soft drinks.

Elevated hair phosphorus in the presence of normal orderly mineral transport suggests evaluating kidney function. Mainstream medical laboratory tests are generally adequate for this purpose. Relevant tests are serum creatinine, serum BUN[131], and a creatinine clearance.

Sodium (Na)

Hair sodium is not a relevant marker of body sodium levels. The relationship of sodium to other elements is relevant and is discussed below.

Potassium(K)

Hair potassium is not a relevant marker of body potassium levels. The relationship of potassium to other elements is informative and is discussed below. For example, very elevated or very low potassium may suggest thyroid problems, but the ratio of potassium to sodium is more informative.

Relation of Calcium, Phosphorus, Sodium, Magnesium, Potassium and Other Elements

Elevation of hair magnesium and calcium suggests the possibility of excess parathyroid activity and osteoporosis.. Supplementation with calcium and magnesium (citrate, malate, aspartate or amino acid chelate forms work best, hydroxyapatite for people with good digestion and adequate stomach acid) helps. Some also believe that eliminating soda pop consumption may improve this situation

An abnormal ratio of calcium to magnesium suggests poor regulation of blood sugar. There are many interventions for this, the most basic ones being a low glycemic index diet and supplementary chromium, vanadium and zinc.

For a low ratio of calcium to phosphorus (fast metabolizer) some believe that a high fat diet with lots of vitamin D is helpful.

For a high ratio of calcium to phosphorus (slow metabolizer) some believe that a goodly intake of salt, B complex vitamins, and protein while reducing fat intake is helpful. It is more widely accepted that these people need to take supplementary acid with meals, e. g. ascorbic acid or betaine hydrochloride. A CO_2 level below 22 on a chemistry panel (blood test) confirms the need for more stomach acid.

If the sodium and potassium lines end at one place and are near each other, and the calcium and magnesium lines go to another place and are near each other, but the ends of the sodium and potassium bars are 2" or more from the ends of the calcium and magnesium bars, then there are adrenal problems. Supplements intended for adrenal support such as adrenal cortex extract or licorice are helpful. There are many combination products available for this purpose and mostly they do work well.

An abnormal ratio of sodium to potassium suggests getting thyroid testing done. See page 116 and following in *Amalgam Illness: Diagnosis and Treatment* for a discussion of thyroid issues. Simply testing TSH and checking whether it is within normal limits or not is irrelevant. Checking the actual hormone levels is essential, with free T3 and free T4 being the most relevant tests.

---

[131] BUN stands for blood urea nitrogen.

Another indicator of possible thyroid problems is the ratio of potassium to calcium. If this is not between 0.03 and 0.35 further thought should be given to thyroid function.

For people with calcium and magnesium higher than sodium and potassium, exercise can be helpful in improving their health and mood. When sodium and potassium are higher than calcium and magnesium exercise often worsens things unless done in very moderate amounts.

When sodium is LOW compared to magnesium (Na/Mg less than 1) the body can't make enough adrenaline and supplementation with tyrosine, trimethlyglycine, folic acid and vitamins B6 and B12 can improve mood and energy. If this ratio is high (more than 5) then supplementary tyrosine and phenylalanine should be avoided.

When Ca/K and Na/Mg are elevated and Ca/P is low then supplementary thymus glandular is helpful as it promotes normal healthy thymus function and cellular immunity

Having a high ratio of Ca/P along with a low ratio of Na/Mg suggests the need for supplementary vitamin B6.

Digestive enzymes are often helpful when either Ca/P is low, or when both Na/Mg and Ca/K are elevated.

An elevated ratio of calcium to phosphorus and a reduced ratio of sodium to magnesium are both signs that more vitamin B-6 is needed. Low hair magnesium and high hair zinc also indicate this, as does a result below 50 for alkaline phosphatase in a blood test.

B-6 supplementation should be undertaken with some care as high levels do not agree with everyone. For an adult sized person, 100-250 mg of B-6 with meals and at bedtime is usually adequate. Excessive B-6 can cause tingling and numbness in the hands and feet which may occur in some adults at levels of supplementation greater than 500 mg 4 times a day. It is claimed that this is occasionally seen at levels as low as 250 mg in a day, though these claims are from sources that categorically state nobody should ever take any nutritional supplement and thus must be viewed with caution. If numbness or tingling in the hands and feet[132] occur discontinuation of the B-6 leads to its cessation in a few days. The relevant amounts for children need to be adjusted for weight based on an average adult weight of 135#. Generally if there is some noticeable response at a reasonable dose (e. g. 100-250 mg 3-4 times a day for an adult) then it is worth experimenting with amounts. If absolutely nothing happens, it is unlikely that jacking the dosage up and up and up will result in benefits and it will surely result in the numbness and tingling at some adequately high level.

Sulfur (S)

There are no interventions that should be undertaken based on hair sulfur levels.

Iron (Fe)

Since hair iron is not an indicator of body iron stores the interventions I discuss here should not be used based on hair test results. The relevant tests are ferritin, hemoglobin and hematocrit.

The ideal range for ferritin is 30-70. People who have other heavy metal toxicities may improve somewhat on having ferritin brought down into this range if it is higher but still within normal limits.

[132] Referred to as peripheral neuropathy in medical terminology.

Low ferritin is more common in women, high ferritin in men.

Iron absorption can be increased with acid supplements such as betaine hydrochloride or the acid form of vitamin C (ascorbic acid). Iron from red or dark meat is more easily absorbed and utilized than that from vegetable sources. People with heavy metal toxicity often have low iron because their body needs to keep it that way in order to reduce oxidative stress. In this case supplementary iron may lead to symptomatic worsening. There are injectable forms of iron if it is not absorbed from oral supplements, but before using these careful thought should be given to the idea that the body is trying to keep iron levels down for a good reason. The usual good reason is heavy metal toxicity.

When the body is holding iron (ferritin) levels down for a good physiologic reason the person has a symptomatic worsening on getting iron injections, and after a few weeks the body has dumped the iron and their ferritin is back down where it started. In these cases further iron therapy is not helpful, but increased antioxidant intake (with the water solube supplements being given with meals and at bedtime) will often result in symptomatic improvement and a rise in ferritin levels without the need for iron administration.

The most effective interventions for high iron are eliminate red meat and high iron foods from the diet, and give blood frequently. Desferrioxamine is an injectable medication that can remove iron from the body, but it is only relevant in cases of severe iron toxicity. It must be used properly (as with any chelating agent) or it makes the problem worse, not better. This means infusion every 6 hours or so over a period of days.

While not relevant to iron status, hair iron is relevant to other conditions. Low hair iron suggests the possibility of lead toxicity. Recall that old lead exposures may not show up as elevated hair lead levels.

Very elevated hair iron in the presence of normal mineral transport suggests a ***very low*** probability of cancer. Routine cancer screening may be appropriate if it is otherwise relevant to the person's condition.

Essential and Other Element Hair Test Results - Trace Minerals and Other Elements

Barium (Ba)

If hair barium is elevated in the presence of normal orderly mineral transport and it is believed to be significant, an increased intake of calcium and magnesium along with decreased intake of vitamin D may reduce body levels of barium and a high intake of potassium opposes its effects. Also use heart supporting supplements like carnitine, magnesium, hawthorne extract or other relevant herbals, and coenzyme Q-10. Any source of barium exposure should be identified and removed.

Boron (B)

In the presence of normal orderly mineral transport, if hair boron is low, take 3 mg/day (adult dose), and if it is high take riboflavin with meals and at bedtime. Riboflavin is most conveniently available in B complex vitamins such as "B-50" or "B-100" which can be found at most grocery and drugstores.

Chromium (Cr)

If hair chromium is in the yellow or red low range in the presence of normal orderly mineral transport, take supplement with meals. At least 200 mcg of the picolinate or similar form is appropriate for adults. Also exclude high glycemic index foods from the diet (sugar, easily digested carbohydrates).

If chromium is in the red high range in the presence of normal orderly mineral transport, restrict intake and take vanadium supplements. If the level of chromium may be causing symptoms, then a blood count and liver function test are relevant things to check.

Cobalt (Co)

Hair cobalt is not coming from vitamin B-12. If it is very elevated in the presence of normal orderly mineral transport, identify the source of exposure and eliminate it. Also use heart supporting supplements like carnitine, magnesium, hawthorne extract or other relevant herbals, and coenzyme Q-10.

Copper (Cu)

When mineral transport is normal and orderly, the ratio of zinc to copper should be kept within the normal range by supplementing each if low and reducing supplementation if high. This ratio is irrelevant if mineral transport is deranged.

Whether mineral transport is orderly or deranged, body copper levels should be reduced by the use of zinc with every meal and possibly molybdenum as well if the hair copper levels are in the red high range.

If hair copper levels are over 75 then chelation with ALA for other metals should be restricted somewhat. Copper levels of 75-125, average one day on for every two off chelation. Copper levels over 125, average one day on for every three off chelation.

Generally people with mixed copper plus other heavy metal toxicity respond more rapidly to interventions directed at reducing copper levels than they do to interventions directed at the other metal.

To reduce copper levels, give zinc with every meal, molybdenum 1,000-3,000 mcg a day (or 10-30 mcg per pound per day for children under 100 pounds), and nutritional supplements that promote bile flow such as taurine, glycine, milk thistle extract and phosphatidylcholine – all with meals.

High copper levels suggests the possibility of gall bladder or biliary tract disease which should be explored.

The prescription drug trientine hydrochloride[133] is used for Wilson's disease and can be used in other conditions where there is very excessive body copper. Penicillamine is often listed as an older therapy for Wilson's disease but it invariably has horrific side effects in toxic people and should never be used for detoxification.

In copper toxicity it is a good idea to check cysteine, sulfate and other things that often are affected by the heavy metals and use appropriate interventions based on test results.

When there is not an obvious cause for high hair copper it is a good idea to check hair copper for family members of the

[133] Trientine is di-bis 2 aminoethyl ethylenediamine. It is the ammonia equivalent of the acetic acid based EDTA in an overall chemical sense. Chemically copper has a special affinity for ammonia.

toxic person as a tendency to accumulate copper can run in the family.

For red range low hair copper in the presence of normal orderly mineral transport copper metabolism should be investigated by determining red blood cell copper, serum copper or urine copper in a timed sample, and serum ceruloplasmin. If copper supplementation is tried without this testing it needs to be kept in mind that a lack of prompt improvement suggests the possibility of copper overload disease rather than copper deficiency.

Germanium (Ge)

If hair germanium is elevated in the presence of normal orderly mineral transport and there are problems suggestive of germanium toxicity the it is worthwhile to try to identify any potential sources of exposure and eliminate them.

If hair germanium is very low, mineral transport is normal and orderly, and the person has fibromyalgia or a similar condition then it may be worth considering supplementing with organic germanium nutritional supplements, though careful consideration of the unknown but significant risks of toxicity of this supplement versus the potential benefits would be wise.

High or low germanium should not be considered significant in the absence of clearly relevant health problems that can not reasonably be explained in some other way.

Iodine (I)

Hair iodine is very often abnormal in the presence of deranged mineral transport so very high or very low levels should not be considered significant unless mineral transport is pretty clearly normal and orderly (as opposed to simply doesn't quite meet the counting rules).

If hair iodine is in the red low range and this is believed to be significant (e. g. there is goiter) then using iodized salt or eating some seaweed is appropriate. If hair iodine is vastly elevated and there is reason to believe it is significant then the source of exposure needs to be identified and eliminated.

In cases of either very low or very high iodine that are believed to be significant, having an experienced health care provider do an examination for goiter[134] and doing appropriate blood testing[135] may be useful.

If red low take some or eat seaweed. If red high identify exposure and end it, eliminate iodized salt, check thyroid.

Lithium (Li)

If hair lithium is below 0.04 (the *Doctor's Data* detection limit) take supplementary lithium (orotate or aspartate form). The typical supplement contains about 5 mg of vanadium and should be taken with meals.

If hair lithium is wildly high (all the way to the end of the red zone) in the presence of normal, orderly mineral transport and the person is not taking prescription lithium, try to figure out where it is coming from, and check thyroid function.

---

[134] Where the examiner stands behind you and feels the middle and lower portion of your neck.

[135] See the discussion of thyroid function and testing on pages 115-117 of *Amalgam Illness: Diagnosis and Treatment*. See www.noamalgam.com for book information.

Manganese (Mn)

Very elevated hair manganese in the presence of normal orderly mineral transport is a sign of manganese toxicity, and low hair manganese in the presence of normal orderly mineral transport is a sign of manganese deficiency.

Red blood cell/erythrocyte manganese is a reasonable confirming test.

For manganese toxicity it is important to identify the source of exposure and end it. If no source of exposure can be identified and manganese levels are vastly elevated then family members should also be checked to see if it is a genetic susceptibility to accumulating manganese.

Supplementary calcium, phosphorus and iron if appropriate can reduce the absorption or manganese from the diet. Antioxidants oppose the toxic action of manganese and essential fatty acids replace the materials destroyed by manganese toxicity. Bile flow promoting supplements such as taurine, glycine, phosphatidylcholine or lecithin, and milk thistle extract or related products may increase excretion

In cases of very high manganese, thyroid function needs to be investigated. Free T3 and free T4 (both tests, not one or the other) are the most informative. For a discussion of thyroid issues see page 115-118 of *Amalgam Illness: Diagnosis and Treatment*[136].

Manganese may be chelated using EDTA. It is best to try oral therapy every 8 hours as described in the chapter General Comments on Chelation. Amalgam fillings must be removed before any chelation can be done safely.

Drug therapy may be needed in severe cases of manganese toxicity to control symptoms until manganese levels can be reduced adequately. The antiparkinson's drug l-dopa can help with the Parkinsonian like symptoms. Manganese victims often tolerate more of it than Parkinson's victims do. Trihexylphenidyl (another antiparkinson's drug) may relieve some of the psychiatric disturbances but does not help with the movement disorders or other neurological issues. In milder cases use of the drug selegiline (trade names Deprenyl or Eldepryl) 5 to 40 mg a day can be quite helpful.

For low manganese, simple supplementation with 5-15 mg a day of manganese generally resolves the problem. Hair or red blood cell manganese are appropriate tracking tests. Hair is usually more convenient and the test is less expensive. Supplement level can be adjusted based on test results, 2 to 3 months should be allowed to let the body come into equilibrium with the new level of supplementation before getting a blood test, and 3-6 months for a hair test.

People with very low manganese may not handle ammonia properly. In this case they will feel better on a low protein diet as well as if they avoid consuming large amounts of alkalinizing materials such as citrate, malate or amino acid chelate based mineral supplements. Using acid form vitamin C or betaine hydrochloride with meals may also improve well being by helping remove ammonia from the body.

Molybdenum (Mo)

In the presence of normal orderly mineral transport hair molybdenum may be significant. When mineral transport is deranged it is not significant. Red blood cell molybdenum levels are usually significant and are a useful guide to body inventory.

With normal orderly mineral transport and yellow low molybdenum, take 1,000 mcg a day of it. With normal orderly

[136] *Amalgam Illness: Diagnosis and Treatment* may be obtained at www.noamalgam.com.

mineral transport and red low molybdenum, take 2,000 to 3,000 mcg a day of it. With normal orderly mineral transport and red high molybdenum, exclude it from supplements and supplement with copper and zinc.

For children, calculate amounts as follows:

For children over 100 pounds, use adult amounts. For children less than 100 pounds, use the following formulae:

1,000 mcg a day for an adult turns into a children's dose of weight in pounds * 10 = number of micrograms.

2,000-3,000 mcg a day for an adult turns into a children's dose of weight in pounds * 20-30 = number of micrograms.

Rubidium (Rb)

If hair rubidium is very elevated in the presence of normal orderly mineral transport AND it is not tracking sodium and potassium levels, then it is appropriate to identify and remove the source of exposure unless the rubidium is being taken for some specific purpose.

Selenium (Se)

Contamination of hair samples by dandruff shampoos containing selenium or sulfur – which is sometimes contaminated with selenium – is quite common and accounts for most observed very high hair selenium levels. Note that while in theory pubic hair is not contaminated, in practice enough shampoo likely washes over the entire body that contamination of any and all hair is an issue for someone who uses such products.

With very elevated hair selenium in the presence of normal orderly mineral transport that is believed not to be due to contamination, selenium toxicity may in fact be present as selenium is toxic at levels not too far about the essential daily intake. A red blood cell/erythrocyte element test for selenium is useful for confirmation.

One possible source of toxicity is excessive supplementation. Single doses of supplementary selenium in the 2,000-5,000 mcg range are not toxic, however chronic daily dosing over a long period can lead to toxicity in most people at 1,500 mcg a day, and in some sensitive people at as low as 750 mcg a day. In cases of toxicity from excessive intake (the amount in all supplements and estimated amount in food must be added up to get the daily total) discontinuing selenium supplementation is the single most important intervention.

Supplementary copper, 1-3 mg a day, may increase selenium clearance somewhat via the bile. Antioxidants are protective against the toxic effect of selenium. For example, vitamin C 1 gram buffered form plus milk thistle extract with meals and at bedtime, and vitamin E once a day.

Chelation probably does not reduce selenium levels. Sweating (see cautions regarding Sauna) is mildly effective.

Selenium deficiency may be reasonably suspected with very low hair selenium in the presence of normal orderly mineral transport. In that case supplementary selenium and vitamin E are helpful. In a pregnant woman or an infant or child, free T3[137] should be

[137] T3 is the more active form of thyroid hormone and is necessary for brain development in fetuses, infants and children. Free T3 is the active form of T3. The normal range is not suitable for a given individual since some people need more and some need less. There is minimal risk from extra T3, it is T4 that causes the

checked and T3 immediately prescribed to ensure it is in the upper half of the normal range to prevent any risk of impaired brain development for the child.

Silicon (Si)

If hair silicon is very low in the presence of normal orderly mineral transport, taking a nutritional supplement may improve collagen and bone strength.

Strontium (Sr)

If strontium is high but calcium and magnesium are also similarly elevated, don't be concerned about the strontium levels.

If hair strontium is in the high red zone the presence of normal orderly mineral transport, take more calcium and magnesium, reduce vitamin D intake, and identify any source of exposure so as to reduce intake of strontium.

Vanadium (V)

If hair vanadium is in the red low range in the presence of normal orderly mineral transport, take supplementary vanadium and avoid high glycemic index foods (those containing sugar or easily digested carbohydrates).

If hair vanadium is in the red high range in the presence of normal orderly mineral transport, take supplementary chromium and ensure no vanadium is in

---

negative effects of too much thyroid hormone and these are primarily heart risks that are only relevant to adults.

Zinc (Zn)

Zinc is particularly confusing for most people because you take more zinc whether it is high or low. The only time you don't take more zinc is if it is normal AND mineral transport is orderly.

In the presence of very obviously normal and orderly mineral transport (not just that it doesn't meet the counting rules but that it really is very normal) there is some possibility of elevated hair zinc really meaning that the body levels of zinc are elevated. In this situation either a 24 hour urine test or a red blood cell zinc (or red blood cell or erythrocyte element test) will resolve the ambiguity between high and low.

In the rare cases of high body zinc (as confirmed by red blood cell testing), it may be reduced by an increased intake of copper, iron, phosphate, calcium, and fiber.

In mercury toxicity (which means when mineral transport is deranged or mercury is otherwise believed to be a problem) absorption of zinc is severely impaired and very high amounts are needed. For children the amount to use is weight in pounds + 20 = number of milligrams to start with, in divided doses. For adults, use 100-200 mg a day in divided doses. Also do not include any copper with this. Supplementation at this level would result in copper deficiency and zinc toxicity in a normal healthy person, but will not do so in someone with mercury deranged mineral transport and in fact is necessary to get their zinc up and their copper down to somewhere near average. There is some inaccurate information regarding this in the "natural medicine" community based on the always risky enterprise of extrapolating results of scientific studies on normal healthy volunteers to presumptions regarding what occurs in sick people. Any practitioner with such beliefs (the need for copper with zinc, the need to limit zinc intake) can easily dispel them by ordering some red blood cell/erythrocyte element tests on their patients with deranged mineral transport.

If mineral transport is deranged zinc should be supplemented regardless of where the zinc result lies on the report – deranged mineral transport invariably reduces zinc absorption and leads to deficiency.

High hair zinc is a sign that more vitamin B-6 may be needed. Other signs of this are low hair magnesium, a high ratio of calcium to phosphorus and a low ratio of sodium to magnesium, as well as a result below 50 for alkaline phosphatase in a blood test.

B-6 supplementation should be undertaken with some care as high levels do not agree with everyone. For an adult sized person, 100-250 mg of B-6 with meals and at bedtime is usually adequate. Excessive B-6 can cause tingling and numbness in the hands and feet which may occur in some adults at levels of supplementation greater than 500 mg 4 times a day. It is claimed that this is occasionally seen at levels as low as 250 mg in a day, though these claims are from sources that categorically state nobody should ever take any nutritional supplement and thus must be viewed with caution. If numbness or tingling in the hands and feet[138] occur discontinuation of the B-6 leads to its cessation in a few days. The relevant amounts for children need to be adjusted for weight based on an average adult weight of 135#. Generally if there is some noticeable response at a reasonable dose (e. g. 100-250 mg 3-4 times a day for an adult) then it is worth experimenting with amounts. If absolutely nothing happens, it is unlikely that jacking the dosage up and up and up will result in benefits and it will surely result in the numbness and tingling at some adequately high level.

Zirconium (Zr)

If hair Zirconium is vastly elevated in the presence of normal orderly mineral transport I would suggest that sources of exposure be identified and eliminated (this is almost always a deodorant). The conventional wisdom is that zirconium is completely nontoxic and safe. Conventional wisdom often proves wrong later and it seems unwise to maintain a physiologically abnormal level of some mineral without a good reason to do so.

In someone with sarcoidosis or a related disease which zirconium is a possible trigger of, eliminating exposure if there is elevated hair zirconium is essential.

[138] This is referred to as peripheral neuropathy in medical terminology.

Appendix: Statistical Tables and Calculations

The statistical calculations behind the counting rules are quite simple, if a bit brain numbing for those who aren't scientific or engineering techno-geeks who love caressing every button on their calculator.

In these calculations, I take the number of essential and other elements, ignoring those that are considered toxic or unnatural to come into contact with, or which are widely varying in people from place to place, and call that n.

I take the number of results that fall into some specific category I am counting and call that r.

The probability that a result falls in the category I am counting is P(counted).

The probability that a result does not fall in the category I am counting is 1 - P(counted), since the result has to either be counted or not and thus the total probability has to add up to 1.

I assume that there is no correlation between the levels of the different elements in hair. This is an approximation and makes the statistics not be a perfect reflection of reality, but then statistics never actually are a perfect reflection of reality. The experience of seeing a large number of hair tests has shown that the statistics I derive below are an adequately good approximation of reality to be of practical utility. These simple calculations do in fact produce rules useful for identifying deranged mineral transport.

For any given counting criteria, the probability that a specific result with r counted and n-r not counted occurs is of course

P = nCr * P(counted) ^ r * { 1 - P(counted) } ^ (n - r)

This is simply the binomial probability distribution. There are many good discussions of it available. Two properties of some interest are that for n large enough it approximates a Gaussian or normal distribution if P is close to 0.5, and it approximates a Poisson distribution if P is less than about 0.1[a].

In this formula, n is the number of elements tested that are considered in "counting," and r is the number of those elements that satisfy the particular criteria counted. E. g. for counting rule #3 on a doctor's data test n is 23 and r is 4 on a test that just meets the rule. On a *Doctor's Data* Hair Element Profile, with n = 23, this means that there are 24 different probabilities for each of the 24 different values of r ($0 \leq r \leq 23$). None of these probabilities are very large so it would be silly to pick specific results and say something like "if there are exactly 15 bars going to the right, there is a mineral transport problem, but for 14 or 16 there isn't."

[a] One good source is Schaum's Outline of Theory and Problems of Probability and Statistics, by Murray R. Spiegel, 1975, published by McGraw-Hill. See pages 111-112.

Instead of saying something silly like that, I introduce a threshold. That is, for r from zero to some number (or from some number to 23) the total probability of the result happening is X, where X is the sum of all the probabilities of the individual results.

This leads to rules like: "there is a high probability that mineral transport is deranged if 5 or fewer bars extent to the right."

It doesn't matter if it is no bars, one bar, two, three, four or five.

I have selected the rules based on seeing a lot of tests and knowing what happened with a lot of the people. These are the rules that accurately sort tests into two groups. One having the kind of deranged mineral transport characteristic of mercury problems, the other having less than a high probability of this.

Tables and graphs of the relevant statistics follow.

It is useful to recall that nCr = nC(n-r)

For the question of whether the bars go right or left, where P = 0.5 (with equal likelihood either way),

P = (0.5) ^ r * (0.5) ^ (n-r) * nCr

and of course since the probabilities of going right or left are equal at 0.5, this reduces to

P = (0.5) ^ 23 * nCr

Probability table for counting rules #1 and #2

Number of bars one way	P for this exact number	P for 0 to this number
0	$1.1921x10^{-7}$	$1.1921x10^{-7}$
1	$2.7418\ x10^{-6}$	$2.8610\ x10^{-6}$
2	$3.0160\ x10^{-5}$	$3.3021\ x10^{-5}$
3	$2.1112\ x10^{-4}$	$2.4414\ x10^{-4}$
4	$1.0556\ x10^{-3}$	$1.2997\ x10^{-3}$
5	$4.0113\ x10^{-3}$	$5.3110\ x10^{-3}$
6	$1.2034\ x10^{-2}$	$1.7345\ x10^{-2}$
7	$2.9225\ x10^{-2}$	$4.6570\ x10^{-2}$
8	$5.8450\ x10^{-2}$	$1.0502\ x10^{-1}$
9	$9.7417\ x10^{-2}$	$2.0244\ x10^{-1}$
10	$1.3638\ x10^{-1}$	$3.3882\ x10^{-1}$
11	$1.6118\ x10^{-1}$	$5.0000\ x10^{-1}$
12	$1.6118\ x10^{-1}$	$6.6118\ x10^{-1}$
13	$1.3638\ x10^{-1}$	$7.97656\ x10^{-1}$
14	$9.7417\ x10^{-2}$	$8.9498x10^{-1}$
15	$5.8450\ x10^{-2}$	$9.5343x10^{-1}$
16	$2.9225\ x10^{-2}$	$0.982655x10^{-1}$
17	$1.2034\ x10^{-2}$	$0.994689x10^{-1}$
18	$4.0113\ x10^{-3}$	$0.9989x10^{-1}$
19	$1.0556\ x10^{-3}$	$0.999756x10^{-1}$
20	$2.1112\ x10^{-4}$	$0.999967x10^{-1}$
21	$3.0160\ x10^{-5}$	$0.999997x10^{-1}$
22	$2.7418\ x10^{-6}$	$0.99999988x10^{-1}$
23	$1.1921x10^{-7}$	1.0000000000000

Do note that this is symmetric, with P (0) = P (23), P (1) = P (22), etc.

Also note that this is a nice binomial that approaches the gaussian distribution of form e^{-x^2} in the limit of a large number of elements checked since P is 0.5. With an n of 23 this is a reasonable approximation of a gaussian distribution.

The probability that a certain number of bars, n, extend into the red zone is not gaussian. It is

P = (0.95) ^ (n-r) * (0.05) ^ r * nCr

In fact, with an n of 23 and P of 0.05, this approximates a Poisson distribution.

Table of probabilities for counting rule #3

Number of bars in red	P for this exact number	cumulative P for 0 to r bars in red	cumulative P for r or more bars in red
0	0.30736	0.30736	1.00000
1	0.37206	0.67942	0.69264
2	0.21541	0.89483	0.32058
3	0.07936	0.97419	0.10517
4	0.020884	0.9951	0.02581
5	0.004177	0.9992	0.00490
6	0.0006595	0.99991	0.00080
7	0.0000843	0.99999	0.00009
8	0.000008873	0.999999	0.00001

As you can see, the probability of 8 or more bars making it into the red zone by chance is negligible.

You can also see that the probability curve is quite steep. The chance of 3 reds happening at random is high enough it really isn't that suggestive. Four is not very likely by chance, and someone just happening to have five reds without some factor influencing it is really unlikely. This does mean that with five or more reds it is very clear that the overwhelming probability is that the person has deranged mineral transport, but with four reds you have to be suspicious about contamination issues (selenium from shampoo being the major offender here) and look at the other nonstatistical elements of the test to decide what nonstatistical common sense judgment you are going to make.

The probability that a certain number of bars, r, stay within the middle green + white band is:

P = (0.67) ^ r * (0.33) ^ (n-r) * nCr

This also approximates a Gaussian distribution reasonably well since 0.67 and 0.33 are not too different from 0.5.

Table of probabilities for counting rule #4

Number of bars in middle band	P for this exact number	Cumulative P for 23 down to r bars in the middle band	Cumulative P for 0 to r bars in the middle band
0	8.4299×10^{-12}	1.00000	8.4299×10^{-12}
1	3.93649x10-10	0.9999999999916	4.0208×10^{-10}
2	8.7915x10-9	0.99999999960	9.1936×10^{-9}
3	1.24946x10-7	0.9999999908	1.3414×10^{-7}
4	1.2683895x10-6	0.99999987	1.4025×10^{-6}
5	9.785817x10-6	0.9999986	1.1188×10^{-5}
6	5.960452x10-5	0.9999888	7.0793×10^{-5}
7	2.9389416x10-4	0.9999292	3.6469×10^{-4}
8	1.1933884x10-3	0.999635	0.001558
9	4.03823349x10-3	0.99844	0.005596
10	1.1478372x10-2	0.99440	0.01708
11	2.754177x10-2	0.98293	0.04462
12	5.591814x10-2	0.95538	0.10053
13	9.6064496x10-2	0.89947	0.19660
14	1.39314313x10-1	0.80340	0.33591
15	1.697101641x10-1	0.66409	0.50562
16	1.7228153x10-1	0.49438	0.67791
17	1.44028587x10-1	0.32209	0.82193
18	9.7473892x10-2	0.17807	0.91941
19	5.2079352x10-2	0.085092	0.97149
20	2.1147373x10-2	0.028513	0.99263
21	6.1336538x10-3	0.0073667	0.99877
22	1.132107x10-3	0.0012320	0.99999
23	9.9935655x10-5	0.000099936	1.00000

Note that some of the cumulative probabilities should satisfy relations where they add up to one. The probabilities that should do this are the cumulative probability of 0 to n bars in the middle band plus the cumulative probability of n+1 to 23 bars in the middle band must be exactly one, and also the probability of 0 to n red results plus the probability of n+1 to 23 red results must be exactly one. The sum of the probabilities for an exact result from 0 to 23 must also be exactly one for each specific type of result.

Some plots of these probability distributions are presented on the next two pages as an aid in understanding them.

Probability of r bars to the right (or left) of the midline

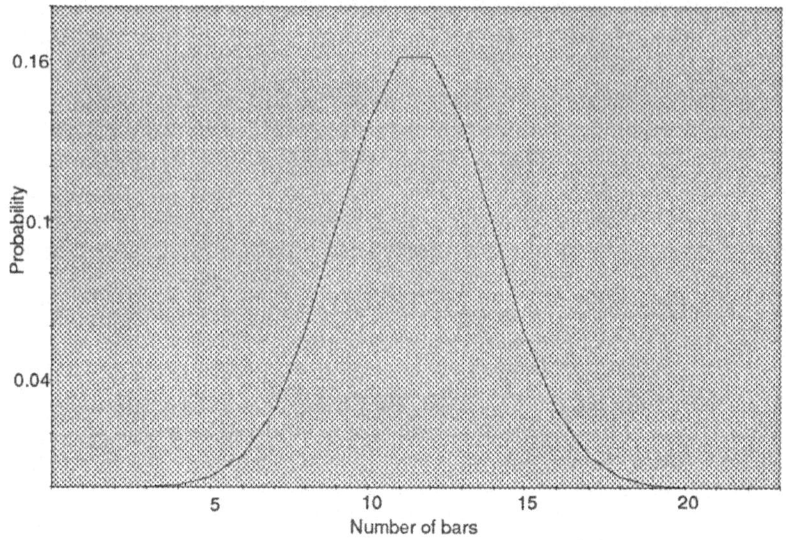

Cumulative probability of zero to r bars to the right (or left) of the midline

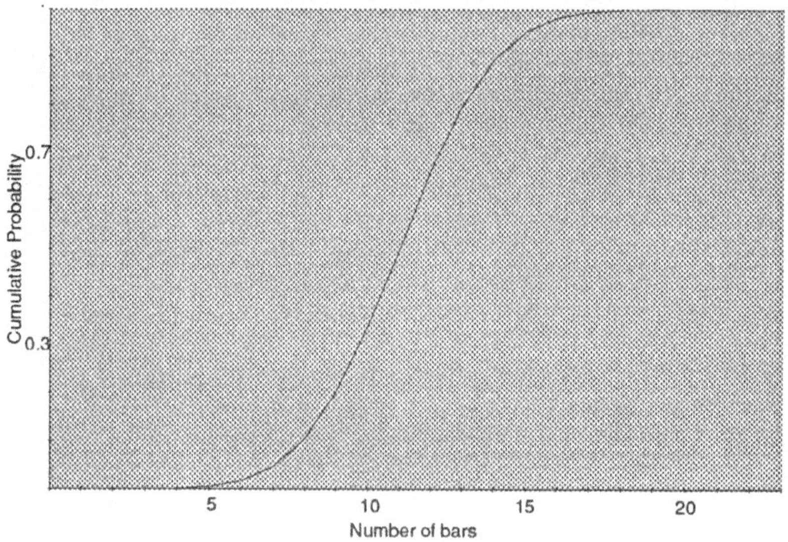

Probability of r bars in the middle band

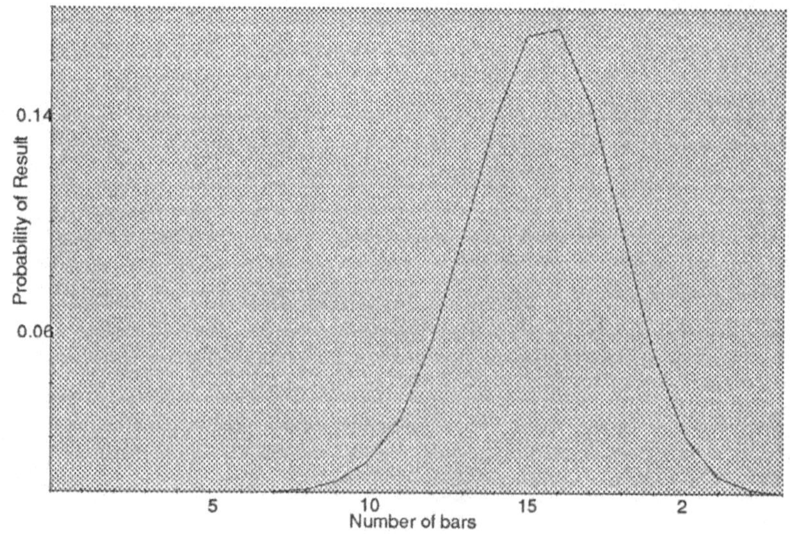

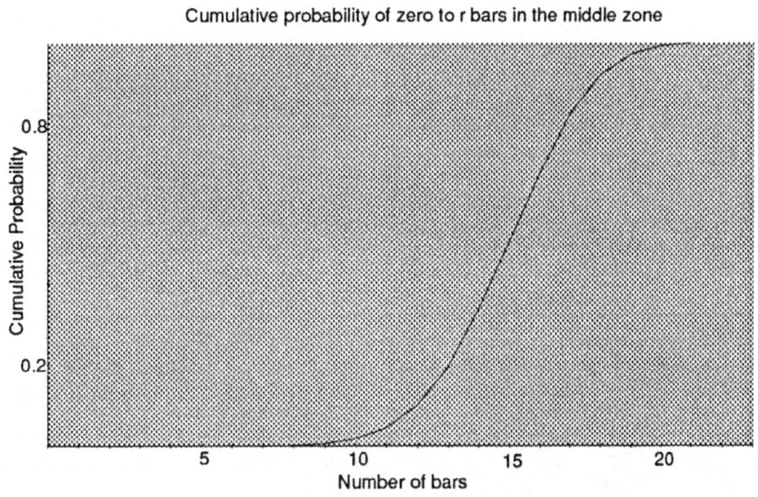
Cumulative probability of zero to r bars in the middle zone
0.8
0.2
Cumulative Probability
5
10
15
20
Number of bars

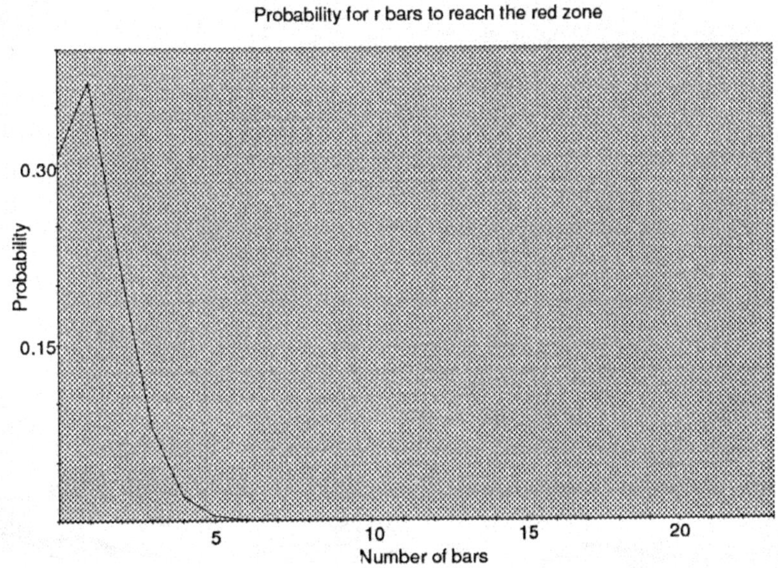
Probability for r bars to reach the red zone
0.30
0.15
Probability
5
10
15
20
Number of bars

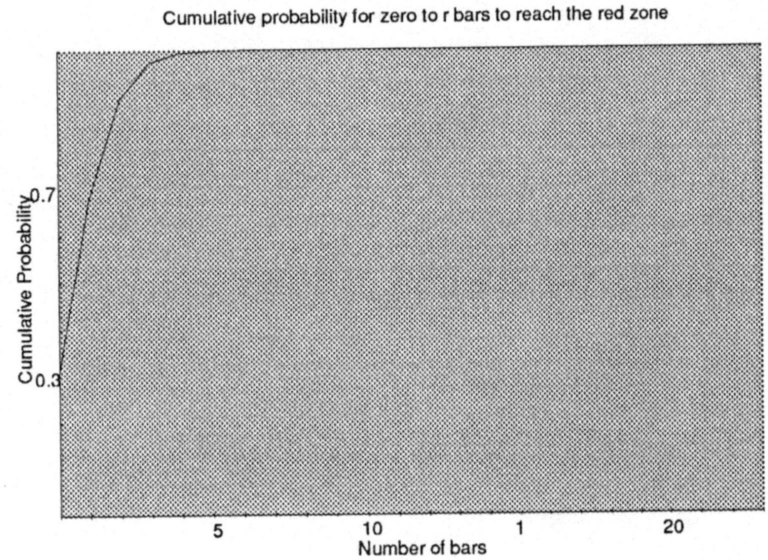
Cumulative probability for zero to r bars to reach the red zone
0.7
0.3
Cumulative Probability
5
10
1
20
Number of bars

Appendix: Various Laboratories' Hair Tests

This appendix lists which elements are determined in hair tests from several laboratories, and gives counting rules relevant to those tests where he counting rules approach to determining deranged mineral transport is possible given the manner in which the report is presented and also which elements are measured.

Two tables are provided for which elements are determined for a particular test. This table is accurate to the best of my knowledge at the time this book was written. If you are getting a test to check a specific element or elements you may wish to verify what is currently tested for with the lab on the web or by phone. Contact information is provided at the end of this section.

The first table is laid out with the elements grouped so it is easy to compare what the different labs do and don't do. The second table is alphabetical by element.

Counting Rules for Various Laboratories..

	DDI hair element profile	TEI	King James	ARL	GSDL Elemental Analysis	Metametrix hair test	DDI toxic element exposure profile	GSDL Toxic Element Exposure
Number of essential elements	23	22	16	15	17	18	9	3
Number of toxic elements	16	13	6	6	11	6	23	17
CR #1	5	5	3	3	3	3	n/a	n/a
CR #2	5	5	3	3	3	3	n/a	n/a
CR #3	4	4	4	4	4	n/a	n/a	n/a
CR #4	11	11	6	6	6	6	n/a	n/a
CR #5	Any two rules are one short on being met						n/a	n/a
suspicious	Any single rule is one short of being met.						≥4 very elevated	≥3 very elevated
Suitable for use with children?	yes	no	no	no	yes	no	yes	yes

Elements tested for grouped for easy comparison of laboratories, essential and other elements

Name	DDI hair element profile	TEI	King James	ARL	GSDL Elemental Analysis	DDI toxic element exposure profile	GSDL Toxic Element Exposure	Metametrix hair test
Silicon				•				
Sodium	•	•	•	•				•
Potassium	•	•	•	•				•
Calcium	•	•	•	•	•			•
Magnesium	•	•	•	•	•			•
Molybdenum	•	•	•	•	•			•
Lithium	•	•	•	•	•			•
Phosphorus	•	•	•	•	•			•
Iron	•	•	•	•	•			•
Zinc	•	•	•	•	•	•		•
Manganese	•	•	•	•	•	•		•
Chromium	•	•	•	•	•	•		•
Cobalt	•	•	•	•	•	•		•
Copper	•	•	•	•	•	•	•	•
Selenium	•	•	•	•		•		•
Germanium	•	•	•			•	•	
Vanadium	•	•	•		•	•		•
Barium	•	•			•	•	•	•
Strontium	•	•			•			•
Boron	•	•			•			•
Sulfur	•	•			•			
Rubidium	•	•			•			
Zirconium	•	•						
Iodine	•							

Elements tested for grouped for easy comparison of laboratories, toxic elements

Name	DDI hair element profile	TEI	King James	ARL	GSDL Elemental Analysis	DDI toxic element exposure profile	GSDL Toxic Element Exposure	Metametrix hair test
Cerium						•		
Cesium						•		
Gadolinium						•		
Gold						•		
Silver	•					•		
Beryllium	•	•				•		
Titanium	•	•				•		
Platinum	•	•				•	•	
Aluminum	•	•	•	•	•	•	•	•
Arsenic	•	•	•	•	•	•	•	•
Cadmium	•	•	•	•	•	•	•	•
Lead	•	•	•	•	•	•	•	•
Mercury	•	•	•	•	•	•	•	•
Nickel	•	•	•	•	•	•	•	•
Bismuth	•	•			•	•	•	
Thallium	•	•			•	•	•	
Tin	•	•			•	•	•	
Uranium	•	•			•	•	•	
Antimony	•				•	•	•	
Thorium	•					•	•	
Tungsten		•				•	•	
Palladium						•	•	
Tellurium						•	•	
Gallium							•	

Elements tested for by different laboratories listed in alphabetical order

Name	DDI hair element profile	TEI	King James	ARL	GSDL Elemental Analysis	DDI toxic element exposure profile	GSDL Toxic Element Exposure	Metametrix hair test	Toxic or essential /other
Aluminum	•	•	•	•	•	•	•	•	T
Antimony	•				•	•	•		T
Arsenic	•	•	•	•	•	•	•	•	T
Barium	•	•			•	•	•	•	E/O
Beryllium	•	•				•			T
Bismuth	•	•			•	•	•		T
Boron	•	•			•			•	E/O
Cadmium	•	•	•	•	•	•	•	•	T
Calcium	•	•	•	•	•			•	E/O
Cerium						•			T
Cesium						•			T
Chromium	•	•	•	•	•	•		•	E/O
Cobalt	•	•	•	•	•	•		•	E/O
Copper	•	•	•	•	•	•	•	•	E/O
Gadolinium						•			T
Gallium							•		T
Germanium	•	•	•			•	•		E/O
Gold						•			T
Iodine	•								E/O
Iron	•	•	•	•	•			•	E/O
Lead	•	•	•	•	•	•	•	•	T
Lithium	•	•	•	•	•			•	E/O
Magnesium	•	•	•	•	•			•	E/O
Manganese	•	•	•	•	•	•		•	E/O
Mercury	•	•	•	•	•	•	•	•	T
Molybdenum	•	•	•	•	•			•	E/O
Nickel	•	•	•	•	•	•	•	•	T
Palladium						•	•		T
Phosphorus	•	•	•	•	•			•	E/O
Platinum	•	•				•	•		T
Potassium	•	•	•	•				•	E
Rubidium	•	•			•				E/O
Selenium	•	•	•	•		•		•	E/O
Silicon				•					E/O
Silver	•					•			T
Sodium	•	•	•	•				•	E/O
Strontium	•	•			•			•	E/O
Sulfur	•	•			•				E/O
Tellurium						•	•		T
Thallium	•	•			•	•	•		T
Thorium	•					•	•		T
Tin	•	•			•	•	•		T
Titanium	•	•				•			T
Tungsten		•				•	•		T
Uranium	•	•			•	•	•		T
Vanadium	•	•	•		•	•		•	E/O
Zinc	•	•	•	•	•	•		•	E/O
Zirconium	•	•							E/O

Please note that this "counting rules" approach to hair tests is relatively new. The laboratories may not be familiar with it, or if familiar with it may not be supportive of it at the present time.

I am mentioning how to interpret a variety of laboratories' tests not because I endorse these tests or the laboratories endorse the approach presented in this book, but simply because a lot of people will already have one of these tests in hand, and many health care practitioners will already have an established relationship with a specific laboratory and be familiar with the interpretation of that particular test. Any laboratories omitted from the list were left off due to my not knowing about them.

Trace Elements, Inc. (*TEI*) test

Of these tests, the *TEI* test is most suitable for this counting rules approach to detecting mercury induced derangement of mineral transport in adults. It is completely useless for children for this purpose since TEI does not use age dependent normal ranges.

For the *TEI* test, the middle band is the two lighter colored stripes horizontally across the report page that is labeled the "reference range." It is between the larger of the two sets of markings on the left hand side of the test. This corresponds to the *Doctor's Data* green + white band.

The bands marked high and low correspond to the *Doctor's Data* yellow bands. Bars not given at all (<< on the bottom of the page) or going all the way to the top of the page are considered "red zone" results that are counted up for rule #3.

I suggest only counting those 22 elements that are used as "essential and other" elements on the *Doctor's Data* test and listed as such in the table above. It would also be perfectly reasonable to consider titanium an "essential plus other" element in the *TEI* test and count it so that 23 total elements are counted. Bismuth, nickel, platinum, thallium, tin and tungsten should not be counted.

All of the other tests are less statistically sensitive than the *Doctor's Data* and *TEI* tests for the detection of mercury induced mineral transport derangement and can reasonably be expected to generate more false positives and false negatives. This is simply a function of the number of elements determined – statistics work better the larger a sample they are applied to. It doesn't mean there is anything inherently wrong with these tests when used for other purposes, such as the ones the laboratories suggest. Be especially suspicious of rule #5 results from *King James Laboratory, Inc., ARL* and *GSDL* tests.

King James Laboratory, Inc./Omegatech test

King James Laboratory, Inc., has normal ranges that do not depend on age and sex, but which do depend on what type of hair was used – head, pubic or body hair. This test is suitable for detecting mercury induced mineral transport derangement in adults, but not children. People who are concerned about the issue of element content variation when using different types of hair this test may be preferable when pubic or body hair must be used.

You count the elements listed as Essential Elements, Essential Macroelements and Other Nonessential Elements. The "acceptable range" corresponds to the middle band and is used for counting rule #4. Simply draw a line down the middle of the acceptable range to apply counting rules #1 and #2. Elements where there results show as < on the left or the bar goes all the way to the end of "high" on the right count as "red zone" results.

Due to reference range issues I personally would be somewhat suspicious of the accuracy of tests meeting counting rule #3 with 4 low "red zone" results mostly involving the elements cobalt, germanium, lithium, molybdenum and vanadium. This is a matter of opinion based on me not being terribly familiar with the test, other reasonable people who use the test routinely may well differ on this.

Analytical Research Labs, Inc., (*ARL*) test

ARL does not break down their reference ranges by age and sex so their test is only suitable for the counting rules approach to detecting mercury induced mineral transport derangement in adults.

For counting rules #1 and #2, use the dark bar in the middle of the "IDEAL" band.

For counting rule #3, consider "red zone" results to be those at either the top or the bottom of the range, if they are not all the way to the edge, they don't count for rule #3.

For counting rule #4, count the number of bars that end in the shaded zone marked as "IDEAL."

Great Smokies Diagnostic Laboratories, Inc. (*GSDL*) test

GSDL does break down their reference ranges by age and sex so the test is suitable for detecting mercury induced mineral transport derangement children.

The *GSDL* Toxic Element Exposure hair test does not test enough essential elements to be useful for detecting deranged mineral transport, its utility is in people with known orderly mineral transport who need to be screened for exposure to elements not on other tests. If three or more of the elements tested for (toxic or essential) are all the way to the right then the test is suspicious for deranged mineral transport and mineral transport should be studied using a test suitable for that purpose before using the Toxic Element Exposure test results to select interventions.

The *GSDL* Elemental Analysis hair test is suitable for detecting deranged mineral transport in adults and children, though with reduced statistical sensitivity. However since sodium and potassium are not checked, the indicators of adrenal and thyroid problems can't be investigated using the hair test result. Thus I suggest using some other test unless these are understood through other means.

For rules #1 and #2, draw a line carefully down the middle of the test results plot. Consider the center of the results ellipse, marked with a hash mark, to be where the result lies for counting whether it is right or left.

For rule #3, consider the "red range" results to be those that are all the way to the edge of the page, with the hash marks on the ellipse lined up with the edge of the results plot. The red zone shown on a *GSDL* test corresponds to the yellow zone on a *DDI* test, NOT the red zone!

For rule #4, count the results where the center of the ellipse lies within the band marked on the results page that contains both the green and the yellow results ranges. This corresponds to the *DDI* green + white middle stripe.

Doctor's Data Laboratories, Inc. (*DDI*) Hair Toxic Element Exposure Profile test

The entire book is filled with examples of how to interpret the *DDI* Hair Element Profile. *DDI* also offers another test, the Hair Toxic Element Exposure Profile. This test is not suitable for the detection of deranged mineral transport. In fact, as is shown in the section *You Need to be Sure to get the Right Test*, this test can be quite misleading and falsely suggest toxicities when mineral transport is deranged.

If you end up with a Hair Toxic Element Exposure Profile and want to check for orderly mineral transport you have to get the Hair Element Profile done. You will also need to pay for both tests. If more than a month elapses you will also need to submit a second hair sample.

If you have a Hair Toxic Element Exposure Profile and four or more of the elements tested for (toxic or essential) are in the red band then the test is suspicious for deranged mineral transport and mineral transport should be studied using a test suitable for that purpose before using the Hair Toxic Element Exposure Profile results to select interventions.

Great Plains Laboratory, Inc. tests

Great Plains Laboratory offers the same two hair tests as Doctor's Data. In certain states and foreign countries Great Plains Laboratories can offer the test directly to consumers without the signature of a licensed health care provider, in other jurisdictions a signature is required.

Metametrix, Inc., test

Metametrix also offers a hair test. They do not break their reference range down by age and sex so their report is only suitable for use with adults. Due to the report format and the way the results are presented no equivalent of counting rule #3 can be derived. For rules #1 and #2, carefully draw a line down the exact middle of the results bars. Count both the Essential Elements, and also barium, boron, strontium and lithium from the Potentially Toxic Elements. The statistical definition of the middle band for this test is P = 0.60, which makes the counting rule #4 cutoff come out at 6.

Direct Laboratory Services, Inc.

The *Doctor's Data* Hair Element Profile (and a variety of other laboratory tests) can be obtained through Direct Laboratory Services without the need for a doctor's visit.

Contact information for the Laboratories

Trace Elements, Inc.
4501 Sunbelt Drive
Addison, Texas 75001 USA
Phone: (972) 250-6410
www.traceelements.com/

Analytical Research Laboratories, Inc.
P.O. Box 37964
Phoenix, Arizona 85069
(602) 995-1580.

Great Smokies Diagnostic Laboratory, Inc.
63 Zillicoa Street
Asheville, NC 28801
(828)253-0621
www.gsdl.com

King James Laboratory/Omegatech
24700 Center Ridge Road
Cleveland, OH 44145
800-437-1404
www.kingjamesomegatech-lab.com/

Doctor's Data, Inc.
P.O. Box 111
West Chicago, Illinois 60186
630.377.8139
800.323.2784
www.doctorsdata.com

Metametrix Clinical Laboratory
4855 Peachtree Industrial Blvd., suite 201
Norcross, GA 30092
(770) 446-5483
(800) 221-4640

Great Plains Laboratory, Inc.
11813 West 77th
Lenexa, KS 66214
913-341-8949
www.glp4u.com

Direct Laboratory Services, Inc.
P.O. Box 601
Mandeville, LA 70470-0601
(800) 908-0000
support@directlabs.com
http://directlab.com/HairElements.php

Appendix: Reference Ranges vary with Age and Sex

I have mentioned the need to use age and sex specific reference ranges to have a test suitable for children and also to have the greatest statistical accuracy in avoiding false positives (test says healthy person is sick) and false negatives (test says sick person is healthy). I have made plots of a few reference ranges from different labs to illustrate this, as well as compared a reference range for males and females for an element where it varies quite a bit.

The reference range is pretty independent of age, sex, and which lab has determined it for selenium. However, for strontium, the reference range is dependent on age and sex and as you can see many children with pathologically high selenium would show as falsely normal on a test with one normal range for human beings, while some normal women would score as pathologically high (or if the range were set to prevent this, men with problems would score falsely normal).

Elements which vary substantially with age are: calcium, magnesium, molybdenum, strontium, barium, rubidium and zirconium.

Elements which don't vary much with age until people are over 50 are: sodium, potassium, lithium and phosphorus.

Elements which do not vary much with age are: copper, zinc, manganese, chromium, vanadium, boron, iodine, selenium, sulfur, cobalt, iron and germanium.

Elements where there is a substantial difference between the male and female ranges are calcium, magnesium, phosphorus, barium and strontium.

A chemically sophistiated person will realize that the alkaline earths – magnesium, calcium, strontium and barium – vary substantially with age and sex. Calcium and magnesium are also involved in many of the relations among elements that are informative in the interpretation of hair tests.

The toxic elements don't vary that much with age as a rule, though older people generally have more than young people, with bismuth and silver appearing to be an exception to this.

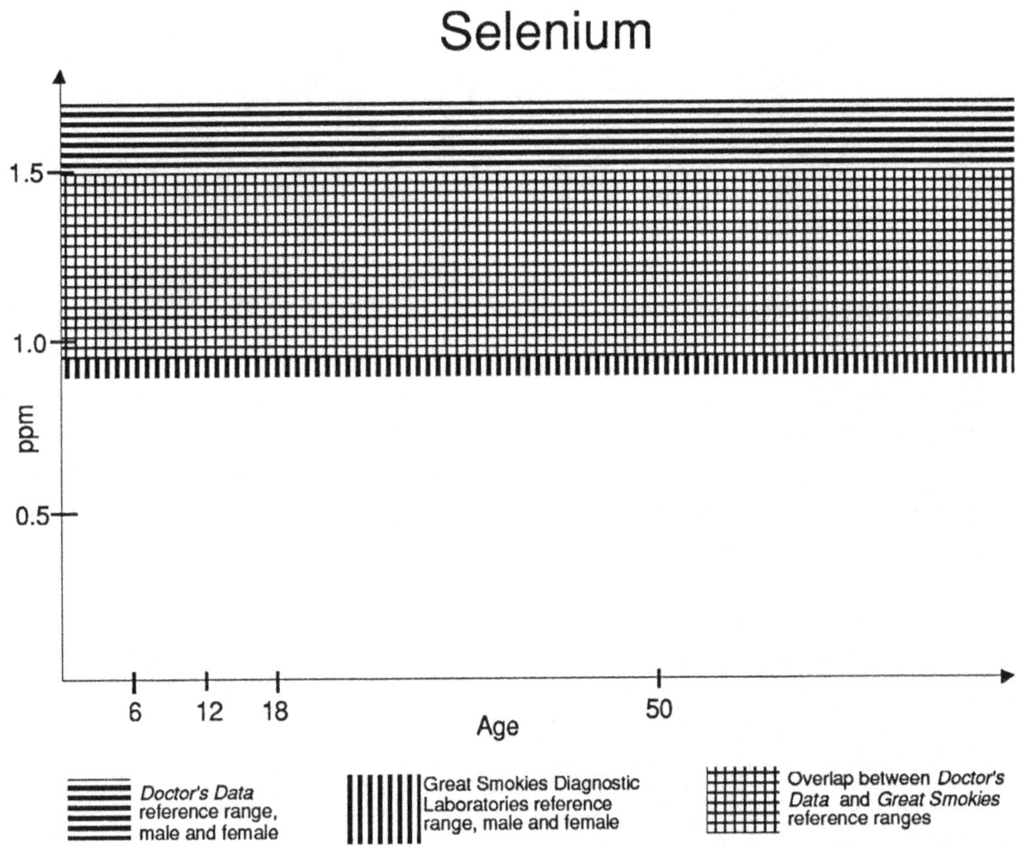
Selenium
1.5
1.0
0.5
ppm
6
12
18
50
Age
Doctor's Data reference range, male and female
Great Smokies Diagnostic Laboratories reference range, male and female
Overlap between Doctor's Data and Great Smokies reference ranges

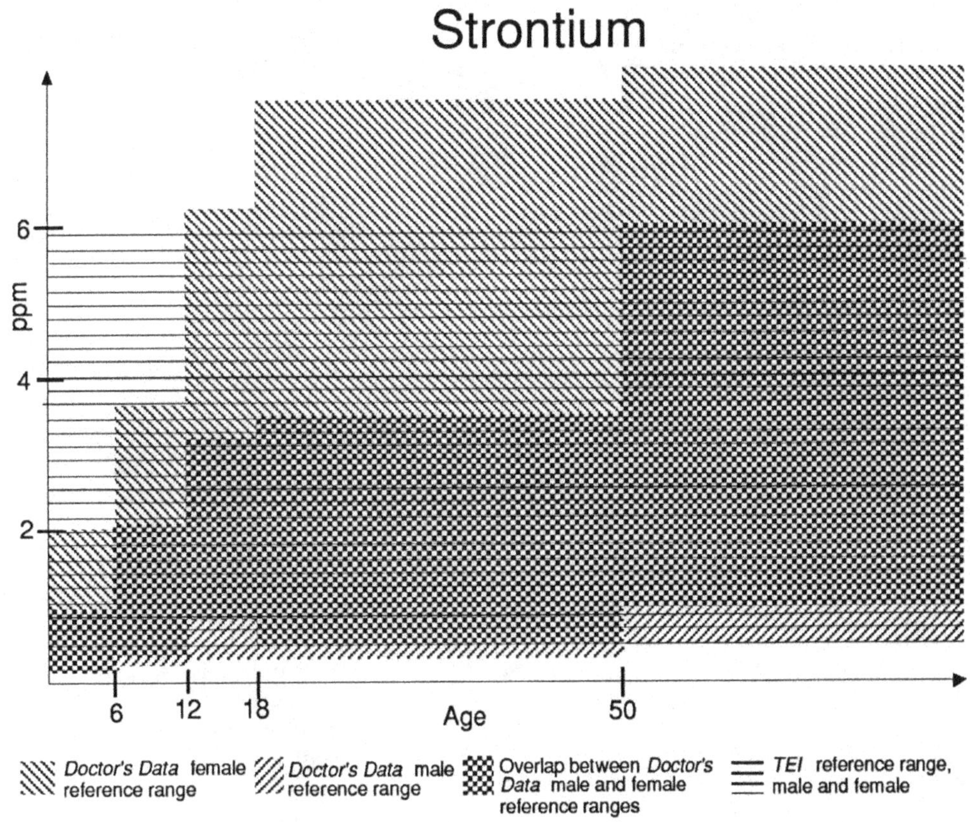
Strontium
6
4
2
ppm
6
12
18
50
Age
Doctor's Data female reference range
Doctor's Data male reference range
Overlap between Doctor's Data male and female reference ranges
TEI reference range, male and female

Index

—B—

—D—

—E—

—F—

—G—

—H—

—I—

—J—

—K—

—L—

—M—

—N—

—O—

—P—

—Q—

—R—

—S—

—T—

—U—

—V—

—W—

—X—

—Y—

—Z—

CPSIA information can be obtained
at www.ICGtesting.com
Printed in the USA
JSHW021121140421
13571JS00007B/8

9 780967 616827